Life in Media

# Life in Media

A Global Introduction to Media Studies

Mark Deuze

The MIT Press
Cambridge, Massachusetts
London, England

© 2023 Massachusetts Institute of Technology

Original photographs © 2023 Piet Hermans

This project has been made possible with support from the Amsterdam University Fund.

All rights reserved. No part of this book may be reproduced in any form by any electronic or mechanical means (including photocopying, recording, or information storage and retrieval) without permission in writing from the publisher.

The MIT Press would like to thank the anonymous peer reviewers who provided comments on drafts of this book. The generous work of academic experts is essential for establishing the authority and quality of our publications. We acknowledge with gratitude the contributions of these otherwise uncredited readers.

This book was set in Times New Roman by Westchester Publishing Services. Printed and bound in the United States of America.

Library of Congress Cataloging-in-Publication Data

Names: Deuze, Mark, author.
Title: Life in media : a global introduction to media studies / Mark Deuze.
Description: Cambridge : The MIT Press, 2023. | Includes bibliographical references and index.
Identifiers: LCCN 2022038286 (print) | LCCN 2022038287 (ebook) | ISBN 9780262545587 (paperback) | ISBN 9780262374613 (epub) | ISBN 9780262374620 (pdf)
Subjects: LCSH: Mass media—Study and teaching. | Mass media—Social aspects—Study and teaching.
Classification: LCC P91.3 D485 2023  (print) | LCC P91.3  (ebook) | DDC 302.23/071/1—dc23/eng/20221031
LC record available at https://lccn.loc.gov/2022038286
LC ebook record available at https://lccn.loc.gov/2022038287

10  9  8  7  6  5  4  3  2  1

Dedicated to students, sharing stories and experiences of their lives in media

# Contents

1    Media Life    1

2    Your Life    25

3    Public Life    51

4    Real Life    85

5    Love Life    127

6    Change Life    149

7    Make Life    183

8    Life in Media    219

Appendix 1: Annotated Sources    253
Appendix 2: Key Concepts in Tweets    297
Appendix 3: Ten Tips for a Life in Media    305
Index    313

# 1

## Media Life

> We do not live *with* media but *in* media. Media are everywhere and cannot be switched off. This is not a new situation; nor is a life in media unique to our time. Media are fun and complicated, and they offer plenty of ethical dilemmas—critical debates that are crucial elements of media studies for a life in media.

Media are part of every single aspect of our lives, including our most intimate feelings and experiences. How we grow up and learn, love, work, and play—it is all to some extent circumscribed by our media. In the process, media influence and structure our lives in all kinds of subtle ways—for better and for worse. At the same time, we are not without agency when it comes to using media to experience, understand, and express ourselves about the world around us. In media, we create worlds: networks, communities, groups, teams, clans, and countless relationships. These worlds spill over, affect us deeply, and complicate as much as inspire life. We get much of our information and insight through social media and news providers; images and ideas are endlessly supplied through films and television; developing tactics and strategies for everyday life get animated by how-to videos or role-playing games online; and navigating public space has its own soundtrack of cool tunes on our headphones. None of these tools and resources is neutral, enabling as much as constraining certain ways of doing things. In other words: we use media to figure out what to do, while media in turn structure our thoughts and actions. It is up to all of us—as we are together alone in media, superconnected to each other and the world yet also stuck inside our own personal information space—to figure out the best way forward.

At the heart of this book and its approach to the role media play in society and everyday life is the fairly straightforward notion that we do not live *with* media but *in* media. Media are to us like water is to fish. Such a view on media as all-encompassing and inevitable is not new; nor is it particular to our current digital environment. Media and communication scholars from around the world—from Canada to Nigeria, Germany to Sri Lanka, China to Colombia (please see appendix 1 for an annotated bibliography for all arguments in this book)—have suggested throughout the twentieth century that it is most fruitful to

think of all kinds of media, including the "mass" media of television, radio, books, and magazines as well as such traditional forms of shared narratives in society as poetry, plays, song, and dance, as a deeply interdependent ensemble of devices, technologies, contents, communicative relations, and interactions that together (and in conjunction) make up people's experience of media life. Our current digital environment integrates all of this into a more or less seamless lived experience of media.

Studying media from a perspective of *life in media* presupposes three essential qualities of people's experience in and with media:

1. *Media disappear*, in that the ways people use and think of media tends to be highly intuitive, ritualistic, and blissfully unaware of the inner workings of media (both as technologies and how media act as industries).

2. *Media are what people do*, in that everything people do in life either directly or indirectly involves media of some kind, subtly structuring, stimulating, and specifying what we observe, feel, process, and perform.

3. *People love media*, in that our primary relation with media is both intimate and emotional, as people take their media intensely personally, having a range of strong feelings about media—ranging from passion and excitement to frustration and fear.

Throughout all chapters of this book, we unpack and explore these fundamental elements of life in media, all the time combining an appreciation of effects, things, and what happens with a deliberate focus on processes, practices, and what can be done. Media need us as much as we need media, a give-and-take that opens up opportunities for intervention, for ways of doing (and designing) things and experiences differently. While it is crucial for media studies to be critical of its object, I would very much encourage us to commit to a hopeful perspective on (our) media. Hope in this context means to actively look for and act on chances and opportunities to change things for the better, based on the assumption that the world is fluid, plastic, and capable of being molded by people's willingness, motivation, and agency. This is not a naive hope! It is a hopeful recognition of the permanently unfinished (and imperfect) nature of society as realized in media and technology, waiting for us to participate and cocreate, to take action, to imagine differently, and to fight whatever evil can be found—no matter how daunting the prospect or task ahead.

Changing the world in media (and with media) necessarily means making a difference in the world outside of media. A life in media inevitably means that the worlds of media and life are inexorably entwined. While such arguments have been made in the scholarly literature since as far back as the 1960s, real-life examples from the 2020s onward show how a life in media has truly become the basis of lived experience of people all over the world. Consider, for example, the global coronavirus crisis. Early in 2020, the World Health Organization (WHO) described the disinformation swirling across all media amid the global coronavirus crisis as a massive infodemic—a major driver of the pandemic. To the WHO and many other experts, organizations, and governments around the world, stopping the spread of misinformation is just as important as providing accurate information about the crisis. As WHO director general Tedros Adhanom Ghebreyesus stated at a conference in Munich, Germany, on February 15, 2020: "We're not just fighting an epidemic; we're fighting an infodemic. Fake news spreads faster and more easily than this virus, and is just as dangerous."[1] This was a significant public recognition of how the realities inside and outside of media have collapsed, how there is no outside to media anymore. Getting people protected from infection and disease during the pandemic would be as crucial as fighting the spread of rumor, disinformation campaigns, and so-called fake news about the virus, its causes, and potential treatments. With the acknowledgment of a parallel pandemic and infodemic having equal consequences for the health and safety of us all, our speculations about the role and impact of technology and media in society move beyond the realm of science fiction or academic debate into the humdrum of everyday life. Such recognition of the porous, disintegrating, or even disappearing boundaries between the real and the virtual—between media and life—makes us realize to what extent our lives have become enmeshed with media and how intimate our relations with and through media are.

---

1. Tedros Adhanom Ghebreyesus, remarks, Munich Security Conference, February 15, 2020. https://www.who.int/director-general/speeches/detail/munich-security-conference

In this chapter, you will find an outline of the key ideas and discussions inspiring contemporary media studies for a life in media. To be sure, this is not an exhaustive survey of the field—it is intended as a starting point for discussion and your own exploration of what a life in media possibly entails. Every chapter in this book can be read on its own, and there is no necessary order to the arguments developed in each. While this opening chapter ostensibly deals with *life*, a comprehensive discussion and definition of *media* can be found in the next chapter. The first and final chapters are attempts at making overarching statements about the elements and prospects of the study and scholarly understanding of media, whereas all in-between chapters tackle life in media through different experiences: how we define and use media in everyday life, how we experience our private lives in public, how we draw distinctions between what is real and what is fake, how we love, how we (can) make a difference in the world, and how media as industries work (as well as how media professionals "make it work").

Before diving in, a brief note about the interchangeable use of the terms *people*, *we*, and *you* in these pages. It is clear that people's experiences of, with, and in media in different parts of the world, coming from different backgrounds, with different motivations and skills, and opting for different kinds of media can be quite different. Something as straightforward as watching television or posting something on social media can mean something vastly different depending on whether you are using a mobile device on the road or settling down in front of a television set at home, whether you are part of a large family or are used to using media on your own, whether you are a white man or a Black woman, whether you seek out media to battle feelings of loneliness or just to get away from it all. It is important to disentangle such differences and compound inequalities, aspects of life and media that serve to either include or exclude, ways in which media can both emancipate and disempower.

In this book, I mostly talk in general terms about life in media, using examples and cases from a variety of peoples and societies, offering a narrative supported by findings from research around the world in an attempt to discuss what we have in common and what connects us rather than focusing on all the circumstances that divide us. Of course, in all of this we must stay mindful of a straightforward result from many, if not most, studies on media influence and effects, (digital) inequality, and exclusion: media have a tendency to *amplify* and *accelerate* existing phenomena in society. Unless people act deliberately—as designers and developers, manufacturers, policy makers, makers and audiences—media in general, and digital products and services in particular, tend to exacerbate such social issues as misogyny, racial and gender discrimination, poverty, war and conflict, and even climate and refugee crises. Yet media can also contribute to meaningfully addressing such issues. It is exactly for this reason that I have opted for a hopeful perspective in this book, grounded in people's efforts and actions in their efforts to reduce harm and improve their lives and those of others.

## Living in Media

The first time I remember directly experiencing an almost complete collapse of the realities of media and life was early in 2003. I had just moved from Amsterdam to Los Angeles to do research at the University of Southern California, when in March the war in Iraq started with the American, British, and Australian invasion of the country. Of course, the *Los Angeles Times* reported extensively on the proceedings, filling most of its pages with news and backgrounds about Operation Iraqi Freedom (as the US government called the war). However, the neighborhood I was living in, University Park, was mostly Latino—and in this part of the sprawling city the second-most-read newspaper in Los Angeles was dominant: the Spanish-language *La Opinión*. In that paper, the war was only mentioned briefly on page 5. On television, networks like Fox News and CBS provided nonstop live coverage of the invasion of Iraq, illustrated with a waving American flag in a corner of the screen. Yet television ratings figures of those early weeks of the war showed that more Americans watched televised reruns of the popular comedy series *Friends* than tuning in to news coverage of the war. All of this made me realize how my neighbors were living in a parallel world, how the media we choose to use does not simply act like neutral providers of news, information, and entertainment—through the interaction of production, content, and usage, people and their media cocreate certain worlds, worlds we all live in and move through daily. While those were the early days of our digital environment, today's media surroundings are much more seamless, intuitive, and interconnected—at the same time, we are more likely (and with less effort) to retreat into deeply personalized media enclaves than ever before.

As media evolve, all aspects of our lives play out in media, and it is in media that we find each other and ourselves. Take love, for example. Using popular dating platforms like Badoo, Bumble, and Tinder, Grindr and Lex (for queer lovers), Aisle (India), AfroIntroductions (across Africa), Latin American Cupid (owned by the Australian company Cupid Media), Matchmallows (Middle East), and Tantan (China), finding some kind of love has never been easier, regardless of where or who you are in the world. Hundreds of millions of people have an account on one or more online dating services, where potential partners are available day and night. With a few swipes, clicks, and chats, you could be off on your next amorous adventure. At the same time, it can also be much more difficult to develop and maintain romantic relationships, as there seem to be endless alternatives available, and the supply of potential paramours never stops. Even if you are matched with someone else, once they put your name into a search engine, your entire life history, including experiences and events that are perhaps not so appealing (especially when taken out of context), may be exposed. Furthermore, are you really "matching" with someone, or is the interface of the site or app you are using subtly steering you a certain way? Is this love, or is it the outcome of a program that runs in the background to connect profiles not necessarily based on mutual amorous preference but governed by the formal indifference of a

computer algorithm? What about all the people who do not have online dating profiles, who are less savvy or uncomfortable presenting and promoting themselves in a way that befits the laws of attraction as determined by the affordances of an app or website or who simply did not log on that day?

---

*As media evolve, all aspects of our lives play out in media, and it is in media that we find each other and ourselves.*

---

In these brief examples, many issues related to life in media come together, all of them extensions of the three fundamental elements (of media as disappearing, interwoven with everything we do, and engaging us primarily on an emotional level) of media studies:

- living in media, even our most intimate feelings and experiences somehow involve media;
- media structure our lives in ways that have become rather invisible, their influence difficult to disentangle;
- "the" media includes a variety of industries, including media firms, technology companies, and telecommunications corporations, all working together as well as competing for our time and engagement;
- using media tends to include some kind of production of media (from sharing personal data each time we log on to composing and sharing countless posts, memes, blogs, and vlogs online);
- in media, participating in private also makes your life public to some extent;
- to be certain about what is genuine, authentic, true, or real in media can be complicated and frustrating;
- we can appreciate how we clearly love media, considering how we use media not just for fun, information, education, and work but for anything in life, including to channel our deepest infatuations, passions, and desires; and
- all this sometimes makes us feel a bit uncomfortable, and we even may want to do something about it—without necessarily throwing out our coveted devices or logging off indefinitely.

In this opening chapter, I invite you to explore these and related issues, using as case studies the 2010 and 2020 versions of the crowdsourced *Life in a Day* films on YouTube. These films, composed of video clips of everyday life submitted by people from every continent, offer a unique window on contemporary life in media as documented a decade apart. Using these films, we can consider the collapse of the boundaries between media and (everyday) life and the subsequent mediation of everything. Subsequently, we have a chance to engage the discussion about media in society and everyday life in terms of what people worldwide are actually doing with media—a consideration of which is (or should be) the core of studying and understanding (the) media.

Welcome to your life in media. This is not a revolution; nor is it an emergency. If it was, you would be told where the exits are! The question is not so much how to escape media or stay above it all. The key is to dive, with head and heart, into the ocean of media we all swim in. What we explore is the issue of how we got here and what we can do now that we find ourselves living in media.

### Life in a Day

In 2010, the online video platform YouTube (launched in 2005, since 2006 owned by Google) asked people around the world to chronicle a single day in their lives and share their recording. The platform received over 80,000 submissions, totaling 4,500 hours of footage, from 192 countries. The director Kevin Macdonald and executive producer Ridley Scott turned this into a ninety-minute film, *Life in a Day*, intending to show what it means to be human in the world on July 24, 2010. Ten years later, the team repeated this effort, this time receiving 324,000 videos submitted from 192 countries, documenting life in a day during the worldwide coronavirus crisis on July 25, 2020.

In several interviews about the *Life in a Day* films, the Scottish director talks about how all the contributors were generous enough to share often quite intimate moments from their lives as part of a huge, life-affirming filmmaking experiment. Macdonald also reflects on how people's relationship to filming themselves has changed over a decade, noting how camera-enabled phones were much more common in 2020 than in 2010 and how YouTube has since become a truly global video-sharing platform. The intention behind the films was to give voice to ordinary people and to show the intricacy and strangeness of the seemingly mundane in everyday life. In 2010, about 75 percent of the film's content came from people contacted through YouTube, traditional advertising, TV shows, and newspapers; the remaining 25 percent came from cameras sent out to the developing world. In 2020, everything was autonomously submitted via a dedicated website. None of the individual filmmakers whose clips ended up in the films were compensated for their effort. Reviewing *Life in a Day* for the British news site the Guardian (on February 2, 2021), Peter Bradshaw found the film to be "about everything and nothing," suggesting that the most significant change has been in the areas of technology and technical skill: "A decade on from the first film, smartphones with higher-quality cameras have probably resulted in a technically more sophisticated haul; people have better equipment and are arguably more savvy about curating their content."

Speaking online at the 2021 Sundance Film Festival in the United States, Macdonald noted how he considered both films as amazing levelers. "They really make us feel as a viewer that 'everyone is like me.' I can see they care about the same things as I do whether they are Egyptians or Congolese or Siberian. They care about the same basic things. There

Media Life 9

are very few strong, powerful parts of human life that we don't share."[2] Writing for the *New York Times* (on February 5, 2021), Chris Azzopardi considered the crowdsourced film "a call for empathy with some genuinely moving moments," a sentiment echoed throughout many comments available on the popular review-aggregation sites Rotten Tomatoes and Metacritic.[3] A poignant moment in the 2020 film is when a mother reveals that her teenage son, who appeared in a funny opening sequence of the first *Life in a Day* (where she tries to get him to wake up in the morning), has died of COVID-19—introducing him by way of turning the camera on his funerary urn. The juxtaposition of such genuinely moving moments with more ordinary, contextless snapshots of everyday life for many reviewers was either an exemplary showcase of how kaleidoscopic and confusing life can be or a reason to condemn the film for being scattershot, voyeuristic, and in the end quite conventional given the abundance of "slice of life" content now available on social media and video platforms all over the world.

There is much to say about *Life in a Day*. We can examine the film moving from clip to clip and wonder about the selection and editing process. There are important questions to be asked about the way various people—teenagers and the elderly, men and women, people from different backgrounds and different parts of the world—are represented and used in the film. There is a metaprocess of curation going in this film: people were asked to film and select from their daily lives, while a team of editors—Nse Asuquo, Mdhamiri Á Nkemi, and Sam Rice-Edwards—curated all their submissions, and Macdonald in turn edited their selections for his director's cut. In every instance, crucial decisions were made about who gets a voice, how people are framed, and what stories warrant inclusion. A critical debate furthermore involves the use of all of us as free labor by commercial corporations, specifically Alphabet, Google's American parent company; Ridley Scott's production company, RSA Films; and Flying Object, the British advertising agency hired to coordinate and promote the project.

While all these approaches are perfectly valid in media studies, and a critical reading of a text such as *Life in a Day* would most certainly be a key element in learning to be media literate, an approach to media studies for a life in media compels us to look more closely at what these films may have to say about us and our lived experience in, with, and of media. Overall, carefully considering the *Life in a Day* films is a good example of how we can reflect on our media life and that of others around the world. Using both films as a case study, it is possible to explore key insights and concerns grounding media studies for a life in media on a global level.

---

2. Geoffrey Macnab, "Sundance Interview: Kevin Macdonald on Life in a Day 2020," *Business Doc Europe*, January 28, 2021, https://businessdoceurope.com/sundance-interview-kevin-macdonald-on-life-in-a-day-2020/.

3. Chris Azzopardi, "'Life in a Day 2020' Review: A Video Diary of a Difficult Year," *New York Times*, February 5, 2021, https://www.nytimes.com/2021/02/05/movies/life-in-a-day-2020-review.html.

| | Chapter 1 |
|---|---|

> **Eight Key Insights Informing Media Studies for a Life in Media**
>
> 1. Media are pervasive and ubiquitous.
> 2. Media have long histories that are complicated.
> 3. Media raise all kinds of ethical issues.
> 4. Media are a source of profound pleasure and fun.
> 5. People seem to be increasingly comfortable with media.
> 6. People seem to be relatively confident about their media.
> 7. People seem to accept surveillance as familiar and ordinary.
> 8. People seem to have little concern about doing the work for media.

Half of these key ideas have to do with media; the other half, with life. All these issues may seem redundant, obvious perhaps, even cliché. It is exactly the banality of (life in) media that demands our time and effort to analyze, as it is in their mundane, everyday nature that media disappear and have a significant impact on society and power over our daily lives. With this understanding as a starting point, we can in fact achieve some critical distance from our media, while at the same time never forgetting how deeply we are always already entangled with media.

## Media Are Pervasive and Ubiquitous

Living in media means appreciating that media are *pervasive*—we cannot simply switch media off. Quite literally, it has become rather difficult to switch off any of our devices—especially our smartphones and smart televisions, tablets, laptop and desktop computers, and handheld and console gaming devices. Finding the actual off switch on many devices can be quite difficult sometimes. Even when switched off, many devices remain operational and connected in some way, retrieving messages, running an operating system, updating and upgrading software, checking in with cell towers and Wi-Fi hotspots to report their location and status (or that of the applications running on such devices).

Another way of considering the pervasive nature of media acknowledges how everything is mediated: all our experiences, relationships, and ways of making sense of the world around us involve and are shaped and influenced by media. This is one explanation for the rich and eclectic diversity of scenes, moments, and everyday life sequences present in the films. This does not mean life is determined by media—it just suggests that, whether we like it or not, every aspect of our lives takes place in media. Quite often this "mediation of everything" involves people actively challenging and resisting media—for example, when people demand a different version of the truth than the one received through mainstream media or when we use social media to rally for a cause. Other times we use media to make our own version of events and, in doing so, purposefully create our own reality. Alternatively, we can try to avoid media as much as possible—when going on a romantic date, for example, or when parenting about screen time. On the other hand, it is inevitable

that media play a prominent role in what we talk about, in what makes us laugh or cry, and what we find significant or trustworthy. Seen this way, embracing, resisting, and avoiding media all are part of and contribute to the pervasive nature of media in our lives.

Next to pervasiveness, living in media entails considering media in all their omnipresence—to see media as an environment rather than a series of discrete technologies. Instead of thinking of different kinds of media separately—such as newspapers, magazines, books, radio and television, films, the internet, digital games, apps on your mobile device, and so on—a life in media means that we look at media as *ubiquitous*, where not being exposed to or not using some form of media has become quite rare, and most of our media use involves multiple media simultaneously. Research around the world documenting how people go about their daily lives consistently shows how so many of us spend most of our time with media and how much of this *media exposure is concurrent*, beyond conscious deliberation. We do not "multitask" our media, as this implies deliberate effort on our part to pick and focus on specific media. We also do not generally pay much attention to all the media around us, as we are often in situations involving multiple media—on the phone, with the TV on, perhaps also paging through a magazine, maybe listening to music, walking outside with headphones on yet also surrounded by (electronic or print) billboards and signs on bus stops and park benches, as well as adverts on cars and buildings. Both *Life in a Day* films remind us of the ubiquity of media, especially when showing scenes in places far away from the hustle and bustle of cities around the world.

Living in media is inattentive rather than focused, distracted more so than task oriented. It could be argued that people are generally semi-aware of themselves and their surroundings when using media and consume media while blissfully unaware of how media work. Switching on your smartphone has become a thoughtless habit as much as putting on a dress or a pair of pants in the morning. Clicking your way around a variety of websites, apps, and software programs can be just as comforting a ritual as eating your favorite snack. Listening to music, browsing a newsfeed or someone's timeline—all of this also comes with ignoring other media, disconnecting from devices, turning away from countless other options you may have. The key is to appreciate how all the different media in your life are both unique to you—as your experience of and in them is quite personal and intimate—while this media manifold is also the same for almost everyone, awash as we are with screens, advertisements, news headlines, and seemingly endless entertainment choices.

There is a paradox at work in the way media function in our lives. The digital environment of media is exactly so irresistible to us because it makes all our interactions with it seem special. Website functionalities, viewing options, game play characteristics, and app selections are all unique to the individual media user. No one has the same experience in media. Furthermore, most adverts seem to be talking directly to you, and algorithms running in the background make sure that the next ad you see seems to coincide with what you are thinking about, and the next song appearing in your playlist is something you really feel like listening to. In media, we are all individuals. Yet . . . most people visit the same handful of websites, click on the same top search engine results, binge-watch the same shows everyone else is talking about, listen to the same selection of songs on repeat, and tune in to the same live sports broadcasts. It is quite hard to find something truly uncommon or surprising (let alone original) in the media, despite their ubiquity and potential for personalization. Seen as such, a life in media contributes to making us feel more remarkable and unique while simultaneously contributing to a more superficial, more-of-the-same kind of universal experience. This inspires a reading of *Life in a Day* as rather mundane or not particularly revealing, as the many scenes blend into a rather undifferentiated whole, which was the point of the film to begin with—to underscore what binds us as documented through personal devices. Yet at the same time, being truly exposed to the private lives of people around the world often feels like an exceptional experience. In media, you are (not) special.

## Media Have Long Histories That Are Complicated

It certainly seems as if our lives in media are particular to contemporary technologies, to the rise of digital devices and applications, and to the "right here, right now" culture of being always online. On the other hand, humankind has forever been feverishly documenting its existence, its relation to the world, and what all of this means in media—especially since the days that we traded in our nomadic ways, settled in permanent dwellings, and formed more or less stable communities. It is from this period on that we find, in gradual succession over thousands of years, increasingly widespread media, such as cave paintings, musical instruments, public forums and theaters, writing systems, scrolls, printed books, and so on. Perhaps we have always lived in media, and it is not our media that are new but rather their intimacy, pervasiveness, and ubiquity, as exemplified throughout these two films and the lives of the people featured in it.

All so-called new media invariably contain versions, elements, and uses of media that came before (more about this *remediation* of media in chapter 2). Whether it is Apple's iPhone debuting in the United States in 2007, the invention of printing on paper in ninth-century China, the start of radio broadcasting early in the twentieth century around the world (from Argentina to Sri Lanka, from England to Australia, and elsewhere), or the introduction of a fully functional mobile internet platform in Japan in 1999, all such technologies and devices have components and functionalities that came from and were inspired by older media. Touch-screen technology was invented and developed in the 1960s and 1970s in England and the United States, printing on cloth in China dates as far back as the second century, broadcasting via telephone lines was in place by 1980 in various European countries, and the first mobile phones with internet access were pioneered in Finland in the mid-1990s. Beyond specific technologies and components, the way we use newer media always contains traces of the past, as new media practices tend to follow old media habits. Television news originally consisted of presenters reading the radio news on camera. When the internet was introduced into newspaper newsrooms, journalists primarily used it to look up information they previously used books and archives for—such as addresses and phone numbers. And whenever we get our hands on a gadget—a fancy television set, a shiny new smartphone, the next-generation game console—we tend to replicate what we did with earlier versions of such devices.

Next to the technologies and ways of using media, history also plays a profound role in media content, as every story, song, film, or game contains parts or patterns from previous media. Such sampling occurs through the recurring use of certain narrative structures, closely adhering to genre conventions, copying and pasting elements of earlier media, and through the countless ways of referring to characters, memorable licks and lines, and any other building blocks of media products that came before. Often, such references are overt, such as in the sampling pioneered in modern classical music in the 1940s and 1950s and popularized in rap, hip-hop, house, and dance music. Other examples of the *remix culture* that is particular to media are mash-ups and parodies in film and video production and all the countless gifs and memes circulating online.

More subtle ways of remixing happen when different media texts refer to each other at different levels and across genres. This kind of *intertextual referencing* also occurs in the process by which we, as media audiences, make meaningful connections across the formal boundaries of specific media objects (such as songs, movies, and games). The advertising and marketing industry makes good use of this, for example, by letting well-known actors promote a product as a character of a famous film or using a popular hit song in the soundtrack for an election campaign. Such collaborations become increasingly complex and tend to involve much more than commercial cross-promotion. Global examples include media users, audiences, and fans following (and, through their engagement, collaborating with) artists like the South Korean boy band BTS and American singer-songwriter Taylor Swift—from their award-winning music and decidedly personal lyrics to their sponsored concert tours, to their advocacy against violence (where they partner with such organizations as UNICEF) and support for the lesbian, gay, bisexual, transgender, queer, and questioning (LGBTQ) community, their involvement in numerous philanthropic actions and charitable causes, and across the many brands both artists partner with (including Diet Coke, Apple Music, and Keds footwear for Swift and Puma, Samsung, and Coca-Cola for BTS). Such partnerships go beyond appearances of the artists in commercial campaigns and include becoming ambassadors and spokespersons where a brand associates itself with causes the artist advocates for and involves cocreative design on products as varied as greeting cards, fragrances, fashion apparel, footwear, accessories, and other merchandise. BTS furthermore collaborates with McDonald's with their own meal—the BTS Meal—which was released gradually around the world during 2021 with great fan engagement on social media, to such enthusiasm that some restaurants had to be shut down temporarily. The intertextual relationship between artists, advertisers, and audiences makes everyone a participant in the remix culture of our digital environment.

We could say, as the American filmmaker Kirby Ferguson stated in his 2010 online documentary series of the same title, everything is a remix (see www.everythingisaremix.info). The same goes for the style and approach of the *Life in a Day* projects, as the director deliberately remixed earlier work. In many interviews, Kevin Macdonald acknowledged how original inspiration came from the work of the English television and film director and producer Michael Apted, in particular his *Up* series of documentary films following the lives of fourteen British individuals from 1964 (when they were seven years old) to 2019. Another influence was the English documentary filmmaker Humphrey Jennings, one of the founders of the Mass Observation project, running from 1937 to the mid-1960s (and relaunched in 1981). The Mass Observation organization originally called on people around the United Kingdom via ads in newspapers to record and submit a day-to-day account of their lives in the form of a diary. Later, a panel of around five hundred volunteers regularly maintained diaries or responded to open-ended questionnaires about their everyday lives. The organization also paid investigators to anonymously record British citizens' conversations and behavior at work, on the street, and at various public occasions including public meetings and sporting and religious events—all of which led to complaints about surveillance and invasion of privacy. Another concern was the projects' lack of representativeness, given the fact that only middle-class, educated, and literate people participated. Similar concerns can be expressed about the *Life in a Day* films.

Nothing in media is necessarily new nor original—even if some new device is remarkable (as the iPhone, block printing, broadcasting, and i-mode clearly are) and films, games, and books are often wonderfully creative and innovative in their take on a particular issue. Contemporary collaborations and sponsorships between musicians and brands take their cue from the high-profile and generally mutually profitable relationship between Michael Jackson, Nike, and Pepsi Cola in the 1980s, whereas the practice of celebrity branding and endorsements goes much further back.

All of this serves to remind us that nothing of what is happening now in media is truly, uniquely, and exclusively related to our time, and it is enormously helpful to investigate the past to find answers to what we care about today. Considering media historically is crucial, as it prevents us from being blinded by all the shiny new toys introduced every year in the consumer electronics market and should stop us from becoming overly invested in the novelty of a new initiative, narrative, product, or service. A property of all media technologies is that they seem to persist in defiance of all assumptions to the contrary. Despite its century-old history, radio survives today, is in fact the prominent medium in some parts of the world (such as in local communities across the African continent), and through podcasting remains highly relevant worldwide. The same can be said for books, newspapers, and television, as well as for the various ways in which media are used in society and everyday life. Newer media therefore do not replace older media—they adopt and remix features of older media while becoming increasingly intuitive to operate and use, which in turn contributes to their ubiquity.

## Media Raise All Kinds of Ethical Issues

As media over time gradually nestle themselves in all aspects and experiences of everyday life and in the various ways in which society functions, we naturally invest much into our media—our emotions, our relationships, our memories, what we believe in, what we think the truth is, what we feel is right and wrong. At the same time, the media we surround ourselves with shape us too, as each device, each message, and each way of using media has a limited range of characteristics and actions we can take, all contributing to a specific way of understanding and comprehending the world. Our existence as human beings—and humanity as a whole—therefore cannot be seen separate from media. It follows that living in media is neither a "good" or "bad" thing for us—it is simply our reality; it is who we are and what we do. Importantly, this position does not negate critical debates about the media; nor does it dissolve us from the responsibility to carefully reflect on the role media play in society and everyday life. It is, however, really challenging to critically reflect on "the" media when our lives and our common humanity are so much conjoined with media. Perhaps the only time we see media abundantly clear is when they break down and when they fail to do what we expect them to. On the other hand, most of

us do not use media as we are intended to. We generally do not read user manuals carefully, and as we overwhelmingly use newer media in ways that we have grown accustomed to from older media, we miss out on a lot of new options and functionalities. The very fact that we all have our own idiosyncratic ways of using (and understanding) media complicates claims we can make about their supposed impact or influence. It is good to acknowledge, right from the start, that our media (like all relationships) are both particular and really complicated.

When making sense of media in society and everyday life, it is key to consider context—what is problematic for some can be helpful for others, what clearly does not work in some situations is of excellent use somewhere else. The material context of media reminds us how the technologies and devices we often take for granted are prospective for a possible future for many. The number of people around the world without access to the internet is about the same as those who do not have access to safe drinking water—about one-third of the global population. Whereas most households around the world own one or more TV sets (and pay for their television), household sizes vary sharply by region and country, with many sharing the viewing experience with six or more family members (especially in sub-Saharan Africa, North Africa, and the Middle East), while some have much more individualized access (mostly in Europe and North America). Such geographic awareness of media distribution, access, and use is important to prevent all too easy generalizations. For example, the *Life in a Day* films exclude submissions from Cuba, Iran, North Korea, Sudan, Syria, and the Crimea due to American export controls and sanctions governing the production process.

The context of media is more than just their materiality and includes people's feelings and attitudes toward media. Despite our lives in media, many people are not all that interested in going online; some fear computers or are overwhelmed if not intimidated by new technologies. This contributes to a lack of (digital) experience and skills to use complex and insufficiently user-friendly media devices. Digital literacy and media education programs are still relatively rare and often inadequate in many parts of the world, especially when it comes to adult learning and the experience of seniors. At home and in private, despite our unique ways of configuring and using media devices, people from all over the world tend look for similar things—games and entertainment, some news and information, and mostly to connect with other people (via email, chat apps, and online social networks). All these contexts—material, psychological, experiential, and differential—of how people access media combine in their impact on *digital inequality* around the world. Someone who does not have access to certain technologies is less likely to develop advanced skills and experiences; a person who does not engage in a particular way online is also less likely to engage in other ways, all which results in a less than ideal media experience. Any kind of statement about a certain consequence or implied impact of media in society immediately involves critical questions: Who is involved, and which people get excluded? What kind of uses and outcomes are we talking about? A (perspective on) life in media therefore inevitably calls for an ethical stance: What do we want from this mediated life, and how are we going to take responsibility for it? Is our digital environment and economy solely about (the extraction and monetization of) data, energy, and attention, or can we expect more from the various media industries, technology companies, and telecommunications corporations that provide us with their products and services? Ethics are a fundamental aspect of a life in media, as it centers our attention on what can or should be done as much as what is happening to us.

### Media Are a Source of Profound Pleasure and Fun

Next to these critical considerations for media studies, it is crucial that we do not forget that media primarily are a source of pleasure, joy, and fun. It is hard to watch these two films without smiling now and then, without being moved. Such affect sometimes makes us forget what we are doing with media; it makes us vulnerable to commercial exploitation and can lead us down a rabbit hole of never-ending YouTube playlists and countless hours spent on completing a quest in a digital game. Yet all these sensations, desires, and thrills are important to keep in mind when making sense of media—if only because it reminds us that our entire body plays a part in the media experience. A life in media is not just the images in our head or the thoughts we have about bits of news and information; nor is it reducible to a set of cognitive skills. It involves our innermost feelings, wants, and desires. It encompasses how we move physically through the world—preferring to travel through areas with wireless connectivity, customizing the experience of our surroundings by wearing (noise-canceling) headphones, changing our posture while we are absorbed in a screen or paging through a magazine. Our whole being is involved, body and soul, from head to toe.

---

*We have always used and relied on media to experience, express, and comprehend our humanity, and it is up to us to take responsibility for the world we want in media.*

---

In short, if we want to dig deeper into the role, meaning, and influence of media in society and everyday life in the context of our lives as lived in media, we need to be mindful of their pervasiveness and ubiquity, historicity and complexity, ethics, and pleasure. We have always used and relied on media to experience, express, and comprehend our humanity, and it is up to us to take responsibility for the world we want in media.

### People Seem to Be Increasingly Comfortable with Media

It is not always easy to own up to our entanglement with media, especially as our media often mean so much to us, and their use has become second nature. The enormous success of the worldwide *crowdsourcing* approach used by the team behind the films suggests the relative ease with which people show themselves in media. Almost no aspect of life is left out of the many submissions—from heartfelt celebrations of pure joy to painful moments of intense sorrow and everything in between. It all gets recorded, sometimes slightly edited, and shared without hesitation for all to see. When people started expressing themselves online (starting in the late 1970s, accelerating globally from the late 1980s) in newsgroups, within virtual worlds, via personal websites, blogs, and vlogs, observers were quick to note the unique and uneasy balance between mass media and self-communication. As people were talking about their personal lives, feelings, and experiences, they were sharing these stories not just with family and friends but with the world online, adding a distinct public element to otherwise private messages. Such blurring of the boundaries between public and private life in media was met with concern and inspired much debate about the nature of online communication. Fast forward well into the twenty-first century, and *mass self-communication* has become the primary way people use media, which in part explains the enormous success of the crowdsourcing approach used in the *Life in a Day* projects. All around the world, social networks and online communities are the most visited and most used places online. It is not that we do not care about privacy anymore; it is just that connecting with others, expressing ourselves, and sharing our lives for many people seems to be more essential—and, importantly, more fun—than any concerns we may have about what is done with our personal information. Furthermore, the privacy rules and controls in place at all these sites, apps, and communities are generally so dense and complex that users report feeling overwhelmed, intimidated, or just powerless to do anything about it.

### People Seem to Be Relatively Confident about Their Media

A further observation must be that many of us are apparently quite comfortable with all the (personal) media in our lives—wielding all kinds of smartphones and video cameras, effortlessly composing compelling narratives and making skillful shots, and documenting, editing, and curating our lives with abandon using a wide range of electronics, hardware, and software applications. Although in theory everyone was invited to participate, those who submitted high-quality videos are not a cross-section of the global population, and their work was furthermore selected and compiled by various editors, adapting the final rendition of the films to fit the expectations and professional standards of the producers at RSA Films. On the other hand, it is quite remarkable how even the most basic personal media devices can quite efficiently record, store, and share our experiences to the extent that these can be featured in a professional motion picture. This is not just an option available and used by those in privileged parts of the world, as global smartphone sales and market reach continue to grow on all continents. Several nonprofit organizations, such as Free Press Unlimited (based in Amsterdam, overseeing media projects in more than forty countries worldwide) and OpenUp (a civic technology organization based in Cape Town, South Africa), offer free-to-use smartphone applications—called StoryMaker and PocketReporter—with the purpose to enable and educate people to produce their own multimedia stories, specifically in places and communities underserved by existing media or without access to a free press. Community, alternative, independent, and indigenous media initiatives thrive all over the world—from Australia to Mexico, the favelas in Brazil to shantytowns in Pakistan, within neglected neighborhoods in the Netherlands as much as by rural associations in countries like Ethiopia and Uganda (see chapter 6 for more detail on how people use media to change their world). People are becoming more media literate and more comfortable in using their devices not just as users but also as makers of media.

### People Seem to Accept Surveillance as Familiar and Common

Another important inference we can make about the two editions of *Life in a Day* is how commonplace, perhaps even unremarkable surveillance is in contemporary society. Traditionally we think of surveillance as something that is happening to us—*a panoptic form of surveillance*, where the many are systematically monitored by the few (such as companies and government agencies). Although few people would be surprised to learn how widespread such types of surveillance have become, the global coronavirus crisis has accelerated the introduction and use of increasingly sophisticated digital surveillance mechanisms—including health-check applications, online vaccination passes, thermal cameras and facial-recognition software to screen for access to public places, digital contact tracing (using Bluetooth and geolocation data to track individuals and their proximity to others), and a wide variety of telemedicine (using video conferencing and other tools to provide care remotely) and e-health (platforms for electronic patient records) protocols. Numerous organizations continue to raise significant concerns about privacy, civil liberties, and the rights of already vulnerable groups in society, releasing reports and documenting instances of widespread surveillance without clear norms or ethical considerations in place (see chapter 3 for much more detail on this). Similarly, people worldwide protest and resist unbridled tracking and monitoring, sharing their uneasiness and misgivings via social media and using hashtags (such as #antisurveillance for Twitter or #instasurveillance for Instagram) to publicize their efforts and gather support. Such protest campaigns are a form of surveillance as well: *synoptic surveillance*, where the many—organizations, researchers, and networks of individual citizens—scrutinize the few.

It is remarkable that during the pandemic the *Life in a Day* project was able to collect detailed digital recordings of people's private lives, willingly shared by the tens of thousands of participants around the world. In these clips, people carefully monitor and record themselves and each other, suggesting a third form of surveillance at play: *omnoptic surveillance*, as in a life in media everyone is (or can be) monitoring everyone else, using readily and widespread available technologies such as smartphones and mobile internet connectivity. With more than half of the world's population using mobile internet and having access to smartphones, and mobile ownership and internet usage forecast to increase in the future, so will the opportunities for all kinds of surveillance. The two films showcase the everydayness of massive mutual monitoring via media, making us aware of our own participation in global systems of surveillance.

### People Seem to Have Little Concern about Doing the Work for Media

With so many people participating in the *Life in a Day* experiment by submitting clips of their carefully recorded and edited life, the project had a wealth of material to curate. Every minute of every day, hundreds of hours of video are uploaded to YouTube alone, all becoming the exclusive property of Google. This does not consider the additional millions of posts, apps, emails, tweets, clicks, and scrolls and all the other images, video, audio, text, and data people share on any of the hundreds, if not thousands, of online platforms and social networks around the world. With over half of the world's population using the internet and social media being among the most visited and used places online, all of us have become part of a commercial enterprise that spans the globe. The part we play is that of providing free labor: we do the hard work—in terms of our activities and engagement, sharing of personal information, and leaving a digital shadow of every place we visit online—that provides value to the platforms offering us their services, generally for free. *Life in a Day* is a film that attracts attention and online traffic, thereby generating value for YouTube (and its parent company, Google), while every clip is provided for free.

The phenomenon of free labor is not new—it forms the basis of the advertising business model of mass media, where audiences (for books and newspapers, shows on radio and television, and titles in magazines) are generated to be sold to advertisers. Back in 1973, the American video artists Richard Serra and Carlota Fey Schoolman created a short film titled *Television Delivers People*, consisting of a scrolling text while generic music plays in the

background. A repeated memorable line in the film reads: "You are the product of television," a sentiment that fuels much of the critique against media as an industry. A similar argument is made in the docudrama *The Social Dilemma*, directed by Jeff Orlowski and released on Netflix on September 9, 2020, to record viewing numbers around the world. *The Social Dilemma* focuses on how social media companies like Facebook, Google, and Twitter use algorithms to encourage people to spend all their online time on their platforms, in the process collecting enormous amounts of personal data to be repackaged and sold to advertisers. In 1973, the product of television was the viewing audience. Today, it is all of us, clicking and swiping our way through the digital environment, recording, archiving, and sharing our lives online while making money for the platforms and services involved.

## Media Life

Of course, there can be many more topics for discussion relevant for media studies distilled from the two *Life in a Day* films. The issues introduced here address the main reasons why studying the media is both important and almost impossible. Important, because our experiences and ways of making sense of the world are deeply entangled with media, suggesting that they have considerable power and influence over our lives. Impossible, because how media saturate our lives makes it very difficult to reflect on media at a distance. The examples and cases offered here (and throughout this book) serve to unpack phenomena typical of a life in media (or *media life*), showing how we can perhaps reclaim analytical potential for study and critical reflection in a context where media underpin the experience of everyday life.

Looking back at the various issues, concerns, and debates about media life raised in this opening chapter, it is possible to identify some fundamental properties of media studies for a life in media. When we study the media (and the digital environment in particular), we must start from their sheer pervasiveness and ubiquity. This comes into clear focus when

- examining our own media use, finding how every single aspect of our lives is enmeshed in media;

- acknowledging how we feel about (our) media and how much we enjoy and love media;

- considering the research on how people around the world spend their time, showing that we use media more than doing anything else in life (including sleeping); and

- exploring the "black box" of technology and production techniques that make media work, seeing how they are designed to keep us engaged and to monetize that engagement.

Although it seems obvious to appreciate the pervasiveness and ubiquity of media in our lives, it needs constant reminding as the digital environment has become so natural to us. For sure, we need to thoroughly question any causal claims about the effects media have on us, yet it is safe to say that the power of media exponentially increases when we become oblivious to their omnipresence. In other words, it takes real effort, all the time, to bring forth the media around us. When doing so, we are not just aware of all the different ways in which media are really involved in almost anything we do and in what happens in society but also how every act of using media changes our experience and understanding of the world around us.

The second analytical step we can take is to deal with their past, present, and (possible) futures while recognizing how complex media are. We can establish this by

- looking beyond media as shiny new toys to see what people are actually doing with (their) media;
- documenting how all media are versions of each other, both in terms of their design and content and relating to the various ways in which people use media;
- acknowledging that media, if anything because of their combined everywhereness and everydayness, are hard to see and think about critically; and
- respecting the fact that it is therefore never easy to use, understand, and make sense of media—neither as a scholar or student of media nor in the practice of everyday life.

Media are so much part of everything we do, from such an early age onward, that it is sometimes easy to forget how little we know about how exactly they operate, how complicated all their functionalities are, how differently people engage and make sense of media, and how intricate the story lines and narrative structures are of every single media message—whether it is a chat message, a novel, a television show, a motion picture, a level in a digital game, or just a single photograph. All media are amazing and, in many ways, quite magical in how they convey messages and meaning, how they produce certain ways of seeing the world, and how they invite (or limit) possible understandings. At the same time, media are inherently messy: they tend to break or do not do what they are supposed to do without clear cause; the way people understand them can differ widely depending on even the slightest variations in how media operate and function; and many of the effects of media are unintentional, counterintuitive, and unpredictable.

A final step in the study of media involves developing and taking up ethical positions vis-à-vis the phenomena and experiences people have in media, based on research past and present. Recurring themes and issues here include

- all the different ways in which people use, accommodate, or resist the media;
- struggles over access and participation in the media;
- which people, values, and ideas get represented and promoted via media and which identities and voices are marginalized or unheard;
- the ownership, business models, and control of the media (including the technology and telecommunications sectors);

- how the key institutions of society, such as politics and the state, security forces and the police, health care, education and schools, families and households, and the media as an industry function under the influence of media;
- how the media, telecommunications, and technology sectors can or should be regulated; and
- what the role is of (increasing) automation in the inner workings of media devices and industries, specifically when it comes to processes governed by algorithms (i.e., automated instructions) or more complex sets of algorithms working largely independent of human intervention (as in the case of applications involving machine learning and artificial intelligence).

Media are important to study and understand as, in media, we are drawn into all kinds of struggles—over representation, identity, access, community, belonging, and so on. In a media life, it matters who gets to use the media and who does not, what kind of values and ideas circulate in the media and which are ignored, sidelined, or marginalized, and how people and institutions get represented and present themselves in the media.

What is of additional significance here is the role of nonhuman actors in the media: the hardware, software, technological infrastructure, and programming that makes everything in our digital environment work. In the media industries, algorithms are used for producing such media content as written news stories and adverts, generating subtitles and any other additional information, filtering out specific content (like hate speech), and making recommendations and suggestions, up to and including advanced uses of such technologies to enhance or replace humans (e.g., news anchors, actors in film and television, and the people who work in media manufacturing and production; see chapters 7 and 8) with digital copies. Society's key institutions—including the financial sector, health care and medicine, police and security forces, and the government—rely on algorithms, automation, and machine-to-machine communication to make decisions that affect people in their everyday life. All of this suggests that the study of our life in media goes well beyond people's concurrent exposure to all kinds of media and includes the various technological and computational processes that run in the background, shaping what kind of world we visit when we enter the media.

In the next chapter, we will carefully unpack what it is that we are talking about when we talk about media. We will see that media are not just the devices we surround ourselves with, not just the things we do with all these technologies, and not just the ways in which we organize and arrange our social environment, including the industries that media are part of. Media are all of this.

# 2

## Your Life

> Media are the devices and channels we use to communicate—including what we communicate, the industries that are in the business of communication, and how all of this fits into our everyday lives.

To understand and ask meaningful questions about what it means to live in media, it is necessary first to thoroughly define media. When people talk about media, they usually refer to information carriers—such as newspapers, magazines, television, radio, computers, tablets, and mobile phones. What such a definition does is to reduce media to devices (or channels that transport information) that are external to us—which we can use to communicate with each other; to inform, entertain, and educate ourselves; and to connect with others and the world around us. It is a promising first step toward a definition of media. What it lacks is a critical awareness of how the relations we have with our devices are much more intimate and personal than simply switching them on or off to communicate, to be informed, or to be entertained. A definition of media based on technological specificities also ignores the fact that we generally are not all that aware of how we use media, and most of us do not understand how media work. In other words, a definition of media strictly in terms of devices and technologies runs the risk of rendering media imperceptible—or perhaps only observable when media reveal their "thingness," that is, when they break down.

Another way of defining media is to look at content: What do we say in the messages we send to each other? How do people respond to a particular news item? What kind of advertisement persuades us to make a purchasing decision? What is a director trying to achieve with their new film? How did the designers decide to depict the female characters in their game? Considering the content of media allows us to dig deeper into what kind of values and ideas they convey, whose voices are heard and who gets excluded, and where the biases and assumptions of media makers lie, making us aware of the *polysemy* of mediated communication and exploring all the different ways we can make sense of the media that surround us. If we combine a focus on media as content with a definition of media as devices, we get to appreciate that the same message can look and feel quite different when seen on

another device or when created or accessed through a particular platform and interface. A music clip on TikTok is not the same nor does it have the same meaning and impact as a message via WhatsApp, a video on YouTube, or a scene from a big-budget movie—even though all of these media can transmit the same message. Having a loved one break up via a phone call, a text message, or a Facebook status update ("It's complicated") affects us differently. The materiality of media matters, just as media content does, and much can be gained if we consider these two aspects of media together when we define media.

A complication in defining media in terms of their materiality and content is the notion that media are felt and experienced differently for everyone. For some, a scary Japanese horror film that pushes the boundaries of man-machine relations—like director Shinya Tsukamoto's 1989 cult hit *Tetsuo*, in which people turn into metal killing machines—is the ultimate form of relaxation. Someone else (like me) would rather take refuge in an adventurous computer game like *Heroes of Might and Magic II* (published in 1996 by the now defunct 3DO Company), a turn-based strategy game filled with angels, unicorns, and castles. Others would turn to the news, to their favorite book, to pornography, or to scrolling through social media updates, listening to a specific playlist on a music streaming service, or tapping away happily on their phone with immensely popular games like *PlayerUnknown's Battlegrounds* (PUBG; published in 2017 by the South Korean video game company Bluehole) and *Honor of Kings* (published in 2015 by the Chinese media giant Tencent). All of this has to do with taste, with what we are used to, with our social context, with access to particular technologies and having the ability to use them, with different levels of media literacy, and so forth.

Media are mixed with our everyday life and social context to such an extent that it is exceedingly difficult to divine precisely which device carries what kind of information, what is communicated, and what people do with all this. A fully operational definition of media must therefore take into account the various ways in which media are integrated into our lives: how we fit routines and rituals around them, both as individuals and institutions; how media, even if we are not using them, continue to shape our conversations, affect what we do, and provide a meaningful backdrop to our daily lives. The downside of such an expansive definition of media is that it can be problematic to put to practical use. On the other hand, reducing one's definition to a single device, industry, or piece of content runs the risk of making any analyses of the role of media in someone's life impossible to sustain. A definition of media must be both precise and applicable to a variety of contexts. It must consider the materiality of media, the various ways in which people use media, and how media as devices, channels, and uses fit into everyday life. Whenever we talk about media, we are therefore talking about three things at once:

1. Media as *artifacts* or devices, incorporating distinct devices and technologies

2. Media as *activities* or practices, considering all the different ways of using media

3. Media as different social *arrangements* and organizational forms in society and in everyday life

In this chapter (as in the rest of this book), we deploy these three dimensions of the definition of media—as artifacts, activities, and (social) arrangements—to explore how media shape and influence people's daily lives. The focus in each instance of media is on an evolutionary perspective: showing how different media evolve, how our uses and behaviors unfold in the process, and what we can extrapolate from these developments in terms of what to expect of our future life in media. The combination of these three approaches to a definition of media enriches our understanding and, most of all, makes us mindful of the different ways in which we can appreciate and understand media in our life.

## Media as Artifacts

The history of media as artifacts suggests a gradual convergence of formerly distinct devices and technologies, increased user control and personalization, as well as a development toward ubiquitous computing, rendering media increasingly intuitive and imperceptible.

When we consider media as artifacts, we have to look carefully at our devices—whether digital or analog, used for work or just for fun, both from the outside and on the inside. A material perspective on media additionally calls for an exploration of their family history: where all the constituent parts came from, how different *functionalities* and *affordances* emerged, and when different people, companies, and institutions in different parts of the world introduced us to the technologies that play such a profound role in the world. When doing so, we first realize that media history is evolutionary and nonlinear. Second, as people become increasingly intimate and affectionate about (and with) their media, devices converge and become more complex, turning into magical black boxes. Finally, our media even seem to disappear as distinct technologies. At the same time, and perhaps somewhat paradoxically, an exploration of some of the most common and widely used media artifacts—such as the remote control, the video recorder, the joystick, the computer mouse, the mobile phone, and the computer interface—leads to the conclusion that we, as individual users, throughout history are increasingly put in control of our media experience. This apparent tension between intuitive use and deliberate choice—between disappearance and reappearance—is at the core of any definition and discussion of media.

All media have long material and social histories. This can be hard to see, blinded as we sometimes are by every new technological invention and every subsequent iteration of an exciting device. For example, the introduction of Apple's iPhone in 2007 seemed like a stunning revolution compared to the standard key phones available up to that time, but the device was an updated version of the Simon from 1994: a mobile phone with a touch screen, offering various applications (such as a calendar, address book, and notebook) and featuring mobile internet access, which was patented and designed by the American technology company IBM and manufactured by the Japanese electronics company Mitsubishi Electric. The Simon was not revolutionary either, as it combined two existing artifacts: a cellular phone and a personal digital assistant. In fact, technologies for the touch screen, a mobile communication protocol, and the internet have been around since the 1960s and were imagined and conceptualized well before that time. Even Apple itself came up with a concept prototype of a tablet computer featuring a gesture-based interface and a virtual assistant decades earlier (in 1987), calling it the Knowledge Navigator. Combining this conceptual device with the successful iPhone led to the announcement of the iPad in 2010—a tablet computer envisioned by twentieth-century science fiction writers, such as the Polish author Stanislaw Lem and British futurist Arthur C. Clarke, and developed in many different versions by various companies in Japan, South Korea, the United States, and the United Kingdom since the mid-1980s. The iterative development of media toward ever-greater complexity suggests an evolutionary rather than revolutionary history that is indicative of all media.

The process whereby each medium contains elements of previous media while combining these and adding new options and features is called *remediation*. Especially with the current generation of so-called smart devices, older media get swallowed up whole, becoming part of such digitally connected technologies as televisions, tablets, and phones. It would be logical to expect that this development will eventually lead to one superior kind of device that combines all the elements and functionalities of the media we like to use. One device to rule them all . . . What is certain is that the companies that make and market all kinds of consumer electronics keep trying to sell us their equipment with exactly that promise. Smart televisions, fully integrated game consoles, advanced new phones, any and all wireless technologies—they are pushed out and promoted as a solution to the problem of too many appliances, cables, plugs, and remote controls in the home. This is a *black box fallacy*, as the number of devices in an average household is more likely to increase every year, while people also hold on to the media they love. In many places around the world, various media are shared across families, groups, and communities (based on necessity, motivation, and competence), suggesting that the existence of many different devices is not a problem but rather a solution. Here, too, the hand of evolution can be seen, since Charles Darwin's theory explains not only that all species are related and gradually change over time but that this process does not develop neatly and linearly.

In terms of history, the road forks endlessly, making the future increasingly complex and manifold. It is therefore necessary to recognize that media, as devices, are elaborate and difficult to master—if anything because the underlying technologies change all the time. A funny side effect of this complex material history of media is how our bodies also evolve to adapt to new ways of operating media. Consider for example making a landline phone call on a key phone: pressing a receiver button, putting your finger into a rotary dial and twisting it around many times, waiting patiently for a connection to be made. Such physical gestures are hard to imagine for those who have grown up with touch screen devices, for whom tapping and swiping is second nature. Every device comes with its own embodied routines that become familiar and even inspire feelings of nostalgia, such as winding an audio cassette with a pencil or splicing a videotape with a razor blade.

The gradual development of media is messy rather than revolutionary. What appears above all is that every new information carrier—from scrolls, pamphlets, and books to newspapers and glossy magazines, via radio and television or personal computers, smartphones, and tablets, to the cloud—has managed to conquer a permanent place in our living environment more quickly throughout history. It has taken each new medium less time to be adopted and embraced by more than half of households in any given society. Remediation to some extent explains this speeding up of media, as every new medium has characteristics that we already know from the devices we used before so that it becomes more plausible and easier to try out the next shiny new toy. The messiness of media history is also at work when it comes to the fallacy of thinking that newer media are somehow "better" than older media simply by virtue of being new. Video standards such as VHS, operating systems such as Windows, and screen resolutions such as Ultra HD are not necessarily better than what was already available before. In fact, new technologies often cause all kinds of sentimental reactions in media users—in part because an older technology (V2000 or Betamax, a previous version of Windows, or just Full HD) worked better, felt more comfortable, and was simpler to use. Such warm feelings stem from our affectionate relations with media, as we bond strongly to certain devices (hardware as well as software), especially those that we used or needed in an earlier phase of our life. Nearly every family has a box somewhere—usually in the attic or stored away in a basement—sprawling with old cables, recording equipment, tapes, adapters, and other more or less obsolete media that people are genuinely fond of. This is because media connect us with other people, with communities and shared narratives, with certain places and events, and with memories that are meaningful to us, offering us something to hold on to.

We associate media with some of the most intense emotions in our lives. Couples identify their love through a particular tune that played when they met. Lovers draw on various media to express their adoration, when for example one lover says to another, "You are the Jamal to my Latika," referencing the amorous couple in the film *Slumdog Millionaire* (2008), or "the Ennis to my Jack" in a reference to the main characters of the romantic drama *Brokeback Mountain* (2005). Partners furthermore tend to carefully maintain some sort of archive of the first messages that were exchanged (whether through a dating app or with handwritten love letters). Likewise, people grow quite attached to their phones, as much as many of us develop meaningful fandoms of a certain film franchise or television series, a musical genre, or any other media. Throughout media history, these affectionate relations with and through media have grown stronger, inspiring increasingly reciprocal intimacies. In the heyday of broadcast media and cinema, people would sometimes develop strong feelings for certain characters and celebrities in the media. These relationships were rather unidirectional, of course. When social media came around, such connections intensified, as we can now follow, befriend, like, and favorite celebrities—and sometimes, albeit rarely, they respond. I remember the pleasant surprise one day when my students alerted me to the fact that U.S. president Barack Obama was following me on Twitter. Although that is fun, usually such high-profile accounts are managed by professional public relations professionals. Lesser-known media figures—such as journalists, musicians, practitioners in film and television, and especially social media influencers—tend to put a lot of work into maintaining profiles across several online social networks, promoting their work, interacting with audiences, and managing their persona. Many media scholars do this hard work of managing and maintaining relations (with colleagues, students, and the wider audience), too. This kind of *relational labor* builds on and gets amplified by the progressively intimate connection people feel with media and through media with each other.

Our love for media tends to grow stronger the longer we use any particular medium. As a result, such an artifact begins to fulfill an almost structural role in life, and it becomes harder and harder to imagine that you could do without a particular device. This explains, for example, the fact that today newspapers are still printed and enjoyed by so many people, while for the daily news there are so many other offerings online. For sure, many people find it important to follow the news. But the newspaper as a medium has less and less to do with that—for most readers, the printed paper provides an anchor in everyday life, just as the daily scheduled evening news on television can fulfill such a role.

The affective, intimate, and personal connections we have in and through media increase the likelihood of us becoming quite attached to (our) media, embracing newer versions while keeping the older ones around. The same goes for the content of the media—as every new piece of music, every film, and every game contains components, characters, story lines, or other references to older media. Even when media as technologies and content are constantly changing, they also stay the same and can even be said to be nostalgic in some way. What is more, as consumers of media and generally eager to find something new and fun, people keep coming back to the media of their youth, of their early adult years. To rephrase it: in how we consume, use, make, and feel about media, we keep media timeless or, at the very least, complicate any notion of a straightforward linear history of media. All of this contributes to the enormous and ever-growing popularity of media, reminding us how the distinction between "old" versus "new" media is often not particularly useful nor helpful in explaining the role of media, which makes it harder for us to stay mindful of all the media artifacts swirling around us at any given moment.

Assessing the long history of media as artifacts, something fascinating is going on: they seem to be disappearing. Consider, for example, the two most common consumer electronics on our planet: the television and the telephone. Virtually all households in the developed world own a TV set, while a majority owns at least one set in developing countries. About half of the world's population owns a mobile phone, and many more use mobile devices. Until the early 1990s, televisions around the world remained about the same size, with an average screen diameter of about thirty centimeters. From 1992, the Dutch

electronics firm Philips introduced wide-screen televisions to the European market, and not much later the first (flat) plasma televisions appeared, resulting in an ever-increasing screen size. At the end of the nineties, the diameter could increase to over one meter. People owning such large screens at the time reported feeling the viewing practice to be more intense and more "real," almost like a nonmediated experience. Manufacturers subsequently switched to somewhat more energy-efficient and cheaper LED televisions and, since the early 2000s, introduced a series of variations on this technology. With each subsequent step, the average available screen size of the TV screen increased further—with some brands launching slightly curved screens whose main purpose is to make the viewing experience even more immersive. The average diameter increased further to almost one and a half meters, with some models available up to well over three meters.

People around the world have quickly become accustomed to a huge screen in their living environment—whether at home, a neighborhood bar, in the classroom, a nearby community center. Contemporary television almost completely consumes the viewer. Additionally (and consequently), TV makers pull out all the stops to transport us even more: deploying canny special effects, surround sound, complex story lines that often run over into other media (such as dedicated websites and apps), clever narrative structures, audience feedback options (including live broadcast voting), and social media integration (e.g., stimulating additional conversation and engagement online via show-specific hashtags). We are, as it were, wrapped up in media, and the distinctive thingness of television disappears into an enveloping experience of image, sound, story, interaction, and participation. The device also quite literally disappears, as people replace televisions much faster than they used to.

The telephone has followed a somewhat parallel development, changing in size over time, integrating other technologies and functionalities, and steadily disappearing from our active awareness of the device as it becomes a truly universal apparatus. When the landline telephone with a rotary dial was introduced in the late nineteenth century, becoming a staple of most households by the mid-twentieth century, the device usually could be found in a dedicated spot in the middle-class home—such as on a small table in the hallway. Calling someone would require memorizing their phone number and rotating the dial for every digit. Even when the dial got replaced by a push-button system (from the 1960s), people still had to be quite deliberate in making calls—for example, having to remember phone numbers. Until the end of the 1990s, most people owned a fixed telephone, as mobile telephony only broke through with the rapid spread of GSM (global system for mobile communications) networks and the rollout of fast mobile internet from the early twenty-first century onward. These mobile telephones gradually became smaller and lighter—the former American telecommunications company Motorola's first mobile phone, the DynaTAC 8000X (introduced to the market in 1984 with an astronomical price tag), weighed nearly a kilo and was referred to as "the brick" by fans and critics alike. The average weight quickly dropped to around two hundred grams for the current generation of smartphones. During this time, the devices did not just get lighter, smaller, and more affordable—they also absorbed other media gadgets, components, and capabilities in a process of extreme remediation. The contemporary smartphone is a composite product containing the most popular consumer electronics of the twentieth century: the telephone, newspapers and magazines, books, a television, radio, remote control, computer mouse, game console, and joystick all in one. In fact, we can no longer call the mobile phone a "telephone" as making calls is one of the things people are perhaps least likely to do with the device. The particular telephone, taking up space at a dedicated place in the home, has disappeared, making way for a warm, thin black screen nestled close to our body and incessantly pinging with the promise of a new message, experience, or encounter. Some people say, "My phone is my life" (especially when they just lost the device), which may be a little dramatic but can be seen as indicative of the all-encompassing experience mobile media and communications have come to provide.

It is striking to notice how media as standalone machines and technologies, each with a specific functionality—such as reading, watching, calling, or listening—are gradually disappearing, even though their existence as devices continues. This is often a literal disappearance, because in the process of remediation they are partially or wholly absorbed by other, newer media. Specific applications of media also become more or less invisible—consider, for example, the keyboard on a smartphone, which remains as a haptic illusion, or how our bodies have taken over the process of a remote control and joystick by means of motion sensors built into TV sets and game consoles. With all this device convergence, it is exceedingly difficult to understand exactly what is going on inside. For example, the computing power of a modern smartphone is millions of times greater than the mainframe that controlled flights to the moon in the 1960s, and even the best computers of the 1990s are nowhere near a brand-new smartphone in terms of processor speed. As modern media are integrated with digital technologies, their already imperceptible operations become altogether opaque.

The touch screen, as the common interface of contemporary mobile media such as smartphones, tablets, and laptop computers, is another good example of media attempting to fade into the background of everyday life. An interface is the program that links our actions to specific functions of a digital device. It conjures specific ways of handling devices that have consequences for how we appropriate and understand media. The interface is a bridge between humans (sometimes referred to as "wetware"), computer hardware, and software and therefore a crucial element in our use and making sense of (digital) media.

The original way to get computers to perform certain functions was via a *command line interface* (CLI), where you had to type in specific commands on a screen to get the machine to do anything. It required some understanding of computer programming and, above all, purposiveness—just clicking around made no sense because nothing would happen. The computer literally had to be instructed what to do: go there, open that folder, run this program. The next evolutionary step was the *graphic user interface* (GUI)—popularized with the first Apple Macintosh computer in 1984 and the first Windows computer in 1985—where you can activate all kinds of functions and options by clicking on small images and icons. This made our use of digital media a bit more chaotic and unpredictable, as the rules for using media via the GUI are less strict. With the introduction of the GUI, computers also moved from places of work into people's homes. Although different icons—a trashcan to delete files, a calendar to see the date and time, a camera roll to play different types of video—referred to dedicated functions, it was possible for people to customize the look and appearance of it all, so everyone organized their desktop differently. We do this all the time—changing icons, backgrounds, and folder structures as if we are decorating our home. Replacing written commands with visual metaphors thus contributed to making the functions that a medium such as a computer fulfilled both more abstract and more personal.

The touch screen of a smartphone, tablet, or laptop computer is an example of the next generation of a so-called *natural user interface* (NUI), designed around the idea that we should be able to use all kinds of media interchangeably without being immediately aware of the various technologies or devices involved. A first step in this direction—of what the computer engineering community refers to as the "calm" technology of *ubiquitous computing*—is the emergence of computers with sensors that respond to our hand and body movements. The next step is to integrate these types of devices wirelessly with all objects in our environment: walls, desks, windows, lamps, refrigerators, streetlamps, bicycles, and so on. The third phase of predictions about the natural user interface is to turn the human body itself into an interface—for example, through the use of biocompatible pieces of electric skin, epidermal electronics (applying skin-friendly wearable patches), and digital tattoos. These wireless devices are linked to the internet, providing us, the companies that provide these services, and anyone else who has access with endless information about who we are, how we are (monitoring our heartbeat and sleep patterns, for instance), where we are, and what we do. All of this happens while making our experiences in media more intuitive, personal, and seamless.

In this gradual development of interfaces, the same process is at work that also applies to the history of televisions and (mobile) telephony: media slowly but surely disappear from our view, as artifacts settle in the background and become deeply entangled within the living environment, working everywhere around us and trying to anticipate our actions. Media become, as it were, invisible. When Mark Zuckerberg announced in October 2021 that his company was "going to be metaverse-first, not Facebook-first,"[1] and rebranded Facebook as Meta, he was in fact hoping to deliver on the promise of ubiquitous computing introduced in the 1980s. It could be argued that worldwide computer industries, even though various companies are in fierce competition with each other, are

---

1. Mark Zuckerberg, "Founder's Letter, 2021," Facebook, October 28, 2021, https://about.fb.com/news/2021/10/founders-letter/.

unified in their quest to develop technologies that disappear, weaving themselves into the fabric of everyday life until they are indistinguishable from it. The metaverse idea of a seamlessly integrated virtual or augmented reality experience is simply an extension of the historical trend in all media artifacts, becoming so embedded, fitting, and natural that we use media without even thinking about it. While this may be the grand vision and design expectation of many within the technology industry, it is safe to say that such a future will neither be untroubled nor uniform, will generally not work the way it is envisioned, and will affect people in different ways dependent on context. What does seem rather straightforward considering the genealogy of media as artifacts is the gradual disappearance of artifacts.

Media impose themselves on us in all kinds of ways, claiming a prominent place in all aspects of our lives, and in the operation of any social institution to such an extent that people can't stop talking about media. Their existence as technologies requires constant care and maintenance, they are upgraded and reinvented at an accelerating pace, and their ubiquity means that more and more processes come to rely on their internal (and invisible) operations: shopping, learning, keeping in touch with friends and family, maintaining relationships, working, policing, governing, and so on. Now that we have become less aware of media in our digital environment, the question is how great their influence exactly is—since what is invisible to us usually has the most profound consequences for our behavior.

**Media as Activities**

Media as what we do comprise all the activities that make society and everyday life work. Media use in a digital environment has become synonymous with media making, drawing us even further in. In the process, our active awareness of media activities disappears, as media envelop and enfold our lives.

In addition to media as artifacts, it is important to consider what we all do with these devices to get to a comprehensive definition of media. The activities involving media are just as diverse and complex as media are. First of all, let us have a look at the time we spend on media. Although the results of time use research differ per method used, it can generally be concluded that people around the world on average spend most of their time with media: at least eight hours per day. That number does not differ much from our media time five, ten, or even twenty years ago, nor does this average differ wildly from one part of the world to the next. Every year people acquire more media than the year before— mainly televisions, mobile phones and tablets, and game consoles. If the sale of these types of devices is slightly less one year, this is not due to our saturated desires but

because there is no major international sporting event planned that year or because it is a year in which, for example, the game industry is in a period of technological transition (from one generation console to the next), when a major new hardware or software standard is on the way.

    The typical household, in Iran as much as in Mexico, in Malaysia as well as Denmark, slowly but surely fills up with media. Internationally, it is not uncommon for an average of two televisions, two computers or laptops, two tablets, and two smartphones to circulate in a household, next to a wide array of books, magazines, and the odd newspaper. Oftentimes these types of devices lead different lives in a family, with a computer migrating, for example, from an office-related function on a working parent's desk to a game function in a child's room to a backup for family photos and financial documents stored in a basement or attic. Throughout the life cycle of media, devices and technologies acquire new meanings through our use, and we continually assign new feelings, memories, and functions to them. It is possible to argue that our media, just like all the members of the household, grow up together and in that process never remain completely the same—but also do not fundamentally change. This increasingly intimate and ever-evolving relationship between people and their media as growing out of everyday use suggests that media are *charismatic technologies*, in that they contribute to processes of personal transformation, identity formation, and expression. A good example of a smartphone as a charismatic device is how it moves through a typical household or community, changing primary purpose and identity as it is handed from one person to the next:

- the phone starts its life as a business purchase, to be used by a parent for work;
- after a while, the phone is replaced by a newer model, at which time it gets passed on to one of the children;
- the child primarily uses the phone for games, and as she grows up, it becomes more important as a way to connect to her peers;
- soon she acquires a phone of her own, and the device moves on to the youngest family member;
- as phones accumulate in a household, they also get shared with other members of the family or community, each time getting new affordances and meanings;
- this cycle continues until the phone is either recycled or stored in a desk drawer for future reference.

It is perhaps somewhat surprising that the time people spend on media is not only quite similar around the world, but it also remains relatively constant. This similarity is also paramount in what people do with media, as all of us primarily turn to our media for entertainment, social networking, or just to kill some time—regardless of whether we live in China, Brazil, Scotland, South Africa, the United States, or almost anywhere else in the world. Digital inequalities persist, and it would be a mistake to ignore local, cultural, and material specificities when making claims about the role and impact of "the" media anywhere in the world based only on your own situation and frame of reference. However, simply because someone is (economically) poor or lives in a so-called developing country does not mean they do not have access to newer media, do not want to play games, do not find love online, or do not watch fun movies. In every society, those who tend to be already marginalized in society experience different forms of digital exclusion. Furthermore, a distinct aspect of digital inequality can be the deliberate choice not to engage with some kinds of media.

If we expand time spent with digital media to include all media (including press and broadcast), people's media use today is perhaps not all that different from several decades ago. It is not the case that we suddenly spend more hours using media because of the introduction of smartphones or the popularity of social media. Perhaps the most striking thing about our "mediatime" is how much we forget that we use media. This is abundantly clear when considering the validity of people's self-reported media use, as generally measured via (phone or online) surveys or asking people to keep diaries of their time spent throughout the day. Research consistently shows that self-reports are not an accurate reflection of actual media use—in contrast to, for example, as measured through direct observation or digital logging. In the past, people tended to grossly underestimate their media use, whereas in contemporary studies people often assume they spend more time with media than they actually do.

The key to understanding media in terms of what we do with media is first to appreciate that we generally engage with media without much deliberation, not giving much thought to their omnipresence and remaining unfazed about the *supersaturation* of media and mediated messages in our environment. Second, the discrepancy between the mediatime we think we spend and the many hours we actually spend with media can partly be explained by our *concurrent exposure to media*, in that the majority of mediatime consists of being exposed to multiple devices at the same time. For example, a television can be on in the background while we thoughtlessly flip through a newspaper or magazine and every now and then grab our smartphone for a message or to check a notification from an app. As a consequence, it becomes almost impossible to remember, let alone distinguish between, all the different media in our environment at any given moment.

The aforementioned evolution of media as artifacts also helps to explain our obliviousness regarding mediatime, as media become increasingly successful in melding into the background of our lives—making us less aware of what we are specifically doing with media. Different media blend together in a single device, making it a lot harder to figure out exactly what we are doing with it. A computer or laptop does not excel in unambiguous behavior either: people often have multiple tabs, documents, and programs open, such as a text document, a game, and a browser for email, a search engine, and an online community. This does not alter the fact that a radio, stereo, or television is often on in the background and that there are also all kinds of papers lying around. Since we are constantly shuffling back and forth between all these media, mindlessly assembling and reassembling a meaningful mosaic of media, messages, and meanings, it is difficult to keep track of exactly what we are doing.

By defining media in terms of what we do with media, a significant expansion of the verbs that apply to a description of our media use occurs. Until the end of the last century, there were generally only a handful of verbs for media as activities: *reading, listening, watching*, and *calling*. Gradually, some new words could be added: *paging, tuning, zapping, tapping, button pushing*, and *pressing* or *clicking* (on a keyboard, computer mouse, remote control, or joystick). Especially since the start of this century, the number of words we need to map media activities increased exponentially, each with its own function, rhythm, and context: *checking, scanning, searching, linking, sharing, liking, recommending, commenting, up voting* and *down voting, uploading* and *downloading, posting, swiping, snapping, chatting, emailing, streaming, spamming, trolling*, and so on. This proliferation of verbs stems in part from advancing insight and nuance among researchers studying these types of activities, but it is also due to a fundamental shift in how we engage our bodies when using media.

Where in the past we consumed media by literally sitting in front of it—paging through a newspaper or watching a television program, for example—we tend to use newer media while leaning forward, working hard to keep track of things happening on the screen in front of us (whether in our hands, on our lap, or located further away in the room). As most media become part of an always online digital environment, we have gotten used to them constantly asking for our attention through notifications and updates. Even when we watch television or listen to music, we are often still bent over to keep an eye on a smartphone. Our body is actively involved in media use, and not only through touch screens and exposure to multiple media at the same time. Sometimes the body is involved from head to toe when we hop around in front of a screen via motion sensors to play the popular *Just Dance* game on the Nintendo Wii (or *Just Dance Now* on mobile devices) like my goddaughter and I used to do, despite my loud protests that this series of games never included any extreme metal music. As we grow accustomed to using specific media, our bodies also grow used to them, taking on specific shapes and quite literally wrapping themselves around media. Consider, for example, the slouching position for watching television, the smartphone slump, leaning forward and into the screen while playing a first-person shooter game, the way we stretch our arm to check messages on a mobile phone under a table during a meeting, a jump scare in a horror movie, and so forth. It is a reminder to always consider the body when (making sense of) using media, as it is intimately involved in all our media activities.

All of the many hours people spend using media consist of a bewildering variety of activities, often including a mix between consuming and producing information. On the one hand, we consume a large number of media products every day: we install, appropriate, and customize a wide range of gadgets and technologies for use at home or on and around our bodies. In the process, people domesticate such devices—finding a place for them, using them to connect to other people, developing specific routines and rituals around them. We binge-watch television series (which in turn can lead to binge bonding as a shared love for a series brings people together); broadcast live sports and popular events continue to attract massive audiences; we spend countless hours playing mobile games on the smartphone, and hundreds of millions of people around the world still regularly buy a newspaper or magazine. During the day (and especially before going to sleep), we click endlessly through YouTube playlists, scrolling through updates from the TikTok community or checking the latest on Instagram. Music, radio, and television furthermore provide the semipermanent backdrop and soundtrack to much of our daily being.

As we are consuming media, we are also producing media. To some extent, this is done consciously—by maintaining one or more profiles on social media; by sharing updates, posts, and audio and video uploads on platforms; and by participating in discussions or leaving comments online. Other forms of deliberately making media are equally common in our digital environment: people create podcasts, keep weblogs or vlogs, help with the editing of entries on Wikipedia, maintain an email newsletter about their hobby or expertise, and manage a clan in a massively multiplayer online game. The lists containing these kinds of productive media activities go on and on, signaling a profound shift in what we do with media. The relation people have with media has become much more reciprocal than in the past and, in the process, more interdependent. Yet much of our media making is not entirely intentional nor necessarily voluntary at all. In a digital context, we produce and share an enormous amount of information about ourselves while using media. Every login, online search term, email, app or text message, and sometimes even every keystroke becomes part of the databases of telecommunications companies and media firms that facilitate and provide our media, feeding our digital shadow into *algorithms* that determine what kinds of advertisements we see and influence the recommendations, menu options, and content choices we have online.

Much of the production implied in our media activities involves (personal) data that feeds a wide variety of algorithms—which are basically a series of instructions telling a computer how to transform data into useful knowledge. In our digital environment, given the extraordinary amount of data people generate on a daily basis, just about everything involves algorithms, including but not limited to

- how news gets selected and ordered;
- the order in which information appears on social media;
- who gets what kind of bank loan;
- who gets flagged as a risk by mortgage lenders, tax authorities, and security services;
- what kind of prices you get to see when checking for products (such as flights) online;
- the places where police officers get dispatched to;
- how sentences are calculated for people serving prison time;
- how self-driving cars navigate cities;
- the recommendations and suggestions of any kind of digital media (from smart home speakers to video and game streaming services);
- how dating sites and apps match potential lovers;
- the sequence of when medicine gets administered (in hospitals and care homes); and
- how certain weapons could be deployed in warfare.

Much more can and should be said about the role of algorithms and *artificial intelligence* (AI, which can be defined as a self-learning group of algorithms; see particularly chapters 4 and 8 for more discussion on the role of AI). For our definition of media as activities, it is important to note that pervasive algorithmic intervention suggests that what we do in media to some extent is coupled with, if not determined by, processes and programs that run underneath, operating in the background and often obscure to even the most skilled coders, data scientists, or statisticians given their complexity, opacity, and (in the case of AI) self-programming capability. We should hasten to add that this does not mean algorithms, fed by the data of our media use, are in control. That would be a story too seamless, too neat, and much too simple. Everyday life in media is always messy and imperfect, and if we consider only for a moment all the wildly differing ways in which people appropriate, resist, hack, repurpose, and otherwise more or less skillfully use media, straightforward conclusions about seemingly all-powerful algorithms can be rejected outright. Even so, what is abundantly clear is that through our actions in and with media, the coupling between who we are, what kind of social structures we create and participate in, and the role of media becomes progressively inseparable.

Considering all the things people do with media, it may seem that media come at us from all sides, aided and abetted by huge corporations that provide and promote their hardware, software, products, and services on a global scale. On the other hand, we are the main source of our media activities. Especially in a digital environment, the potential for individual customization, for our own agency when constructing a practically singular information space for ourselves, is profound. People create all kinds of personally meaningful, playful, and practical rituals and routines with their media. Every computer user—whether a desktop, laptop, tablet, or smartphone—has their own habitual "click-round" or checking cycle along a range of sites, apps, platforms, and places online, which we visit at various moments during the day. We modify and personalize our media, decorating and rearranging their appearance and functionalities and filtering the things we read, see, and hear based on our personal preferences and settings. In turn, the algorithms that determine the inner workings of most services online try very hard to fit us into mathematically determined categories—white gay men in their thirties, Polish students living abroad, dog owners from Spanish-speaking regions, and endlessly on, as indicated by the companies that purchased our data—while we aim to make media fit our idiosyncrasies. This is a delicate dance, where the individual user may not be powerful when compared to companies like Amazon, Ali Baba, or MercadoLibre but certainly is not powerless.

The starting point for every media choice is yourself. This prompted US-based *Time* magazine to choose "You" as its "Person of the Year" in 2006, featuring a front cover with a YouTube screen functioning as a mirror. The person holding up the magazine would be looking at herself. The centrality of ourselves as having to take responsibility for reconstructing the world and our lives in it through (the way we use) media cannot be underestimated. Although this would suggest that people make well-considered choices about what to do with media, usually this is not the case: our choices for media activities are generally anything but careful or deliberate. Media are what we do—switching on the smartphone in the morning is as basic an activity as putting on clothes before heading out the door. If you are the source of almost everything you do in media, and most of your media-time is spent without much active awareness, the time you spend in media has a tendency to speed up, seeming to fly by. Our intensely embodied engagement with media amplifies this experience, as we feel strongly about our media, and the body responds to whatever we are doing with media—watching TV or a movie, reading a book, listening to music, or using social media—as if the event were unmediated. Getting a bit lost in media can therefore be seen as both a natural response to a fun, engaging, and often quite routinized experience, while it is also the goal pretty much all media, technology, and telecommunications industries would like to achieve with their products and services.

Beyond the use of media in everyday life, activities with media also have come to play a profound role in the functioning of all institutions in society. Whether it is politicians

using social media to connect to (prospective) voters, police and security forces collecting data, brands commissioning expensive advertising campaigns, schools switching to hybrid teaching and requiring students to use a laptop for their homework, doctors offering online consultations, companies offering employees options to work from home using dedicated teleconferencing software, or parents installing remote monitoring software on their children's smartphones, the inner workings of all social institutions involve media. At the same time, institutional settings and practices change and transform because of their increased orientation toward media. It is likely that anything a politician, a celebrity, a corporate spokesperson, and even a friend or acquaintance posts online is both authentic as well as conforming to the rules of the medium involved—in order to get noticed, to acquire likes and shares, to engender empathy. A good example is the "thirst trap": intentionally posting something provocative to trick people into responding, liking, and sharing your message. Although thirst traps usually involve sexy photographs, the phenomenon (common in social media) can also include politically charged statements or attention-grabbing news headlines. This kind of clickbait is not particular to our digital environment—as human beings, we are social animals and naturally crave validation, and the media as an industry have always sought people's attention. The form and format of clickbait online is both a function of a person's genuine need for recognition and approval, a more or less deliberate ploy to get attention for a particular product or cause, and a specific type of message that is preferred and accordingly receives preferential treatment by the algorithms governing online social networks. In conjunction, these elements work to further cloud our awareness of the time we spend in media.

---

*We are absorbed in our media to the extent that we can be said to be living in media rather than with media.*

---

The conclusion about seeing media in terms of what we do is similar to a definition of media as devices: using media is an intimate, personalized, self-preferential, and ultimately social process, whereby media come to play a part in all activities of everyday life and in the way institutions function in society. Media activities envelop and engulf us, making it harder to maintain an active awareness of all the different ways in which we use media (and how the media use us). We consume and produce so much news, information, and data that media seem to slowly but surely disappear as discrete devices and actions. We are absorbed in our media to the extent that we can be said to be living *in* media rather than *with* media.

## Media as (Social) Arrangements

> Media play an important role in society and everyday life, to the extent that they seem to disappear—as distinct devices as well as particular uses. In the process, we organize almost everything in our lives in one way or another with media and for media.

In media we find love (and break up), work (and go in search of something new), knowledge, fun, and relaxation; we are part of a community, and we can really be alone. We no longer experience a public event—a concert, wedding, demonstration, or competition—without many people sharing what happened via media. This "mediation of everything" suggests that simply looking at media as devices, technologies, uses, and activities is not enough—we need to dig a little deeper and consider media as social arrangements as a crucial third element of our comprehensive definition of media.

Although the design of contemporary media prefers our devices to fade into the background of our lives, and our media use likewise lacks due diligence, there is still plenty of debate in society about the omnipresence of media. Publications about how to limit your mediatime abound, new phones and computers come with software that keeps track of your time spent with media (and can issue warnings if need be), hotels and resorts sometimes market themselves as media-free zones (whereas others supply free smartphones and Wi-Fi hotspots to customers), professionals advise on media diets, how to tackle "infobesity" and taking a news vacation have become a major industry, and governments and academic institutions worldwide are investing heavily in research on (the consequences and possible prevention of) so-called fake news, disinformation and misinformation campaigns online. Since 2019, the World Health Organization includes gaming disorder in its International Classification of Diseases (ICD), and moral panics about anything related to sex (and, to a lesser extent, violence) in the media occur regularly in countries all over the world. It would not be far-fetched to argue that this collective hand-wringing about media gives expression to a widespread feeling of anxiety about media becoming completely enmeshed with society and everyday life—especially because media as artifacts and activities are gradually disappearing.

Our intensive orientation toward media is not just something that occurs on an individual or community level—it also applies to any and all companies, organizations, and institutions in society. Without an up-to-date website, managing multiple profiles on social media, and a timely response to emails, chat messages, and phone calls, it is difficult for anyone—whether a company, club, political party, or government agency—to achieve their goals. Effectively managing and as far as possible controlling your presence and representation in media can sometimes be just as important as simply doing your job (especially if managing social media is your job). We evolve certain ways of integrating media into our daily lives, such as wrapping devices in dedicated cases and covers, decorating computers with stickers and decals, establishing preferred pathways for zapping across TV channels, developing daily app-checking cycles, microcoordinating appointments via text and app messaging, and uploading memorable moments instantly to photo-sharing and video-sharing platforms. As we arrange media (as artifacts and practices) into our social environment this way, they come to provide the inescapable backdrop to our lives.

Media also give shape and meaning to the way people engage with the world, for example, through our participation with such public events as demonstrations and celebrations. Online expressions like #POIDH (short for "pictures or it didn't happen") or "hashtag or it didn't happen" are a common comment received by people posting about experiences and events they are part of. People who have a bit more experience with managing a public persona—such as popular performers, high-profile reporters, and politicians—find themselves in the current digital environment confronted with increased visibility, necessitating constant vigilance regarding the ways in which they appear in media and how they cultivate and maintain relations with fans, audiences, and voters through media. Numerous news reports on refugee crises in, for example, North Africa, the Middle East, and across Europe and Latin America note how people generally have to leave everything behind—except their phones, which are as important to refugees as water and food. After Russia's invasion of Ukraine in February 2022, it is striking to see how everyone involved makes media about this horrific war—from teenage TikTokers uploading videos from their bombarded apartment buildings to Ukrainian president Volodymyr Zelenskyy (himself a former professional media producer), from military regiments using drones to filmmakers producing full-length documentaries, from crowdsourced freelance war reporters to an army of graphic designers dedicating their art to the defense effort. The social arrangement of media in times of the Russian-Ukrainian war turns all of us into witnesses (rather than just consumers) of what is happening on the ground, raising all kinds of complex ethical questions about our involvement.

For refugees and victims of war alike, smartphones and social media provide a lifeline during their perilous journey—maintaining online communities to help figure out unknown circumstances, enabling regular contact with families and friends (and the people that help them), using digital navigation and communication to find their way, and switching between locatability and invisibility as their profoundly precarious situation dictates. Similarly, decisions by Google Maps to disable geolocation from its service in Ukraine (to prevent Russian troops from tracking the whereabouts of defending forces) or an army of IT volunteers using relatively simple software to hack into Russian radio traffic and share the transcripts online signal the significance of media in the context of war. These widely differing social arrangements are similar in their subtle perspectival shift from using media to organize certain aspects of everyday life simply for fun and social connection to media providing a primary orientation for the way people move about and act in the world.

Looking at media as social arrangements involves not only an appreciation of how interdependent media, society, and everyday life are; it also includes recognition of the influential role media play as a global industry. The media as an industry—from the telecommunications sector to the film world, from publishers to broadcasting companies, from advertising agencies to marketing and public relations firms, from record companies to internet platforms—is a potent and economically powerful player in society. Their power to some extent flows from our love for media and from their provision of the channels and stories that we use to connect to each other and to organize our everyday lives.

The significance of media as an industry is exemplified by the ways in which states all over the world use direct and indirect policies to protect, manipulate, or control the media. As globally operating media corporations such as Netflix and Newscorp offer channels, platforms, and content around the world, governments implement rules to protect and promote local media production, sometimes even trying to prevent "outsiders" from coming in altogether. State or public broadcasting is supported around the world through a variety of taxes and subsidies—in many countries, also aimed at influencing media content to be favorable to ruling parties and politicians. The business of media industries, largely based on charging consumers for access to content (either through subscriptions, direct sales, or indirectly via advertising), is furthermore enabled by strict international copyright legislation protecting vast content libraries. In recent decades, many governments around the world ramped up efforts to support and promote local creative industries (of which media are part), in an effort to redirect their national economies away from agriculture and manufacturing—which tend to be seen as unable to sustain significant employment numbers in the (near) future.

A darker indicator of the media's significance in the governance of entire countries is the ways in which governments seek to both control their national media and curtail the operations of international media. The Russian state's media watchdog Roskomnadzor, for example, has bit by bit dismantled the country's independent media infrastructure and, by the time Russian tanks rolled into Ukraine (in February 2022), all but shut down every single independent journalistic platform, requiring all remaining media organizations to publish President Putin's propaganda. The annual World Press Freedom Index of Reporters without Borders is a stark indicator of how governments all over the world seek to tighten or restrict media pluralism and independence, reduce transparency, and control the infrastructure that supports investigative journalism. In 2021, the index showed that journalism is totally blocked or seriously impeded in 73 countries and constrained in 59 others, which together represent 73 percent of the 180 countries evaluated.

Another deeply troubling indicator of how much the media as an industry—and its professions, such as journalism, film and television production, game development, and social media entertainment—matter is the singling out of media firms and professionals as targets for attacks, abductions, imprisonment, and killings around the world. Journalists in particular have come under threat, by state actors, terrorist networks, and even by ordinary citizens—especially when they cover public events and protests or try to do their work in the world's conflict zones. It is a testament to the media's significance that people can get so worried, upset, and even violently angry about the way reporters do their work. In the context of a life in media, it really matters what the media do.

The process by which the media take a prominent role in society, both as an industry and as the functions that media have in the daily life of people and institutions, is called *mediatization* (see also chapter 5). The mediatization of society and everyday life mainly works in four dovetailing ways:

1. Media *increase* the possibilities we have as humans to connect with the world around us. Without the intervention of media, our human communication is limited to a specific place, a specific moment, and our senses. With media, we can suddenly see further, hear more, and connect with many more people, places, and ideas anywhere in the world.

2. Media can also *replace* things—for example, we can play an online computer game with friends instead of playing football outside, transfer money via internet banking instead of going to the branch, and have fun sexting instead of (or in addition to) writing traditional love letters.

3. Media also *converge*, and our exposure to media tends to be concurrent—that is, we use different media more or less simultaneously throughout the day. This convergence contributes to the mixing of our social reality with the reality of media.

4. Finally, people, organizations, and institutions *adapt* in all kinds of ways to omnipresent media.

Whereas the mediation of everything helps us to appreciate the nature and characteristics of media as artifacts and activities, the concept of mediatization makes us more

acutely aware of how deep the entanglement of media and life goes. Mediatization does not decrease by, for example, killing your internet connection or no longer watching television. Not only is it a luxury to disconnect (as engaging online for many is essential for their livelihood, and networks of families and friends can be scattered all over the globe); it is also unlikely that all your friends, neighbors, and family members will log off too, as society is now almost completely organized in and around media. Our own media use also suggests that we are less and less aware of what exactly we do with media and how much time of the day this consumes. Even the devices themselves disappear into the background of our living environment. In this process, media come to co-organize our social reality. With media, we both have the tools and resources to do something about this; at the same time, it becomes easier to simply ignore our individual and shared responsibility to just get lost in the immersive experience. It takes effort to stay mindful and to truly take responsibility for a life in media.

## Your Life

It is perhaps somewhat paradoxical that, as media become more entangled with everyday life, all institutions in society consider media imperative for their functioning (and even survival), and media as a global industry has become an economic, cultural, and political force to be reckoned with, we struggle to accurately define what media are. The approach here, to consider media as artifacts, activities, and (social) arrangements, is an attempt to do justice to the far-reaching presence and coexistence of media, while at the same time offering a somewhat practical definition. As is hopefully clear from the review, this does not mean that media determine our lives. There is a process of mutual shaping, of codetermination going on, whereby we create, make, and use media, and in turn media form and influence us. This process is never-ending, messy, context-dependent, and complex. The history of media does not seem to have a linear progression; nor does it tend toward any goal other than securing survival—both for us and for our media.

Life in media may not be determined by technologies and machines; nor does it make us all-powerful in shaping a metaverse, where all our mediated experiences are seamless and we are safely in control. It does draw us into the world, whether we like it or not. We are witnesses to the lives of countless others, consumers of events and experiences far and wide, and suppliers of personal information that feeds the algorithmic way our digital environment works. It follows that living in media is life lived in public. In the next chapter, we explore what life is like in a global surveillance society, where most of the surveilling is done by ourselves, to each other.

# 3

## Public Life

A life in media is, to a certain extent, a public life. It used to be difficult to get attention for your life—unless you were a prominent politician, a popular athlete, or some kind of celebrity from the music or film world. In our digital environment, it is difficult to keep something really private.

Mark Zuckerberg, founder and CEO of Meta (formerly Facebook), wants "to give people the power to build community and bring the world closer together"—according to his company's mission statement. Zuckerberg finds evidence for this vision in his "law" of social sharing, which he discussed in a 2008 conversation with Federated Media's John Battell: "Next year, people will share twice as much information as they share this year, and next year, they will be sharing twice as much as they did the year before."[1] As people around the world share and reveal more information about themselves in media every year, it seems that we are all opting into a massively monitored world, where we are being watched by everything and everyone. To some, this is a utopian vision, connecting the world and making us more aware of each other and our shared humanity. To others, a global community based on collective inspection and observation is a deeply disturbing scenario, riddled with issues related to a lack of privacy, the loss of personal freedoms, and significant risks associated with self-censoring, stalking, trolling, and other problematic behaviors.

A life in media makes all of us participants in a global surveillance society, where the sharing, collecting, and repurposing of personal information is paramount. Surveillance in such a society, where the lines between private and public life are less than clear, has a simultaneous *panoptic*, *synoptic*, and *omnoptic* character:

- In a *panoptic* context, a small group—usually the state and a few large corporations—keeps an eye on the rest of the population to gather information, manage, and possibly influence or even control people. This kind of top-down or *vertical surveillance*

---

1. Mark Zuckerberg, speaking to John Battell at the Web 2.0 Summit 08, November 5, 2008.

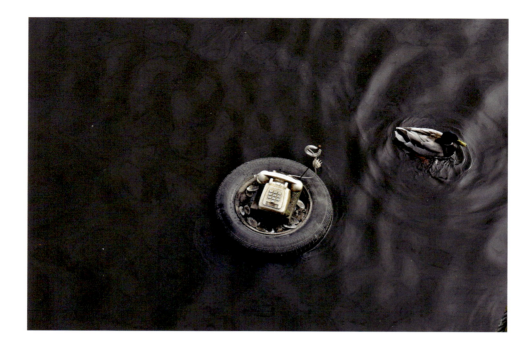

can be obviated by *anti-surveillance*, intended to either avoid monitoring altogether (e.g., internet browsers tend to include options to block companies from tracking you online) or make observation more difficult to achieve. Several companies, for instance, offer sunglasses that stay dark when scanned by facial recognition software (which otherwise would make them transparent); artists around the world create modified jewelry, unique hairstyles, specialized clothing items, and even entire fashion lines that aim to confuse or defeat cameras, trackers, and scanners to raise awareness and critique widespread surveillance.

- *Synoptic* surveillance occurs when the roles are reversed, as people collaboratively monitor, evaluate, and rate the products and services of companies, municipalities, and state actors online. Examples include feedback platforms like Yelp, Trustpilot, and a dating review app called Lulu that asked female users to evaluate the romantic appeal of their male dates. (Lulu launched in 2013 and was sold to Badoo—one of the most-downloaded dating apps in the world, originally launched in Moscow—in 2016 after much controversy.) Another prominent instance of synoptic surveillance (or *coveillance*) is people using their smartphone to record and share footage of police brutality. These kinds of *inverse surveillance* tactics have become part of the operating procedures of many grassroots organizations around the world, deliberately recording their experiences of being watched and using this to analyze and question surveillance practices.

- With *omnoptic* surveillance, everyone keeps an eye on everyone else in a digital culture in which more and more devices are always on, and people can be permanently accessible, which makes every media user a potential target as well as a source of observation and inspection. With this kind of horizontal surveillance—sometimes referred to as *sousveillance*—you, your friends, fellow students or coworkers, family members, and anyone else potentially participate in all kinds of observation and monitoring practices in media. Women in particular experience the dark side of omnoptic surveillance quite regularly, as men (often already at a very young age) stalk them, send or solicit sexual images, bully, and otherwise harass females online (mainly via social media).

In our global surveillance society, all forms of mediated monitoring are intertwined, given the entanglement between us and our devices. It could be argued that a fourth type of surveillance can be added to this list: that of *self-surveillance*, referring to the attention one pays to one's behavior when facing the actuality or virtuality of immediate or mediated monitoring. This observation can be by others whose opinion seems relevant or by someone in the everyday context of caring about themselves. Self-surveillance in media has many forms, including checking up on and maintaining one's profiles across several online social networks, quantifying and tracking all kinds of health-related data (e.g., using step counters, smartphone applications to log food calories, and wearable devices such as a Fitbit to document daily activities), and dutifully recording and sharing various aspects of one's everyday life (also known as "lifelogging" or "lifestreaming" when using a camera to broadcast one's life in real time). In the context of a life in media, omnoptic surveillance tends to subsume practices of self-surveillance, as the information we collect about ourselves becomes part of the archives and databases of network providers, telecommunications services, and platform companies. There has always been some form of monitoring of ourselves and each other in larger communities. Surveillance by states, nobles, and other authorities is a historical feature of diverse societal arrangements all over the world. What is perhaps new or different today is the scope and scale of the monitoring. Our collective lives in media constitute a comprehensive and all-encompassing surveillance society, with everyone involved in the practice of massive mutual monitoring and potentially participating in all kinds of observation practices through sophisticated information and communication technologies.

Before we can estimate the consequences of living in a surveillance society, we need to find out exactly how mediated monitoring works. Surveillance, as almost everyone experiences it, takes place in at least seven basic ways, as enacted by

- the state and security forces,
- health care agencies (including hospitals and insurance companies),
- industries and businesses,
- social institutions (including the workplace, school, and family),

54 Chapter 3

- everyone,

- you (tracking yourself), and

- machines.

In this chapter, we consider all of these forms of surveillance via media to raise awareness of the intensity and complexity of the surveillance society in which we all live and participate. To get us started, I outline and explore where the various expectations and assumptions about surveillance historically come from and how this shapes the debates we have today about the observational context of our digital environment.

**The Panopticon**

When we worry about surveillance, it usually has to do with a top-down, vertical, or *panoptic* (literally, "all-seeing") form of surveillance. The notion of panoptic surveillance originates from a series of letters that the English philosopher Jeremy Bentham wrote to his father in 1786 during a visit to his brother Samuel in Belarus, who was hired by the Russian Prince Grigory Potemkin to oversee the work and workers in the ports and factories on the estate of Krichev. Faced with a large number of complex processes, a largely unskilled workforce, undisciplined (British) migrant laborers, and pressure from Russian nobles to do well quickly, Samuel devised a managerial system that would oversee who issued orders on his behalf. He also designed a new kind of factory building, where the manager could inspect the work from a central office in the complex. The ingenuity of this management system, according to Jeremy Bentham, was its "inspection principle": the workers supposed that his brother was or could be everywhere, doing their work in a disciplined manner under the assumption that they could always be watched. This got the reformist philosopher thinking: Could this imagined omnipresence of a supervising authority be a way to organize and control the whole of society?

In his letters, Bentham elaborated on the idea of indirectly watching everyone using the hypothetical case of a prison complex, where a guard from the center of a circular building could keep an eye on prisoners without being seen. This principle inspired the construction of these types of penitentiary complexes all over the world, in countries as varied as Cuba, the Netherlands, Vietnam, and the United States. Yet Bentham envisioned that this would not only keep prisoners in line but could also be applied to mental hospitals, factories, and schools. The core of a panoptic surveillance system is not so much that everyone is under surveillance but that people feel that they can be monitored at all times and adapt their behavior accordingly. The expectation of surveillance is enough, according to Bentham—as long as the observation would be visible, verification would not be necessary. Control of people in a surveillance society is not carried out by an all-powerful state or some other authority but is embodied and enacted by people themselves in response to omnipresent cameras, scanners, sensors and trackers. Seen as such, our world of ubiquitous and pervasive media would inspire a kind of self-disciplining practice called *panopticism*. Interestingly, an all-seeing surveillance system from which anyone could derive authority in the end depends on the voluntary participation of those it inspects to function effectively. Some therefore consider our lives in media as constituting a participatory Panopticon, given so many people's apparent comfort with sharing (details of) their lives online.

What makes a surveillance society play such a powerful role in people's lives is not so much supervision from above but—as Bentham suggested well over two hundred years ago—more precisely the way in which everyone properly participates in the system. What this origin story also shows is that expectations and concerns about surveillance are not particular to our current life in media. Surveillance has a long history across different countries and cultures, and our current "digital by default" context can be seen as vastly expanding and extending the various ways in which monitoring can take place while traveling, at work, in school, out and about with friends, or simply staying at home.

In the context of our lives as lived in media, two key critical debates about surveillance to engage in some detail involve the role and consequences of *surveillance capitalism* and *technology bias*. First, we can consider the role of surveillance as providing the global digital economy with its main currency: our personal information. Data about the places we visit and people we hang out with, where we are going, how we are feeling, and what we are saying tend to be gathered primarily to identify and promote new commercial opportunities (beyond functioning as entry points for potential oversight and control by state actors). All of this is happening in real time: as you start typing a term into a search engine, software runs in the background anticipating your query, selling top results to the highest bidder. Ultimately, the purpose in surveillance capitalism (for companies large and small) is to directly influence and modify people's behavior for financial gain. Although we certainly have agency in putting together our own media diet, and sometimes people use media to resist prescriptive notions of what we are supposed to be saying or doing, it is important to note that the dominant identity we are ascribed to (by the media, telecommunications, and technology sector) when watching television, surfing the web, or tapping an app is that of a consumer: someone who buys or contemplates buying certain products and services. This passion for personal information for the purpose of profit, coupled with the desire to be able to predict what we want, drives the global push for more channels, services, devices, places, and spaces to become smart: to come online, to be equipped with chips, sensors, and other tracking technologies for collecting people's personal information, and to automate their operations.

Second, the consequences of institutions basing some or all of their operations on the outcome of algorithmic procedures derived from digital data should be carefully examined. This is a widespread concern primarily fueled by the fundamental insight that such technologies as statistical formulas (which form the beating heart of algorithms and artificial intelligence), computational software, and machine learning protocols are neither objective nor neutral. There is an inherent bias in all our technologies, in part as a direct consequence of the particular perspectives and prejudices of the people involved in producing and programming them. This, for instance, results in mass surveillance systems reproducing the racial-profiling practices of security forces, such as the EU policing unit Frontex's use of drones patrolling the Mediterranean (rather than conducting sea patrols that come with the human rights requirement to assist vessels in distress), and automated decision-making in immigration applications regarding refugees and the undocumented. Another outcome of technology bias can be the imposition of specific social values and behaviors on citizens through a social credit system tracking and evaluating citizens for trustworthiness, as developed by the government of the People's Republic of China with the assistance of the country's technology and telecommunications sector. A third example involves computer hardware and software unjustly assisting in data-driven policing efforts by relying on old, often corrupted, and contaminated datasets and databases to predict potential crimes and would-be criminals. This is sometimes referred to as caused by the problem of GIGO, or "garbage in, garbage out," meaning that the quality of information produced by a computer or statistical program can be only as good as that of the data entered into it.

All these examples of technology bias combine the seeming innocence of a computational approach to data gathered through mediated inspection with the subjectivities of the people commissioning and operating such hardware and software, as well as with those who have little choice but to integrate such systems of surveillance into their daily routines. For the perspective of a life in media, where the boundaries between media and life are blurred beyond distinction, it is crucial to additionally note that technology bias is sometimes an actual feature of specific devices and computer programs, well beyond the intentions or control of humans. Facial recognition software for example tends to falsely match people with darker skins with profiles in existing databases, as well as women, to a much greater extent than white males. Explanations for this bias point to structural limitations of the technology: cameras struggle with taking good pictures of people with a dark skin tone, and it is difficult to program a computer to meaningfully distinguish someone's makeup from facial features.

Even if provided with "perfect" data, computers and algorithms still produce outcomes that disproportionately affect anyone who does not fit the "mainstream" or dominant profile as defined by machines. This is the main reason for such organizations as the United Nations calling for a moratorium on the use of certain surveillance technologies, especially those directed at displaced people. Examples include robotic lie detector tests at airports, eye scans for refugees, and voice-imprinting software for use in asylum applications. All of this should remind us that whatever is considered the social norm in any given society is neither a neutral nor an objective indicator, yet it will be one that is inevitably replicated in the technological systems we surround ourselves with, further marginalizing anyone who somehow deviates from the norm. Given the fact that we have manifold selves and our identities, actions, and behaviors always change over time, all of us at some point meander and swerve from the norm. No one truly fits the one-size-fits-all bias in technology.

The same mechanisms that create potential problems can be used to tackle some of these issues. Artificial intelligence can, for example, be used to prevent ageism in corporate hiring practices, and opt-in data gathering methods offered by service providers can assist in making sure people get the kind of products or help they really need. By making procedures, computational protocols, and data public, the organizations involved can contribute to more transparent and principled surveillance processes. A society relying on surveillance to function inherently faces complex and critical questions about the ethics of an expansive data-driven economy and an overreliance on machines and automation to make decisions that affect people's lives.

### How Surveillance Works

In 1637, the French philosopher René Descartes famously wrote ,"Je pense, donc je suis," which was later translated into Latin as "cogito ergo sum," arguing that the only real certainty we have as human beings is the doubt of our own existence. In 2010 the American television comedian Stephen Colbert suggested on his daily talk show *The Colbert Report* (on the Comedy Channel television network) that, in the context of people sharing their entire lives in media, a better variant of this statement would be "cognoscor ergo sum," roughly translated as "I am known, therefore I am." Later that same year, the late Polish philosopher Zygmunt Bauman put forward his update to Descartes's famous statement with specific reference to our life in media: "I am seen, therefore I am." In a TED (short for "technology, entertainment, and design") talk on April 3, 2012, Harvard sociologist Sherry Turkle added her take on this famous phrase: "I share, therefore I am." Turkle, Bauman, and Colbert primarily refer in their respective proposals to the popularity of social media, the process of massive mutual monitoring typical of a surveillance society, and our tendency to share all aspects of our lives with everyone online. It does indeed seem as if life in media is a public life, where it is increasingly difficult to keep separate what is private and what is public. In what follows, let's unpack the basic ways in which surveillance works within and across all aspects of media life.

### Forms of Surveillance: The State and Security Forces

State surveillance is one aspect that has been known to just about everyone worldwide since the revelations of the former CIA and NSA operative Edward Snowden in June 2013. Snowden provided collaborating journalists at *The Guardian*, the *New York Times*, and the US-based nonprofit investigative news organization ProPublica with secret documents, detailing how the American government, working together within the Five Eyes Intelligence Alliance of Australia, Canada, New Zealand, the United Kingdom, and the United States, was in the process of intercepting billions of phone records and emails and

conducting all kinds of other forms of unwarranted surveillance on citizens of countries all over the world. The revelations produced headlines worldwide and caused plenty of political and public debate about the extent and purpose of widespread government surveillance. Snowden became world famous, the subject of the Oscar-winning documentary *Citizenfour* (2014) by Laura Poitras—one of the journalists Snowden originally shared his documents with—and the biographical thriller *Snowden* (2016), written and directed by Oliver Stone; he published his autobiography in 2019.

The government, police, and security services monitor us in all kinds of ways. This type of supervision consists of three main approaches: overt surveillance, covert surveillance, and the processing of data gathered through surveillance.

First of all, the networks that people use to connect with each other can be monitored: telephone lines, internet cables, computer servers, satellite connections. This process, usually governed by laws and legal frameworks, can be done directly—where the state literally listens in on telephone conversations, reads our emails or chat messages, and records us in public spaces using omnipresent closed-circuit television (CCTV) cameras—or indirectly, whereby representatives of a government request this type of data from a technology or telecommunications company (such as some of the corporations mentioned in the Snowden files, including Microsoft, Yahoo, Google, Facebook, and Paltalk). All of this is contingent on such companies cooperating with the state. Sometimes, such collaboration is actively sought by the corporations involved to secure lucrative software-as-a-service (SaaS) contracts, providing the necessary programs, training, and maintenance for automated surveillance systems to operate. For example, facial recognition technology is developed partly through the scraping of images from social media profiles without permission to train the software to recognize people's faces. On the other hand, some of these companies have been reluctant to provide state actors with access to their infrastructure and data, citing privacy concerns or simply being protective of their proprietary software and services. In efforts to combat the spread of disinformation and hate speech online, the Cyberspace Administration of China announced in 2021 that platforms should display each users' unique Internet Protocol (IP) address. This was feared by many as another form of overt state surveillance, yet several companies such as Weibo and Douyin followed suit in 2022.

In addition to public network monitoring, a second method of observation is somewhat more subtle and tactical, in which security services secretly record what we do with media by using dedicated hardware and software operating in the background. Consider, for example, spying on all the connections we make via open Wi-Fi networks (in public spaces, in coffee shops, libraries, and shopping malls) or the installation of spyware on a connected computer. This type of monitoring is covert, difficult to trace, and legally operates in muddy waters.

The third approach to state surveillance refers to what the police and other security services do with the collected data. This involves all kinds of protocols, methods, and techniques for handling, combining, codifying, and further processing information—something virtually no one has either the capacity or capability to adequately oversee.

Although government surveillance has a history that goes back for centuries, our digital environment offers many more opportunities to do this on a grand scale compared to the past, when it depended on the individual tracking of people and covertly opening written letters that suspects sent to each other. Later, technologies were added, such as binoculars (invented by a Dutchman at the beginning of the sixteenth century) and eavesdropping on radio communication (through wiretapping, for example). From the beginning of the twentieth century, governments around the world established their national security services, whereas the two world wars spurred the development of international cooperation and exchange in surveillance efforts. The combination of mail, radio, telephone, and CCTV surveillance dominated state monitoring efforts in the last century. At the start of the twenty-first century, state and security forces have greatly expanded their surveillance operations—a process further accelerating during the global coronavirus crisis of 2020 and 2021, with governments and police forces linking existing efforts to monitor public spaces with facial recognition software to process and prosecute people—with euphemistic titles like "Domain Awareness System" in New York, "Safe City" in Moscow, and the "Ring of Steel" in London and Belfast. In 2021, Amnesty International, with the support of numerous local organizations, launched the #BantheScan campaign to protest the use of facial recognition software, stating that such technologies amplify racially discriminatory policing and threaten the right to protest. This highlights a prob-lematic feature of all surveillance regarding our lives in media: with every new informa-tion and communication technology that state actors use for supervision, the meshwork for data collection gets drawn wider, with the result that everyone can be monitored—not just people who, on the basis of clear and (legally) verifiable criteria, pose a threat to oth-ers' safety. Surveillance begets surveillance.

The overall process of state-directed surveillance tends to be a lot more messy than dystopian visions of an all-encompassing Big Brother–type society suggest. There are many different services that provide surveillance for the state, collecting information on a wide variety of human activity, including but not limited to financial, education, travel and transportation actions, medical, housing, taxes, and communications records. The data that various agencies collect are stored in all kinds of different ways, and in general, different departments do not know who exactly does what; nor are there uniform and trans-parent procedures in place on how the collected data can be archived, collated, and effectively used. The state, security, and police tend to experience *data glut*. Confusion about what data is stored, who is responsible for which files, when data is considered to be obsolete (and on the basis of what criteria), how many copies of certain records exist (and where), how to prevent data hoarding, and poor data management protocols all contribute to an imperfect state of affairs.

The point of all this kind of state surveillance is not so much the fact that citizens are being watched in all kinds of ways. These kinds of monitoring processes have been com-monplace for quite some time, and we have slowly but surely come to accept that someone is almost always watching us. What makes the surveillance society from the perspective of citizens and the state so relevant to consider is, on the one hand, the complex and messy way in which all these data are acquired and archived and, on the other hand, what exactly the state, police, and security services are doing with all the information gathered. Rather than being stored in a database as a unique individual (and a department or agency there-fore looking at us as an indivisible human being), our personal information ends up frag-mented across all kinds of files, digital archives, and databases, housed in different departments, located on computer servers nearby and far away, and resulting in a digital self torn to pieces. As a result, people get reduced to being members of all kinds of statis-tically aggregated groups and population segments associated with different risk levels—such as the traffic-light model authorities around the world use to classify people according to a series of criteria into green (ignore), orange (keep an eye on), and red (check) risk management levels. For example, people from different backgrounds may end up in the same statistical group because they live in a certain neighborhood, travel to certain regions, or shop at certain retailers. In the meantime, all those files get compromised, and the transparency and supervision of the data, the information flow, and the surveillance process become obscured.

Meanwhile, all kinds of commercial entrepreneurs have jumped into this gap in the market of enormous yet polluted data. One of the most prominent is the American data analytics company Palantir, founded in 2003 by the people behind the internet payment system PayPal. Palantir develops software that combines and analyzes data from many different sources—the police, the judiciary, the municipality, and the financial sector—making it easier to detect patterns and connections in the data. Palantir is used by multiple government services in the United States—such as the CIA, FBI, customs, and the military. According to the company, governments and other organizations in over 150 countries also use Palantir's software. One of these is the International Consortium of Investigative Journalists (ICIJ), the international network of investigative journalists responsible for breaking the impactful story of the Panama Papers in 2016—exposing the use of offshore shell corporations by prominent businesses and politicians around the world for illegal purposes, including fraud, tax evasion, and dodging international sanctions. Palantir's cofounder and current chairperson, the German American entrepreneur Peter Thiel, a prominent financial backer of conservative politicians, served on Meta's board of directors from 2005 to 2022. His political influence stretches far and wide and has a particular life in media quality, as the people he financially supports with his "Thielbucks" (in reference to the currency V-Bucks in the popular digital game *Fortnite*) are all connected through what is jokingly called his "Thielverse" (referring to Mark Zuckerberg's plans for a metaverse).

There are also more playful and critical ways of dealing with the practice and technology of state surveillance. For example, the New York–based theater group Surveillance Camera Players has been performing regularly since 1996 in places where a lot of CCTV can be found. The group calls its approach "guerrilla programming" and wants to make the public of its performances, consisting of security personnel and police, aware of the fact that they themselves are being monitored too. In doing so, the actors denounce the indiscriminate supervision of people. Another example is the game Camover, invented in Berlin in 2013. Players had to destroy as many CCTV cameras as possible in a short time while they were being filmed. These videos then ended up on websites like YouTube, adding a meta layer to the game: an action intended to criticize surveillance in the public space was itself captured on video and shared via social media. A third example is the project #NotABugSplat in 2014 by a Pakistani, French, and American art collective. This project involved the reproduction of a young girl's portrait on a vast vinyl poster, displayed in a field outside the city of Peshawar in Pakistan—not far from where the girl's family was killed by a drone strike in 2010. The poster cannot be seen close up and was only fully visible from the air. The artists targeted military drone operators sitting thousands of miles away who refer to kills as "bugsplats" (which is a term used by US authorities for humans killed by drone missiles, a term also commonly used in the game industry to indicate a software crash), as well as the larger public exposed to the many media reports of the work—including on social media using the project's hashtag. Another audience for this installation was the intelligence community, who watch videos of drone strikes on their computers—a kind of horrible "predator porn" (*Predator* was the name of the first-generation remotely piloted aircraft used by, among others, the US, Italian, and Turkish military for observation and offensive capabilities), something Edward Snowden also talked about in his many interviews with reporters. These are but some examples of a rich global genre of surveillance art, using media artifacts and activities intended to record human behavior in a way that offers revealing commentary on the process of surveillance and the technologies used to surveil.

**Forms of Surveillance: Health Care**

From the cradle to the grave, a vast amount of information about us is produced, collected, and stored by the various organizations and agencies involved in health care. Health surveillance warrants separate discussion here, as it pertains to the most intimate information and knowledge about ourselves, affecting not just how we make sense of the world but also our continued embodied existence within it. Public health surveillance is the continuous process of collection, analysis, and interpretation of data and the subsequent dissemination of this information to policy makers and health care providers. From the moment parents go to a hospital or clinic for a pregnancy test, your health data gets collected and archived. Throughout life, a wide variety of people and institutions are involved in health surveillance, including but not limited to primary care clinicians, researchers and technicians in hospital laboratories, consultants in communicable disease control, environmental health officers, health administrators, insurance companies, providers of specialized care, pharmacies, pharmaceutical companies, and software providers for electronic patient dossiers. Since these organizations, companies, and providers generally do not communicate with each other much and in most cases are not allowed to exchange health care data, in the process of health care monitoring we get reduced to an ever-multiplying series of files, without there being any way for either doctor or patient to monitor all this information to check, verify, share, or track.

There are important questions about the most effective use of the data collected through surveillance of our health and care. From 2004 to 2013, I worked for several universities in the United States, where I was privileged to have a decent work contract including health insurance with the company Anthem—one of the largest for-profit managed health care companies in the United States. As part of my package, I was offered a free full medical check-up once a year to see how I was doing. If I could demonstrate that I did not smoke or if the data showed that my BMI (body mass index—a questionable measure of someone's health invented in the early nineteenth century)—had remained steady for another year, my monthly payment would go down, and I would qualify for "premium" care if I ever ended up in hospital. This case provides a direct link between health surveillance, financial incentives, and behavior influence, which can be problematic, as I have to admit it felt quite good to be rewarded for a relatively healthy lifestyle. I think I would otherwise be less likely to actively think about sports, lifestyle, or diet (or to find out what good options would be for my precise circumstances). Yet such rewards also make it unlikely for me to consider critically the consequences of surveillance as applied to people's health and well-being, let alone question the statistics my health care was based on.

Part of these kinds of ethical issues is the question of to whom your data actually belongs. Many medical records cannot be viewed by patients themselves, or its archiving and management are the responsibility of technology companies that are not even located in your country of residence (as the provision of e-health hardware and software tends to be outsourced). An example is the multinational VINCI Energies Group, providing information technology and computing services globally to transform medical centers into "connected hospitals" featuring a wide array of interlinked digital health applications, such as remote access to medical files, dedicated Wi-Fi for patients and virtual consultations, contact tracing apps (which became a global phenomenon during the coronavirus crisis), direct-to-consumer or on-demand telemedicine, automated bots that screen and refer patients, telehealth visits by doctors or nurses, electronic intensive care unit monitoring programs, and so on. Questions of transparency, ownership, and responsibility over all the data such a system collects, processes, and archives are quite pertinent.

Just as with state and security surveillance, the quality of the information collected about us through public health surveillance is a key concern. As with any form of data gathering and archiving in media, there is always information contamination. The hardware and software that process our data become obsolete, falter, or fail completely and are vulnerable to hacking and computer viruses; the maintenance of these systems generally leaves a lot to be desired. And again, if your data somehow skirts the edges of whatever is considered to be the norm—especially in terms of gender, race, age, ability, sexual orientation, or any combination thereof—these issues with health surveillance can become even more problematic. All this (handling of) information has major implications for people, as decisions about medicines, treatments, and insurance depend to some extent on what the computer spews out in a particular office, at a particular part of the vast health care network, in a place somewhere in the world.

Health care is the subject of debates and projects aimed at developing innovative forms of supervision and monitoring of patient information. For example, since 2012 the worldwide Hacking Health movement organizes activities in which a variety of community hospitals, academic medical centers, faculties of medicine, clinics and pharmacies, information technology vendors, pharmaceutical companies, insurers, granting agencies, patient organizations, health care professionals, clinicians, physicians, nongovernmental organizations, and entrepreneurial hubs and networks, as well as various local, state, provincial, and federal agencies, participate to come up with ways to give people more power in the

collection of data and making health information accessible. The network has chapters across all continents, located in such metropolitan centers as Athens, Kuala Lumpur, Lyon, Melbourne, Mexico City, San Francisco, Toronto, and Utrecht.

As with developments in state surveillance, the field of surveillance in health care is complex and evolving. The global coronavirus crisis of 2020 and beyond exposed a related development in public health surveillance, beyond the panoptic data collection efforts of agencies and institutions involved with health care: the synoptic surveillance of people all over the world, collectively monitoring global media—news providers as much as online social networks—for information about all things related to the coronavirus. In some cases, this led to increased anxiety, heightened stress responses, and misplaced health-protective and help-seeking behaviors, all of which was further amplified by exposure to (and participation in the sharing of) disinformation and misinformation about the virus—with equally serious consequences.

**Forms of Surveillance: Business and Industry**
It has become something of a cliché: data is the lubricant of the global information society. The rise of *surveillance capitalism*—of collecting and mining personal data for profit—is a worldwide phenomenon. The economy progressively revolves around data obtained by persistently tracking consumers. This vision for all-encompassing industry surveillance was articulated in detail by Eric Schmidt, chief executive officer of Google

from 2001 to 2011 (and chairperson of Google and its parent company, Alphabet, until 2017), in a keynote for a conference in Berlin, in September 2010, where he stated: "Ultimately, search is not just the web but literally all of your information—your email, the things you care about, with your permission—this is personal search, for you and only for you . . . We can suggest what you should do next, what you care about. Imagine: We know where you are, we know what you like . . . A near-term future in which you don't forget anything, because the computer remembers. You're never lost."[2]

---

### Data are the "digital air" that we breathe in media life.

---

Not only can companies envision such a predictable future because newer media enable all kinds of monitoring, but corporate surveillance is especially straightforward because people seem to generally agree with it or give up trying to prevent or circumvent it despite sometimes having serious misgivings about relinquishing their data. Every time a website or software program asks whether we agree with tracking cookies or a new user agreement, we sign to allow being followed online. A second cliché from our life in media applies here: if a service is free, then its users are the product that is being sold. And this product—our personal information—is extremely valuable. The three largest personal data traders in the world are also among the highest-listed companies in the world based on market value: Apple (which collects information through people's interconnected use of the products and services in its proprietary ecosystem, including iPhones, iPads, Macbooks, the Apple Watch and software such as iOS, Apple Music, the App Store, iTunes, Safari, and iWork), Microsoft (which uses the same principle), and Alphabet (Google's parent company). Data are the "digital air" that we breathe in media life.

The data hunt of companies did not start with the rise of the internet. Market research, surveys, and focus group interviews for many decades have been a staple for the commercial operations of businesses. We have been providing personal details to stores by completing and submitting discount forms and by taking advantage of the benefits offered by numerous customer loyalty and membership programs. In this scenario, a company's insight into our individual behavior is limited to when we buy something. Of much greater value is information about the phase just before that, when we are thinking about a particular purchase, coupled with data that says something about our socioeconomic context: where we live, what other products and services we find interesting, what people like us tend to do in similar circumstances. Loyalty programs have existed at least since the eighteenth century but suffered from a lack of information about the customer. This changed with the rise of membership schemes in the late twentieth century (such as airline frequent flier programs) and in-store loyalty programs (like the plastic cards you can get at supermarkets).

---

2. Eric Schmidt, speech at 2010 IFA Consumer Electronics event in Berlin, September 7, 2010, https://www.youtube.com/watch?v=DtMfdNeGXgM.

Both arrangements require the customer to register their personal details. As electronic commerce (or e-commerce) and teleshopping became possible—consider, for example, the launch of the vast online marketplaces Amazon in 1995 and Alibaba in 1999—the amount of data about who we are, what we want and need, what we think about, and what we like or do not like increased exponentially.

Particular to the discussion about *participatory surveillance* is the insight that much of the monitoring we are subjected to regarding business and industry is the outcome of our active involvement. Next to surrendering our personal information through the *digital shadow* we leave behind when we are online and registering for all kinds of customer loyalty and membership programs, we are regularly asked to fill in or answer a variety of surveys and review questions to assess a product or service of a particular company. Whether it is a board of smileys right after the baggage check at the airport to assess the customs procedure or a reminder to leave a review of a movie, dining out experience, or a repair service, we are constantly asked to evaluate. Each time we share important information about ourselves: our tastes and preferences regarding specific products and services, at a particular place and time. The consequences of these free reviews of products and services are sometimes quite substantial. For example, restaurants get reservations primarily through such review sites as Yelp, OpenTable, Tripadvisor, and Zomato, where everyone can leave evaluations of the places they have been. While this is useful to find out where to have lunch or a drink, establishments have to pay to participate in this system. Tripadvisor, as one of the largest review communities operating around the world, charges a monthly membership rate as well as a fee per booking, while collecting data from each reviewer as well as anyone who visits one of its sites. Other sectors of the digital economy, such as comparison sites for hotel rooms (e.g., Booking.com, a Dutch company that pioneered the business model behind these data-driven systems) and airline tickets, work the same way—as vast surveillance operations (where panoptic and synoptic modes of monitoring are combined) at a cost to participating restaurants, hotels, and airlines.

Since over half of the world's population uses the internet, the databases companies have acquired are colossal in size. This is the age of big data, where data sets with thousands of variables measured among hundreds of thousands if not millions of people are quite common. All these data on their own are of little value. That is why specialized departments and companies perform statistical calculations—generally in formulaic ways, using algorithms—to extract patterns from the data. Statistics enriches data, adding value by linking every button push, click, swipe, and keystroke to arithmetics and databases. As we have seen, the problem is that statistical formulas—just like technologies in general—are never neutral. Each calculation has limiting conditions, excludes certain options, and assumes many factors as given. The patterns that statistical programs extract from data can therefore best be understood as the product of the operations and calculations that have been applied to them, more so than a somehow accurate reflection of the people the data were gathered from.

There is both belief in and fear of these kinds of gigantic datasets and the influence they have on society. The belief, common among programmers and engineers in the technology sector, is based on the assumption that so much data put together make the question of *why* people do what they do redundant as we now know everything about *what* people do. Based on this information, it should be possible to predict how the world can be set up efficiently so that everyone is catered to. In an economic sense, this means that advertisers can predict, based on patterns in the data, when a certain group of people wants a certain product or when they need a specific service, immediately connecting them with companies that are able to provide it. You have experienced this scenario, as in the uncanny sensation of seeing an ad in your social media newsfeed for a product or service you were just talking about with a friend. It is not you that the advertiser is following (or knows about); it is the outcome of the statistical analysis of certain characteristics people like you have in common at a given moment. You are not special! Your data, combined and compiled with the personal information of countless others, are.

On the other hand, the fear of an economy governed by big data tends to be inspired by the fact that decisions about what to do with data and the patterns that are extracted from them are often taken by computers based on artificial intelligence—where the software independently learns from previous calculations and applies these lessons to each new assignment. This loss of human control and oversight is a scary proposition to many. I experienced a telling example of this when I was moving from the United States back to the Netherlands in 2013, after living and working abroad for about ten years. After all this time, I had built up some savings that I wanted to transfer with me—which would have been extraordinarily expensive to do through the regular banking system. Using PayPal, I started monthly transfers of small amounts of money from my American bank account to myself in the Netherlands, thus bypassing the expensive fees and exchange rates that banks charge. Admittedly, I was quite pleased with myself about this little trick—until it failed after a couple of times, and the payment order was refused. No matter what I tried, nothing worked. Finally, I called the company and asked for an explanation. After navigating a maze of computer menus, I finally got through to a customer service representative who patiently overheard my story and offered a solution: to wait a week and then try the transfer again. When asked how this would solve the problem, the man audibly shrugged and told me, "The algorithm does this every now and then." Can't you intervene, I asked—especially if it happens again. "No, we can't," he explained. "No one can touch the algorithm."

An interesting trend among companies that deal with big data is their hiring policy, whereby, in addition to computer programmers, data analysts, and statisticians, positions come up for people trained in ethnography, ethics, and qualitative methods. This is because

Even if you really do not participate (anymore) in all the hustle and bustle online, it is still unlikely you can escape the indirect surveillance of social media. For example, outside of Facebook, you still have to deal with the platform in all kinds of ways. Indirectly, it tracks nonusers through its vast advertising network that follows people as they click around online (via any site or app featuring Facebook like and share buttons), and by *shadow profiling*: using data collected about nonusers to create profiles of people who have never signed up for Facebook. Google's mobile operating system Android does something similar. Increasingly, our experience of online social networks extends to all other aspects of the internet (such as shopping, finding news and information, and playing games), a process called the *platformization* of the World Wide Web. Direct surveillance on platforms, even if you do not operate a personal profile, occurs when acquaintances post about you on their profile. This form of coveillance—where someone close to you records an activity you participate in and shares this online—has a dark side in online bullying (or cyberbullying). This has major consequences for people—when such harassment occurs as part of a networked, more or less coordinated campaign involving group trolling and stalking, for example, as well as doxxing (publicly providing personally identifiable information about an individual or organization) and real-life harassment.

Coveillance can be particularly damaging to people in already vulnerable or otherwise marginalized positions in society. Importantly, this involves children online. For example, teasing and bullying each other in the schoolyard can be magnified in media to such an extent that children see no other way out than suicide. Women, BIPOC (Black, Indigenous, and people of color), and members of the LGBTQ+ community who speak out in their professional role on television and social media are regularly confronted with hateful reactions and forms of harassment online. There is a thin line between visibility and vulnerability for people in the public eye.

An example of a ludic protest against this kind of problematic synopticism is the traveling theater show *Hate Poetry* in Germany (performing from 2011 onward), where prominent journalists with a migrant background read and poke fun onstage at the death threats, threatening emails, and public calls on violence they or the publications they work for regularly receive, shocking and entertaining audiences everywhere. Platforms take measures against bullying and other forms of online misconduct, for example, by hiring curators and offering users more opportunities to report and take action against abuses. Given the overwhelming number of digital communications, the absence of universal standards of what can or cannot be tolerated, and a general lack of transparency about the way protocols about abuse get established, this does not seem to be very effective in the long run. Online misconduct, broadly conceived as *dark participation*, poses a challenge to legislators as countries around the world are trying to put into effect special antibullying laws. Persistent problems surrounding hatred and harassment in media suggest that responsibility does not rest with one person or party but is shared by all participants in the calculations that have been applied to them, more so than a somehow accurate reflection of the people the data were gathered from.

There is both belief in and fear of these kinds of gigantic datasets and the influence they have on society. The belief, common among programmers and engineers in the technology sector, is based on the assumption that so much data put together make the question of *why* people do what they do redundant as we now know everything about *what* people do. Based on this information, it should be possible to predict how the world can be set up efficiently so that everyone is catered to. In an economic sense, this means that advertisers can predict, based on patterns in the data, when a certain group of people wants a certain product or when they need a specific service, immediately connecting them with companies that are able to provide it. You have experienced this scenario, as in the uncanny sensation of seeing an ad in your social media newsfeed for a product or service you were just talking about with a friend. It is not you that the advertiser is following (or knows about); it is the outcome of the statistical analysis of certain characteristics people like you have in common at a given moment. You are not special! Your data, combined and compiled with the personal information of countless others, are.

On the other hand, the fear of an economy governed by big data tends to be inspired by the fact that decisions about what to do with data and the patterns that are extracted from them are often taken by computers based on artificial intelligence—where the software independently learns from previous calculations and applies these lessons to each new assignment. This loss of human control and oversight is a scary proposition to many. I experienced a telling example of this when I was moving from the United States back to the Netherlands in 2013, after living and working abroad for about ten years. After all this time, I had built up some savings that I wanted to transfer with me—which would have been extraordinarily expensive to do through the regular banking system. Using PayPal, I started monthly transfers of small amounts of money from my American bank account to myself in the Netherlands, thus bypassing the expensive fees and exchange rates that banks charge. Admittedly, I was quite pleased with myself about this little trick—until it failed after a couple of times, and the payment order was refused. No matter what I tried, nothing worked. Finally, I called the company and asked for an explanation. After navigating a maze of computer menus, I finally got through to a customer service representative who patiently overheard my story and offered a solution: to wait a week and then try the transfer again. When asked how this would solve the problem, the man audibly shrugged and told me, "The algorithm does this every now and then." Can't you intervene, I asked—especially if it happens again. "No, we can't," he explained. "No one can touch the algorithm."

An interesting trend among companies that deal with big data is their hiring policy, whereby, in addition to computer programmers, data analysts, and statisticians, positions come up for people trained in ethnography, ethics, and qualitative methods. This is because many realize that big data cannot do without small or "thick" data: insights gained from research among a limited number of people with full attention to their unique and specific emotions, stories, and ways of looking at the world. Where large databases provide insight into what applies to most people, small-scale projects offer insight into what all this means for people on an individual level, in a particular context. Making sense of the complex mix of gigantic data files, algorithmic analysis, and small-scale ethnographic research furthermore requires *data empathy*. Om Malik, founder of the technology news website GigaOm, argued back in 2013 that data without a soul is worthless: "As we move towards a fully quantified society, a society shaped by data, we run the risk of ignoring things that are difficult to measure. Empathy, emotion and storytelling—these are as important parts of good business as they are of life."[3]

In concrete terms, data empathy means to enrich the deindividualized statistical analysis of big data with personal stories, backgrounds, and context that give meaning to the material. The somewhat paradoxical consequence of the development of surveillance capitalism may very well be that more attention is paid to the personal, the particularity, and the affective nature of the experiences we have in media. The dark side of this development is the far-reaching commodification of our innermost feelings and desires through the commercial integration of big and thick data.

## Supervision by Institutions

In addition to governments, health care, and business, our lives are steeped in contacts with social institutions, such as political organizations, schools, and universities, numerous charities and nonprofits, and of course the rich nature of social life in cafés, bars, nightclubs, and concert venues. Most of these institutions are mediatizing along with the rest of society, meaning that they pivot increasingly toward publicity and use media to manage and monitor their current and prospective members, constituents, customers, and visitors. International nongovernmental organizations such as Greenpeace constantly develop new media campaigns, asking people to fill in online forms (e.g., to show support for a specific action), to sign up for email updates and newsletters (calling on us to use and follow hashtags on social media), and to post things related to the theme of a campaign. For these types of projects, Greenpeace works together with companies that supply extensive databases of people who are likely to be interested in nature and the environment. This is but one example of how civil society retools its existence to accommodate media, information, and communication technologies across the board.

The field of education is another distinct site of institutional surveillance. Digital media are prominent everywhere in schools, universities, and other settings for teaching and learning—including digital whiteboards in primary schools, laptops for children (that

---

3. Om Malik, "Coffee & Empathy: Why Data without a Soul Is Meaningless," *OM* (blog), March 26, 2013, https://om.co/gigaom/why-data-without-a-soul-is-meaningless/.

may contain secretly installed software to keep track of how the device gets used), the use of Google Workspace for Education and other cloud-based free services, comprehensive learning management systems (such as Blackboard or Canvas) for all coursework, and complete virtual teaching modules (so-called MOOCs, or massive open online courses). Just about everything that pupils, students, teachers, and professors do is digitally recorded and stored. Direct forms of surveillance include the many tests and exams that children have to take from primary school (and sometimes even before that at the day care center) onward, throughout their education career. Parents receive reports at home with statistical analyses of how their child relates to the average line of all other children. Schools and all other stakeholders in the education system use a variety of surveillance-based data that are processed by computer programs to make predictions about how children in certain groups are expected to develop over the following years. In a digital environment, the entire learning process of a child gets mapped out accordingly. Ideally, such a surveillance-data-driven personalization of education provides more tailor-made challenges and can contribute to an engaging education fine-tuned to each individual student. A more critical perspective would note that the more or less exclusive focus on data contributes to the management and direction of education by computer software and administrative systems beyond the particular wants and needs—and, most importantly, voice—of the child.

As in other sectors (such as health care), education professionals around the world tend to complain that they spend an increasing amount of their working time filling out forms online, maintaining data files, and doing administrative work for a range of supervisory bodies. Extensive records are kept of all aspects of education—from programs and subjects to individual teachers and students, whereby it is not always clear what these data are intended for, how they are exactly used, and who ultimately has control and direction over all the information gathered. The administration and practice of surveillance at all levels of education, as throughout other organizations and sectors of society, generate many critical responses. For example, there is an international countermovement that organizes informally around the concept of the "edufactory," since higher education—where more and more students (from all over the world) are taught by fewer and fewer staff—according to these critics looks more like a faceless knowledge factory than a place where autonomous learning is central.

Surveillance in a life in media extends to all sectors and domains of social life, as no aspect of our growing up, learning, working, and living together is left untouched. This also includes the workplace, since many sectors of the economy work with proprietary software packages that streamline the business process while keeping track of what everybody is doing. It is not unusual for CCTV to be installed in the offices of companies and for employees to have to report being at work by swiping a card, using smart tags, or clocking in via any other form of digital check-in system. Contracts often stipulate the possibility that company hardware and software (including electronic correspondence) may be monitored from time to time, while overt workplace surveillance becomes commonplace at corporations around the world—for example, through remote tracking of the location and use of company vehicles, computers, and phones, using facial recognition software to monitor the expression and mood of staff while working, and demanding employees carry handheld or wearable devices with them at all times to document their exact location and movements. Some companies introduce software (with such titles as Impraise, Crewmojo, and Engagedly) to get employees to rate and review each other—just like consumers are asked to do on such sites as Amazon and Tripadvisor. All these initiatives, services, and practices generate detailed data on peoples' whereabouts, movements, actions, and feelings at work. During the global coronavirus crisis, many employees working from home are subject to intensified forms of monitoring through the use of remote network connections, shared software tools (such as Slack or Microsoft Teams), and other dedicated corporate applications, extending the reach of workplace surveillance deep into the home.

An institution most of us have extensive experience with is the family—and it is riddled with surveillance. Numerous companies sell hardware and software enabling parents to track and monitor their children's media use. So-called *teleparenting* styles have become quite common, as parents are in almost constant electronic touch with their children as they grow up—for example, by interacting incessantly via messaging, chat, and video telephony services. My social media newsfeed fills with friends near and far sharing photos and videos of all the funny and quirky things their offspring are doing. I regularly come across babies napping or (not) eating, kids singing and dancing, teenagers driving away in their first car or receiving a diploma, and so on. Despite all our legitimate concerns about the loss of privacy, this type of "sharenting" is widespread. As a result, the lives of children from the cradle (and before, when parents share their first ultrasound online) and well into puberty are preserved and published in all kinds of more or less invasive ways. By the time children become teenagers, it is not uncommon that they have to contend with a vast digital archive of snapshots, videos, sound recordings and many kinds of other data associated with them, without having ever been in a position to give permission to record, archive, or share any of it. At the same time, kids get online at ever earlier ages, with major platforms catering specifically to them—such as Facebook's Messenger Kids and YouTube Kids—and advertisers using "kidtech," such as that provided by the UK-based company SuperAwesome to target children online without violating strict privacy laws that prohibit advertising directly to specific children's profiles. This is why some in the scholarly and legal community argue for the right to be forgotten online, especially for teenagers.

Companies are taking advantage of parental pride by offering devices that make sharing everything your baby or child does even easier. A far-reaching example of this was the Aristotle (a product of the American multinational toy manufacturer Mattel), announced in 2017 as an "all-in-one voice-controlled smart baby monitor," taken off the market only a few months later after protests from consumer organizations, child psychologists, and

privacy advocates. The device had a Wi-Fi connection, a built-in camera, and was running on artificial intelligence programmed to grow up with the child. According to the company, the Aristotle would be able to independently monitor babies and small children, teach them to spell and speak, warn of dangers, read them bedtime stories, soothe them if they cried in the night, and correct inappropriate behavior. Even though this product is no longer available, similar smart devices are popping up in households all over the world, such as the Amazon Echo, Google Home, the Apple Homepod, and Xiaoyu Zaijia (Little Fish) from Baidu. These are all permanently online devices that collect detailed information about us while we use them for everyday household things: asking for a particular song at dinner, dimming the lights for a romantic setting, keeping track of how long our food has been in the oven, contacting a delivery service. The global technology industry hopes that in the near future our homes will be equipped with "social robots," such as the Jibo—developed through a crowdfunding campaign by the American roboticist Cynthia Breazeal and acquired in 2020 by the Japanese telecommunications company NTT. The Jibo closely resembles the Extraterrestrial Vegetation Evaluator (EVE) robot in the Disney movie Wall-E (from 2008)—a film that offers a stark warning about the possible consequences of omnipresent surveillance and control by technology.

Despite this lengthy list of examples of surveillance in the realm of civil society and social life, there are plenty more instances to be found across all social institutions (including religion, sports, and law). Taken together, there are a few trends common across all cases of institutional surveillance as it has developed over time and accelerated in our digital environment:

1.  An ever-growing data hunger that is never satisfied as each subsequent generation of hardware and software demands additional information

2.  A general lack of overview and consensus regarding responsibilities and control over all the different types of monitoring involved

3.  A lack of transparency as to for whom and to what purpose all these data are obtained, how they get archived, and how all this information may be subsequently used

**Observing Each Other**

On April 30, 2008, then senator Barack Obama visited the Indiana University campus in Bloomington—the American university I was associated with at the time. In a packed basketball stadium, he addressed a wildly enthusiastic crowd as part of his campaign for the Democratic Party's presidential nomination. His opponent, Hilary Clinton, visited a few weeks earlier but was significantly less popular among the students and in the region. On the way to the stadium, I was approached by a cheerful young woman, who asked if I would like to take a seat on the stage with the senator, clearly an offer I could not refuse. Everyone onstage was supplied with signs—half of them official "Change We Can Believe" campaign placards, the rest consisting of fake (yet equally professional) makeshift yard signs roughly painted with various slogans: "Barack Rocks," "Fired Up," and "Obama Oh Yeah." I was seated directly behind where Obama would address the crowd—right across from a battery of television cameras from all the major US networks (such as CNN, ABC, NBC) and the BBC. Our role as stage audience was clear: we would be visible to the people at home and served as the decor for the candidate with our supposedly homemade signs and posters. When the senator finally climbed onstage, the stadium erupted with cheers. Caught in the moment, I could not help myself but cheer along—and take photographs. As I stood directly behind Senator Obama, the pictures I took showed his back and the upturned faces of the crowd in front of him. Looking back at these pictures, I realized that every single person in that audience carried some sort of camera. In effect I was taking pictures of other people taking pictures of me. This was the first time that it occurred to me to what extent we experience the reality of the world and the people around us in media, as much as having a direct, immediate experience of it. What was also clear was not just the staged nature of the political performance but more so the reciprocal surveillance and sousveillance that was going on: the television cameras dutifully recorded the stage show, myself and several others providing the backdrop, in front of a roaring crowd in the process of documenting and sharing the event as it unfolded, all of us in various ways constructing or "worlding" a moment in terms of the media we were using at the time.

Since that special afternoon in 2008, this kind of omnoptic surveillance—where everyone is (or can be) monitoring everyone else—has become paramount. In fact, it could be argued that this is the default experience of a life in media. Whenever we engage with the digital, some personal information gets collected, archived, sold, and subsequently acted upon. Although it may seem that the rapid rise of online social networks since the early 2000s (such as Friendster starting in 2002 and MySpace and Facebook in 2004) can be blamed for the grand reciprocal mediated observation that so many of us are part of, there have been crucial coveillance moments in media before. Consider, for example, the history of reality television genres, from programs pulling pranks on regular people and all kinds of contest shows—including cameras installed in the homes of "regular" people via structured reality shows (e.g., *An American Family* in the United States in 1973 and its British spin-off, *The Family*, in 1974)—to combinations and remixes of all these genres, including most famously *Survivor* (debuting in 1997 on Swedish television) and *Big Brother* (first airing in the Netherlands in 1999). Adaptations of these and other shows circle the globe, and producers in countries all over the world introduce innovative variations using local scenery and celebrities to attract audiences, such as *Keep Running* (from China, with contestants participating in outdoor challenges involving a lot of running), *Little Cabin in the Woods* (from South Korea and featuring entertainment stars locked up in a cabin), *La Cenicienta* (made in the United States and aimed at Latinx audiences, a version of the hugely popular *The Bachelorette* franchise), and the Nigerian edition of *Big Brother*, *Big Brother Naija*, which has spawned several reality shows starring its most popular contestants (such as *I Am LAYCON*, *Shoot Your Shot*, and *Mercy & Ike*). We are used to spying on people like us.

Even though the sharing of details about our lives—voluntarily as part of a good conversation, involuntary because others gossip about us in the café or teahouse—can be said to be a historical and even somewhat healthy feature of society, in media such exchanges get supercharged, magnified, and commodified: suddenly, we can enjoy the most intimate details of the lives of others, whenever and wherever we want, and this information gets immediately sold on a global marketplace. This sharing goes so far that people who do not participate or stay offline for a few weeks are approached by concerned friends and family members to see if they are all right. While this may be done with the best intentions, it also makes all of us excellent workers in the service of surveillance capitalism. Media industries contribute to this—games you have not played for a while send messages suggesting your avatar needs you, applications remind you that friends are waiting for you, and most commercial services would love it if you registered your email address with them, so they can regularly send you updates and information about new products and sales. Dedicated insurance policies for your digital life, such as My Life Locker, founded in 2009 by the Canadian mortgage councillor Sandra Tisiot, offer services that assume that you have died after a certain period of inactivity in the digital world, automatically initiating a farewell protocol across all your profiles online so that all your documents and accounts are protected and preserved. Perhaps we never have to die in a digital environment—as there is enough personal information about (people like) us stored in databases to allow a sufficiently developed computer program to endlessly keep posting and sharing a life for us.

Even if you really do not participate (anymore) in all the hustle and bustle online, it is still unlikely you can escape the indirect surveillance of social media. For example, outside of Facebook, you still have to deal with the platform in all kinds of ways. Indirectly, it tracks nonusers through its vast advertising network that follows people as they click around online (via any site or app featuring Facebook like and share buttons), and by *shadow profiling*: using data collected about nonusers to create profiles of people who have never signed up for Facebook. Google's mobile operating system Android does something similar. Increasingly, our experience of online social networks extends to all other aspects of the internet (such as shopping, finding news and information, and playing games), a process called the *platformization* of the World Wide Web. Direct surveillance on platforms, even if you do not operate a personal profile, occurs when acquaintances post about you on their profile. This form of coveillance—where someone close to you records an activity you participate in and shares this online—has a dark side in online bullying (or cyberbullying). This has major consequences for people—when such harassment occurs as part of a networked, more or less coordinated campaign involving group trolling and stalking, for example, as well as doxxing (publicly providing personally identifiable information about an individual or organization) and real-life harassment.

Coveillance can be particularly damaging to people in already vulnerable or otherwise marginalized positions in society. Importantly, this involves children online. For example, teasing and bullying each other in the schoolyard can be magnified in media to such an extent that children see no other way out than suicide. Women, BIPOC (Black, Indigenous, and people of color), and members of the LGBTQ+ community who speak out in their professional role on television and social media are regularly confronted with hateful reactions and forms of harassment online. There is a thin line between visibility and vulnerability for people in the public eye.

An example of a ludic protest against this kind of problematic synopticism is the traveling theater show *Hate Poetry* in Germany (performing from 2011 onward), where prominent journalists with a migrant background read and poke fun onstage at the death threats, threatening emails, and public calls on violence they or the publications they work for regularly receive, shocking and entertaining audiences everywhere. Platforms take measures against bullying and other forms of online misconduct, for example, by hiring curators and offering users more opportunities to report and take action against abuses. Given the overwhelming number of digital communications, the absence of universal standards of what can or cannot be tolerated, and a general lack of transparency about the way protocols about abuse get established, this does not seem to be very effective in the long run. Online misconduct, broadly conceived as *dark participation*, poses a challenge to legislators as countries around the world are trying to put into effect special antibullying laws. Persistent problems surrounding hatred and harassment in media suggest that responsibility does not rest with one person or party but is shared by all participants in the process: the attacker, followers and bystanders, all relevant authorities, the law, platforms, and the material infrastructure of the media used for such dark participation.

A world in which everyone shares everything with each other and where most aspects of everyday life run through platforms forms the basis of the book *The Circle* by American writer Dave Eggers (published in 2013) that was turned into a motion picture for Netflix (in 2017, with Emma Watson and Tom Hanks in the lead roles). Eggers's book and film describe the enormous success of the Circle, a company modeled after a hybrid of Facebook, Google, and Amazon, pressuring everyone to publicly share everything about their lives on its platform. To belong and be popular, it is crucial to keep a constant eye on your reputation score on the network and to never ever turn off your smartphone. Having a memorable personal experience yet failing to share it with the network is, according to the logic of the Circle, selfish. Soon, the characters in the book (and the movie) learn that meeting the demands of the Circle software has all kinds of benefits: discounts, recognition, preferential treatment, and so on. All they give up is their personal privacy, and this goes so far that the protagonist puts a tiny camera on her clothes that permanently broadcasts everything she sees online for the whole world to watch. The only character in the book that is critical of such oversharing—a previous boyfriend of the lead character—in the end commits suicide to escape (and protest) omnipresent surveillance. The 2010 book *Super Sad True Love Story* by the Russian American writer Gary Shteyngart offers a similar, slightly more satirical view of a life in media where everyone can wear a so-called äppärät (similar to smartwatches such as the Apple Watch or a Fitbit), enabling people to publicly live stream all their thoughts, conversations, and experiences. The device also produces a popularity and personal beauty score. Everything in this rather bleak media life revolves around how you are perceived by others—while society collapses in the background.

In the meantime, the reality of our life in media has caught up with the gloomy vision of Eggers and Shteyngart. Tiny cameras—no bigger than a screw—that can record and broadcast razor-sharp images are, for example, commonly available online. Between 2012 and 2016, the Swedish company Narrative marketed the Clip (previously known as Memoto): a stamp-sized camera that records and shares high-definition video online to one or more platforms of your choosing. For some time now, all kinds of websites and applications make it possible to use your computer (with web camera) or smartphone as a personal television channel, such as Stickam (operational between 2005 and 2013), Justin. TV (founded in 2007, spinning off in 2011 into the popular streaming service Twitch, and sold to Amazon in 2014), Ustream and Bambuser (online since 2007; Bambuser got global headlines for being widely used by protesters in Egypt during the Arab Spring in the early 2010s), Livestream (as of 2009), and Periscope (part of Twitter, founded in 2015, discontinued in 2021). Facebook is also committed to this trend and has been allowing its users to broadcast video of their lives via the platform since 2016—a service it was forced to curtail after it was used to live stream a terrorist attack on a mosque in Christchurch, New Zealand, in March 2019.

As with everything in and about media, these developments are neither without controversy nor really new. For example, researchers at Yale University developed software in the mid-1990s that made it possible for people to keep a kind of permanent electronic diary of everything they encountered in their lives—from a digital birth certificate to any other official document, all photos and videos, letters, bills, tickets (for concerts and movies and the like), voicemails, software, and so on. This inspired Microsoft to sponsor the MyLifeBits project (running from 2001 to 2010), aimed at making it easier to live a life in media while your devices would document everything, turning all of these recordings into a searchable and shareable digital "lifetime store of everything," as the researchers involved called their application.

What makes *The Circle* and *Super Sad True Love Story* particularly dystopian in their analyses of life lived in a context of a global surveillance society is the uniform and superficial culture that ensues. In these books, the humans gradually come to perform their lives more or less exclusively as set by the parameters and rules of the media. Media, in terms of hardware and software, come to determine how we live in these narratives, reducing everyone to an individual statistic to compare and contrast yourself with. Greeting someone just a little less pleasantly, occasionally not participating in a social event, or opting out of a party harms your reputation score. Self-disciplining seems to be the only solution. Constantly worrying about your score is also not science fiction anymore. It is common practice for agencies, companies, and platforms to quantify their relationship with us, scores that to some extent come to determine our access to and experience of various services and institutions. Banks maintain credit ratings, tax authorities attribute risk levels to citizens, health care providers rate and rank people's medical needs in later life, customers review your reliability on online marketplaces, and dating applications maintain a desirability score for every user. As the volume of data increases, such scores get cross-tabulated and enhanced through algorithmic calculations, adding an unpredictable yet highly influential element to an already nontransparent system.

Whether all this surveillance by each other necessarily leads to a more uniform, perhaps even depressingly bland global culture is very much the question. In general, it could be argued that massive mutual monitoring amplifies and accelerates both conformity as it does idiosyncrasy, based on a variety of variables including one's personality, local and cultural context, time, and place. The number of databases, digital archives, scores, and algorithmic computations affecting our lives seems to be increasing rather than decreasing, suggesting complexity more than entropy. What seems clear is that (without formal legislation and enforceable protections) the responsibility for surveillance in a synoptic context shifts from top to bottom on the basis of self-tracking—from the state to the citizen, from the company to the consumer, from the public sphere to the private circumstances of the individual.

## Supervision by Ourselves

Ultimately, we are perhaps best at keeping an eye on ourselves. In the past, the art of vetting yourself was generally reserved for a small social elite—people who had the time, resources, and skills necessary to keep a personal or financial diary, for example. So-called egodocuments or self-life writings—such as autobiographies, diaries, memoirs, and personal letters—were traditionally the provenance of famous rulers, politicians, artists, military leaders, and scientists. Writers and poets have been known to keep autobiographical notes from about the seventeenth century onward, followed later by members of the upper class and, in rare cases, their children. Of particular interest for a life in media is Otto van Eck's diary from the period 1791 to 1797, written from his tenth to eighteenth year of life (he died of tuberculosis in 1798). Otto came from a Dutch family of high-ranking public officials and regents. His parents had him keep a diary as a means of self-disciplining and to get to know their son better. What makes this diary special is not only that it is one of the first and most extensive kept by a child but also that Otto entrusted all aspects of his life to the diary in great detail: documenting his relations with all the people in his environment (parents, sisters, friends at school, neighbors), what books he read, details about his personal health and how he was feeling, and what he did day in and day out. Otto van Eck's diary can therefore be seen as a true precursor to the ways in which we report to ourselves and about ourselves in the contemporary digital environment of the internet, smartphones, and wearable media. Otto's example additionally raises a fascinating question about self-surveillance: To what extent does documenting one's life contribute to genuine self-knowledge, and where can we draw the line—if any—between such insight and the practice of self-disciplining? This is a pertinent historical conundrum, especially given the fact that the details of our digital dairies are not necessarily private (either archived online or published via social media), just as Otto's personal archive of his life was not.

With rising literacy rates and media becoming more intimate, people keep track of and report on themselves in all kinds of ways. Sometimes the doctor asks you to monitor a certain aspect of your health, and in media, people use a variety of mobile applications for this purpose. Since everything we do in a digital context is stored and we live in media, just about every aspect of our lives becomes available for self-tracking and quantification. The output of all this self-surveillance can be categorized into passive, active, and affective data. We measure *passive data* by keeping track of where we are (via the GPS function on our mobile devices), what time it is, what our body temperature and heart rate is, logging our weight and counting calories of what we eat and drink, whether we stand or sit down, and so on. *Active data* include the number of steps you take per day, logging any kind of exercise, time spent on a variety of activities (including how much time we spend on our media devices), what we do when using media (e.g., saving the movies and television programs you watch, which games you play, the music you listen to, and everything you read in books, magazines, and newspapers), how many hours you sleep, how many words you type, how many kilometers you swim, cycle, or run, and so on. Third, a distinct trend is to

register *affective data*, such as keeping files on daily routines, cravings and tendencies, personal goals and motivation, moods and feelings, your thoughts, ideas, and dreams.

Data produced by self-surveillance supposedly help us to lead a better life: to watch our weight, keep us moving, make room for new discoveries (e.g., regarding books, films, and music), and even to get enough sleep. Finding out more about yourself, pursuing personal goals (such as losing or gaining weight), a friendly competition with others, keeping an eye on one's health or simply out of curiosity: people have a variety of reasons to dig deep for data about themselves. It is unclear what exactly we do with all that data (and what happens to the vast digital archive of all our self-tracking in the hands of the companies providing the necessary hardware, software, and connectivity), and it remains to be seen whether this self-tracking is really all that effective. For example, calorie trackers and pedometers are notoriously unreliable. Specialist (medical) knowledge and analysis are required to interpret and assess the scores of any individual. The recommendations of an algorithm based on past musical preferences may lead to new discoveries but is equally likely to send you down a rabbit hole of endless recycling and repetition. Companies come up with products such as a Body Battery—a Garmin smartwatch feature combining heart rate, stress, and activity monitoring to score a users' energy level as determined by an algorithm. No one can verify how it works, and legitimate questions can be asked about its reliability across different people and contexts (as with the use of BMI), yet it builds on people's motivation to self-monitor and self-discipline.

And while we engage in self-surveillance, people worldwide consistently express serious concerns about their privacy online. Often those same people survey themselves in media in one way or another. While it is tempting to point to others, the state, or the commercial world when it comes to our worries about how "Big Brother is watching you," it is safe to say that "little brother" surveillance—the kind of mundane tracking and monitoring we do to ourselves and each other—is a prominent omnoptic feature of our everyday lives in media.

### Monitoring by Machines

Finally, we have to consider a form of surveillance that affects everyone but remains invisible: *dataveillance*. In this process, machines record, store, exchange, process, and repurpose the data obtained from all the various forms of surveillance documented in this chapter and more. All parties involved (including the government, police and security services, companies, and organizations) outsource the processing of our digital shadow to machines that extract patterns from this endless data stream on the basis of mathematical formulas and algorithms. At its heart, this process works the same for the police when they engage in predictive policing and for all of us when we let an online social network like Facebook determine whose news and updates we see first when we log in. As the algorithms that collate and translate all our data tend to be self-learning, and the sheer amount of data flowing through the global internet at any given moment is beyond the comprehension of human beings, only other machines are involved in the supervision of machines. Every year machine-to-machine communication grows more rapidly than other types of communication—with the possible exception of our mobile communication. None of these interactions and the algorithms that govern them are neutral. Computational devices tend to reproduce every bias, error, misconception, and inequality in society as expressed in data. Left unchecked, dataveillance tends to exacerbate social inequalities in part because of the GIGO problem (as discussed earlier) and insofar as mathematical formulas are programmed to identify and reproduce patterns in the data. Although this can be corrected for, the volume of data processed at any given moment, the tremendous variety of data sources involved, and the limited capacity of human oversight make for a potentially problematic, ethically challenging, and altogether rather messy situation.

More and more everyday objects wirelessly connect to the internet—generating a worldwide internet of things (IoT) with connections and communication flowing between the servers and routers that make up the internet backbone and our digital televisions, washing machines, and cameras to office photocopiers, self-driving cars, and entire so-called smart cities equipped with a multitude of sensors. The question is not so much whether we are (or can be) monitored by machines but who has access to all the data, what uses are made of the data, and who can be held accountable for actions undertaken based on the information produced by endless data processing (using computational methods, algorithms, and artificial intelligence). It is clear that regulatory issues about responsibility, accountability, and ethics in the context of dataveillance still leave much to be desired and that technological developments move faster than sustained critical reflection on these trends seems to allow for.

There are all kinds of (mostly technological) strategies to discover, avoid, or resist dataveillance—such as only using public or shared facilities like internet cafés or libraries for going online, keeping up to date with the latest information and software regarding privacy and surveillance, opting to encrypt emails and chat messages, paying with cash only, and so on. Tactical approaches to minimize surveillance include anonymized web surfing, sharing accounts with family and friends across different devices, providing fake information when registering for anything online, and other attempts to keep using all that our digital environment has to offer while trying to remain unseen. In fact, all kinds of digital resistance can be found around the world, by individuals as much as communities; institutional networks (such as local organizations, specific groups of citizens, and employees of a certain company) and even national governments can enact tactics to circumvent the surveillance operations of other states. However, none of this seems to be a long-term sustainable solution, as it requires a level of skill and understanding not available to all; the technological context tends to evolve faster than people's discussions and competences can keep up with, and a focus on avoidance and resistance outsources the responsibility for dataveillance to those surveilled. Beyond strategic and tactical resistance, a perhaps more democratic approach involves critical reflection on how to live with surveillance in all its forms and guises—including an appreciation of why we all participate in the rise of a global surveillance society, where panopticism, synopticism, and omnopticism exist side by side.

## Public Life

The question why we keep an eye on ourselves and each other and why so many people continue to make so much effort to share their lives in media can be answered to some extent by looking at possible technological, economic, cultural, psychological, and biological explanations of why people choose to live in public. To conclude this overview of how surveillance works on so many levels of society and everyday life, let us explore the reasons we all have to collaborate with the current global surveillance system and its inspection principle enabled, amplified, and accelerated by our digital environment.

*Technological advances can be deployed to encourage organizations and institutions to be more cooperative, productive, and creative rather than nontransparent, controlling, and disciplining, depending on what people actually want from technology and how we are going to take responsibility for our desires.*

*Technology*  Developments in technology make forms of direct and indirect surveillance possible and in some cases even unavoidable. The abundance of increasingly smart devices, providing all kinds of household or otherwise everyday objects with wireless connections and computing power, speaks to this trend. Even though developments in consumer electronics are somewhat predictable—every new device contains elements and parts of older media, which also applies to how we use newer media—it can feel as if technology is running away from us as people and society. For example, it sometimes seems that surveillance happens to us, while in practice it is the result of a long-term historical development, occurring within and across all levels of society and everyday life, generally based on our more or less voluntary participation in it. From a technological point of view, each new media artifact (hardware and software) contributes to our surveyance. Given this development, growing concerns about the potent role of technology, machines, and media in society align with more hopeful expectations of abundant digital data contributing to better social systems. Technological advances can be deployed to encourage organizations and institutions to be more cooperative, productive, and creative rather than nontransparent, controlling, and disciplining, depending on what people actually want from technology and how we are going to take responsibility for our desires.

*Economy*  Keeping an eye on each other and ourselves offers all kinds of economic benefits. In a best-case scenario, criminals can be caught faster through predictive police work, patient care improves through the effective exchange of medical data, keeping track of your body statistics inspires a healthier lifestyle ensuring long life, and enrolling in a business loyalty program provides helpful discounts in the store and supermarket. Accepting surveillance has its rewards: access, discounts, and preferential treatment—all powerful motivators for participation. At the same time, avoiding or resisting inspection often comes with penalties, such as decisions that may get a mortgage delayed, one's ability to travel curtailed, or an application for a loan or grant denied. Whereas these kinds of outcomes highlight the problematic nature of surveillance, overall people seem to accept less-than-transparent forms of mediated monitoring to benefit from its real or perceived rewards.

*Culture*  From a cultural and sociological perspective, our more or less spirited participation in ubiquitous top-down, bottom-up, and collective espionage can be interpreted as a possible solution to the complexities of being part of a global community. Every time we tag someone in a photo, hand out a like, or leave a kind comment online can be seen as equivalent to giving someone a hug, shaking a hand, or saying hello. These are small yet significant gestures of communication that contribute to the maintenance of social relationships, enhancing our feeling of belonging. Similarly, a key reason for us to tune in to the latest episode of a popular TV show, to check out an exciting new artist on a music streaming service, to follow a live sports broadcast, or to read that book that everyone is talking about is to be able to do exactly that: talk about it, on social media as much as at the office, on the playground, and in the pub or teahouse. It is through such moments of "small" sociability and idle talk that a sense of community gets established. In a way, all the public interactions we perform online—from liking, favoriting, and sharing something to leaving comments, uploading pictures, curating, and cultivating our (multiple) profiles—can be considered as acts of social bonding and grooming when we are part of communities beyond our physical copresence. Surveillance plays a significant role in this display and expression of shared feelings and social connection, as it allows us to sustain a sense of community despite the size of the groups we (would like to) belong to. Media in general and online social networks in particular contribute to the establishment of trust, identity, and cooperation among groups of people—especially those that do not necessarily live in proximity to each other.

**Psychology**   Massive mutual monitoring feeds (and is an expression of) our desire to see and be seen. The human need for recognition is a crucial psychological interpretation of our participation in a surveillance society. Recognition and appreciation can translate into feeling powerful or even superior, but for most people this mainly means that they just want to be seen and acknowledged in some way. With a global rise in prosperity and the gradual spread of education and literacy, being able to express yourself and be heard and seen becomes an important part of life for people around the world. If a roof over your head and food on the table is something you do not have to worry about, postmaterial values such as self-expression and recognition rise in significance. Despite our misgivings about platformization and a loss of privacy, sharing our lives online is a clear-cut way to gain recognition, appreciation, and understanding.

Another uniquely psychological aspect of our media use as it relates to mutual monitoring comes from our knowledge that people are watching us. When online, every post, every upload, every share of a photo, video, or piece of text, and possibly even every keystroke potentially has an audience. In practice, we never know who exactly reads our posts or watches that funny video we just uploaded. This makes our mediated interaction online oriented toward an indefinite range of potential recipients, which in a general way explains how we tend to stretch ourselves to make sure people understand the point we are trying to make—for example, through the use of emojis and emoticons, by making use of all kinds of filters to change the feel and color of our pictures and videos, or simply by expressing ourselves more forcefully than we would do in a face-to-face situation. Our enthusiastic engagement with surveillance in part stems from a deep-seated desire to be seen and to be understood—a quest for authenticity that, at the same time, becomes muddled through our active direction of it.

**Biology**   Every time something or someone interacts with us in a kind or pleasant way, we tend to enjoy this recognition in an embodied way: it just feels good! This feeling extends to life in media. Every thumbs-up on Facebook, every advertisement that seems to address us directly, every time we beat a level in a computer game, and just about any message or notification from all our apps cause our brain to produce a tiny bit of dopamine. Dopamine is sometimes called the "happiness hormone," a substance that our brains produce throughout the day that makes us feel good, especially during activities that require some concentration and focus. The blue light from omnipresent television, computer, tablet, and smartphone screens also influences our body in various ways, contributing to the production of dopamine at times when the body generally needs to relax and wind down from the day. Media companies make good use of this biological insight by appealing to us to stick around for more of that show after the break, to produce narratives (whether for a full-length motion picture, a long-running TV series, or an advertising campaign) with recurring hooks to entice us to stay on, to watch yet another video, to click on one more link. In a reference to the promise of ubiquitous computing (see chapter 2), Mark Zuckerberg at a 2011 developers conference called this the principle of *frictionless sharing*, where people keep on using media, unhindered by due deliberation. Another example of a specific step platforms take to reduce friction and get us to spend more time with them are all the statistics we are being shown: how many friends you have, how many people like your holiday photos, how often others have shared your funny status update (and who clicked on it again), and so on. It really takes conscious, considered effort to wean yourself off media—and this difficulty in part has to do with how our bodies react to living in media.

A life in media seems to come at a price: we live our lives in public. Beyond the plentiful instances where it produces deeply problematic outcomes—especially for minorities; such vulnerable populations as children, refugees and migrants; or anyone who deviates from a dominant social norm—the most profound danger of a global surveillance society is arguably its inspection principle: how people gradually adapt their behavior based on the expectation of being constantly monitored. Self-censorship vis-à-vis invisible, nontransparent, and uncontrollable oversight stifle diversity and dissent. In fact, it is precisely this kind of conforming behavior that makes us less human. Contrastingly, living in a world where everything and everyone is potentially visible in media reminds us that we are all interconnected, that we can use that link for things other than self-disciplining or self-promotion, and that we are not powerless, as the true power of omnipresent surveillance ultimately depends on our willing participation in it.

# 4

**Real Life**

As society and everyday life become more entangled with media, significant questions emerge about the nature of human-machine relations, the usefulness of maintaining distinctions between humanity and technology, and the extent to which we are able to determine the reality (or authenticity) of events, information, and behavior.

There are few things of which the present generation is more justly proud than of the wonderful improvements which are daily taking place in all sorts of mechanical appliances. . . . We refer to the question: What sort of creature man's next successor in the supremacy of the earth is likely to be. It appears to us that we are ourselves creating our own successors; we are daily adding to the beauty and delicacy of their physical organisation; we are daily giving them greater power. . . . We take it that when the state of things shall have arrived which we have been above attempting to describe, man will have become to the machine what the horse and the dog are to man. . . . Day by day, the machines are gaining ground upon us; day by day we are becoming more subservient to them. . . . The upshot is simply a question of time, but that the time will come when the machines will hold the real supremacy over the world and its inhabitants. . . . Our opinion is that war to the death should be instantly proclaimed against them. Every machine of every sort should be destroyed.

These words read like the screenplay for a science fiction movie, closely resembling the scripts of the various *Terminator* movies and television series (between 1984 and 2009) starring the Austrian American actor Arnold Schwarzenegger as a murderous cybernetic organism, half human, half machine. One would be forgiven to look at media this way, given the inseparability of media from society and our daily lives. In the process of this mediation of everything, media permeate the inner workings of society's institutions and the practice of everyday life—up to and including our most intimate processes and feelings. This can be a rather unsettling realization. Media (as artifacts, activities, and arrangements; see the definition of media outlined in chapter 2) are gradually disappearing, nestling in the background and providing an environment within which we find meaning and express ourselves—and where we live our lives increasingly in public, often acting assuming we are being watched (as explored in chapter 3). All of this can be said to obscure and blur the boundaries between life and media, between reality and virtuality, between humanity and technology. This state of confusion, of constantly moving or altogether collapsing frontiers between human life and its mediation, is of key significance to media studies, just as the general anxiety about the increasingly prominent place of media and machines in social affairs is. What is particularly striking about the perspective expressed in this quote is that it is not a critical response to, for example, the rapid rise of artificial intelligence, a global surveillance society, or the ever-increasing intimacy between people and their media. Instead, these words were penned and published much earlier—in the summer of 1863.

In September 1859, Samuel Butler—a young Englishman disillusioned after losing his faith while preparing for the priesthood—boarded the full-rigged ship the *Roman Emperor* to New Zealand,. Upon his arrival in Christchurch (in January 1860), he set out to work as a sheep farmer and started reading a book that was published during the long voyage to widespread discussion in scientific circles: Charles Darwin's *On the Origin of Species* (published November 24, 1859). In this book—which is considered one of the most influential works of scientific literature, inspiring a global "Darwin industry" of dedicated scholarship—Darwin explains in a delicate but determined way that man is not

the measure of all things. On June 13, 1863, the editor of *The Press*—at the time a newly established New Zealand newspaper—published a letter captioned "Darwin among the Machines," written by Butler using the pseudonym *Cellarius* (which is Latin for "butler"), extending Darwin's argument about evolution to what he called "mechanical life." If technology evolves as humanity does, Butler argued, sooner or later they would take over—as machines are better equipped for survival than us fragile human beings. A copy of the paper was sent to Charles Darwin via his publisher, who responded with a letter stating that he considered Butler's take on his theory "remarkable from its spirit."

Darwin's theory of evolution suggests that humans in their present form are no more than a very well-adapted organism. Humanity is therefore neither necessary nor the best outcome of evolution. This was a stark departure from the received wisdom of the time, which was heavily influenced by religious dogma. This dogma included the work of his grandfather Erasmus Darwin, who also acknowledged evolution but saw it as having a purpose, with human life and the progression toward civilized society as its logical outcome. Upon his return to England, Butler paid several visits to Darwin, getting to know the family well. After his initial letter to *The Press*, Butler proceeded with his line of inquiry, anonymously self-publishing a novel in 1872 titled *Erewhon* (to be understood as "nowhere" backward) about a fictional society that deliberately functions without machines, considered by the Erewhonians to be potentially dangerous as they might develop consciousness and thereby render human beings obsolete.

It is in these early intellectual skirmishes about humanity's relationship with technology that we can find debates that strike at the heart of contemporary concerns about the role of pervasive and ubiquitous media in society and everyday life:

- the growing interdependence of humanity and technology tends to be met with concern and uneasiness;

- peoples' hopes and fears about omnipresent media can be traced throughout (and are subsequently deeply shaped by) the history of human-machine relationships;

- discussions and anxieties about the relations between humanity and technology (including about what is real or virtual, true or fake) are about much more than our current news, information, and entertainment landscape, inviting a fundamental consideration of what it means to be human in a comprehensively mediated, algorithmic and to some extent automated context; and

- taking up a critical and ethical stance regarding media life is complex in a situation where the boundaries between humanity, nature, and technology seemingly blur beyond meaningful distinction.

In this chapter we explore these discussions, first by unearthing the origins of media and mass communication theory and research in the late nineteenth century out of a growing concern—among both academics and the public—about the various ways in which mediated communication distorts, manipulates, circumvents, and altogether shapes what people experience and recognize as reality and truth. This leads to the realization that media come, as a strict definition of media suggests, in between us and whatever we perceive "reality" to be and that people's historical interdependency with technologies in general and media in particular is especially controversial because of this. Over time, this inspires people to develop three more or less distinct (and often overlapping) strategies to deal with the ongoing interference of media with society and everyday life, which we consider in some detail in this chapter (and further on in this book) by focusing on the implications of each approach:

1. Following Samuel Butler's call, we can wage war against the machines. This can be done quite literally by, for example, destroying or discarding media, dismantling hardware, uninstalling software, disconnecting from networks, deliberately opting against using certain devices (such as a smartphone), and through active nonparticipation in online social networks. A more indirect way of warfare includes developing a critical, if not suspicious, attitude toward media—for example, by advocating for rigorous media literacy programs and education.

2. We can also submit to a mediated existence, recognizing the inevitable boundary-blurring characteristics of a life in media and trusting ourselves to technology without surrendering to machines. This, for instance, means that people develop more deliberate choices and cultivate some level of critical awareness about their media, while still enjoying the digital environment for what it has to offer.

3. A third option would be to become media, as it were: becoming like the character Neo in *The Matrix* franchise (of motion pictures, animation series, comics, and digital games, beginning in 1999) by either accepting the reality with which we are presented or embracing a notion of reality as open source: malleable, subject to hacking and cocreation. Examples of such an approach would be designing alternative social media that are not built on the exploitation of people's personal information, investing in free and open-source software, and developing a not-for-profit *digital commons*.

Each of these options has its corollary in the study of media and (mass) mediated communication, including quite critical as well as more hopeful perspectives. To investigate these options is important, as this touches on some of the most consequential debates in our time, for example, regarding so-called post-truth or fact-free politics, the role and impact of "fake" news and disinformation campaigns, people's susceptibility to manipulation (by politicians as much as advertisers), and our common quest for authenticity (especially when we are online, which is almost always). The chapter concludes with an argument that perhaps we have already become media and that this rather uncanny sensation provides a fruitful way forward to study and understand our digital environment.

## On the Imperfection of Communication

Whether it is one's love life and romantic relationships, the organizational forms that make work and the functioning of companies and corporations possible, the processes and processing of mass migration and urbanization, or the intricacies of democratic politics and the political system—in all of these areas, media and mediated communication play a formative role. This fundamental realization, dawning in the late nineteenth and early twentieth century with the first mass media of newspapers, magazines, and radio, gave rise to much hand-wringing about the potentially problematic impact media would have on people and institutions. The first studies on the effects and influence of media emerged shortly thereafter—offering inconclusive evidence, which would become the dominant theme of media impact and effects research (see also chapter 8). Media have some effects on some people some of the time—but effects are generally ambiguous, the direction of such effects remains unclear, and individual differences between people turn out to be much more powerful predictors of effects than anything that can be generally ascribed to particular media or certain groups of people. It is important to note that this does not mean that media do not have effects—quite the contrary. Media clearly play a powerful role in all aspects of human existence; it is just that the research shows how complex and contextual such influence always is.

Undeterred by evidence to the contrary, both politicians, researchers, and the wider public have, since the early twentieth century, recurrently projected their fears onto media as a massive unidirectional influencing machine. This, for example, involved historical fears, such as

- the technological threat from more advanced civilizations, as depicted in the *War of the Worlds* serialized stories of the English author H. G. Wells (published between April and December 1897);
- scary predictions about robots rebelling and exterminating human life, as in the 1921 play *Rossum's Universal Robots* by the Czech writer Karel Čapek (which introduced the word *robot* into the English language, derived from the Czech word *robota*, meaning "forced labor," and which narrative, interestingly, ends with robots displacing humans from power, proclaiming, "Mankind is no more. Mankind gave us too little life. We wanted more life");
- the rallying potential of propaganda posters before and during the Second World War;
- investigations and even court cases about popular music (including The Beatles in the 1960s, Led Zeppelin in the 1970s, Queen and Slayer in the 1980s, Motörhead in the 1990s—all among my favorite bands) "backmasking" their music to insert evil messages;
- the widespread panic about violence in video games after the 1999 Columbine high school shooting in the United States; and

- global distress about the supposed power of online social networks to target political advertisements to individual users and thereby sway elections (as claimed in 2017 by the data-mining firm Cambridge Analytica, using Facebook data to create customized propaganda for its clients) and about online platforms acting as amplifiers of disinformation and especially widespread conspiracy theories about COVID-19 (from March 2020 onward).

Time and time again, people's deepest fears and anxieties are marshalled and projected onto the (mass) media of the time, whether it is propaganda, rock 'n' roll, erotic literature, fake news, or artificial intelligence. More often than not, such angst is specifically reserved for children and teenagers, whereby a global field of experts continually seem to fall victim to ephebiphobia: the irrational fear of young people as expressed by adults throughout history. Consider, for example, this statement about media's supposed impact on the young: "The popularity of this new pastime among children has increased rapidly. This new invader of the privacy of the home has brought many a disturbing influence in its wake. Parents have become aware of a puzzling change in the behavior of their children. They are bewildered by a host of new problems; and find themselves unprepared, frightened, resentful, helpless." This statement is not about smartphones or video-sharing platforms like YouTube or TikTok; nor is it about problems caused by the endless hours children spend on games like *Minecraft*, *Roblox*, and *Fortnite*. It comes from a high-profile study on the perceived effects of radio on children, published in 1936.[1] From as early as the late seventeenth century, representatives of religious institutions, learned societies, universities, and other authorities would regularly publish scathing missives about the ruinous consequences of reading, followed up in rapid succession throughout recent history with references to the dangers of television, comic books, film, popular music, video games, and social media. Taken together, all these fearful expectations of what media do to us have much in common, generally centering on the loss of control over one's emotions with a corresponding loss of focus on one's duties and responsibilities in society.

The twentieth century can plausibly be described as the "first age of mass media" in the institutional sense of the concept—as earlier forms of mass media and communication existed in many parts of the world. However, the kind of industrial organization, institutional arrangement, and networked infrastructure available marked the twentieth century as a turning point. As shown, this period inspired both wonder and much alarm at the perceived predominance of various mass media. Despite enormous changes in media institutions and technologies and in society itself and considering the rise of media studies and communication research as a more or less distinct field of scholarly work, the terms of public debate about the potential social significance of the media have changed remarkably little. The early twentieth century formed the backdrop for the first studies

---

1. Azriel L. Eisenberg, *Children and Radio Programs* (New York: Columbia University Press, 1936), 17–18.

into the role and influence of the media on what was generally considered a largely illiterate and anonymous mass of people. Beyond the particulars of the research and its less than straightforward conclusions, what is of interest in the context of a life in media is the underlying unease about what media do when they come between us—that is, when they disrupt and distort what are generally considered more authentic, pure, and real interpersonal communication and experiences.

A consistent trope throughout the history of technology in general and media in particular is the assumption that without or outside of media, our communication, experience, and (mutual) understanding would be better—if not perfect. In Western thought, this kind of thinking can be traced as far back as the work of Plato in the fourth century BC, who documented his teacher, Socrates, explaining to a student outside the walls of Athens how much he loathes the tools of speechmaking, musical instruments, and writing, as they prohibit genuine interaction and dialogue—which was his preferred method of teaching and communicating. Relying on media such as books, Socrates argued:

> will create forgetfulness in the learners' souls, because they will not use their memories; they will trust the external written characters and not remember of themselves. The specific which you have discovered is an aid not to memory, but to reminiscence, and you give your disciples not truth, but only the semblance of truth; they will be hearers of many things and will have learned nothing; they will appear to be omniscient and will generally know nothing; they will be tiresome company, having the show of wisdom without the reality.[2]

Of interest here is, first and foremost, Socrates's reference to reality and real experience as something that inevitably gets lost when media are involved. Second, this anecdote highlights the historically profound assumption that the role of media in communication necessarily makes for less-than-perfect interaction. A critical perspective on this would start with the suggestion that all communication is inevitably impaired, taking place between people who cannot really understand each other as they are intrinsically different. All communication is—or at the very least contains—miscommunication, and we tend to project our pain about this inevitable imperfection onto media.

A second point of critique of the assumption, that face-to-face communication through direct dialogue somehow leads to a "better" form of mutual understanding, is that in conversation two (or more) people ideally achieve consensus through the art of compromise. In the process, one or more of the discussants would have to surrender their own unique way of looking at and understanding the issue, in effect losing their authentic voice. A dialogue can even be seen as a rather aggressive and exclusive form of communication, as it forces people into interaction, often privileging those with certain personality characteristics, better training, capacity for empathy, or a talent to express themselves charismatically.

---

2. Plato, *Phaedrus* (circa 370 BCE), https://www.gutenberg.org/files/1636/1636-h/1636-h.htm.

A final, more fundamental reflection concerns the wistful premise that pure, authentic, and immediate communication is indeed possible and would culminate in perfect understanding—between a romantic couple, within a group of friends, across society and its citizens. It is, for example, no surprise that a key piece of advice for people in a relationship is that partners—romantic or otherwise—just need to communicate. All of this supposes that people, as individuals, are or can be completely true selves (and true to themselves) to the extent that they possess some kind of coherent core or inner self that can be openly, honestly, and effectively communicated to others. Yet such a classic, Platonic notion of people having a true self that can be known, seen, and understood and expressed authentically is subject to much criticism in the literature, as scholars (in such fields as psychology, psychiatry, anthropology, and philosophy) suggest that whomever we think we are can perhaps best be understood as a temporary and situated model, intersubjectively constituted out of the countless experiences, emotions, and interactions we have on a daily basis throughout our life. In other words, we are a complex and contingent mess, just as our media are.

Of course, this critical take on the supposed merits of communication without media does not negate our desire and need to be seen and understood. In fact, it perhaps explains why people are so invested in their media and care so much how they come across to others in media. For our lives in media, it is important to recognize the historical nature of how we think about machines and technology in general and media in particular, as they come between us and reality, ensconcing themselves in the background of our lives, influencing and shaping all possible interaction, and forever frustrating yet also facilitating our deep-seated desire for pure, authentic, and uncorrupted communication and understanding.

Since human society, throughout history, has increasingly relied on the power of all kinds of machines, technologies, and media to function, our dependence on them has expanded. Especially since the late nineteenth century, technological developments have followed each other at an accelerating pace, amplifying anxieties about their role in human affairs. In this process, it is perhaps unsurprising that people project much of their anxieties, frustrations, and especially all the negative emotions associated with the loss of perfect communication in a fast-moving, rapidly expanding, and increasingly technological environment onto the machines and media of their time. In 1919, the famed psychologist Sigmund Freud's former student Victor Tausk—a Slovakian psychiatrist—published his essay "On the Origins of the Influencing Machine," one of the first research-based analyses interrogating our fundamentally ambivalent relationship with technologies, machines, and media. Tausk outlines the case of one of his patients, Natalija A., who was convinced that her body and emotions were being manipulated from afar by a grotesque machine. She felt that all her family and friends were under the influence of similar machines, everyone connected to their own device, living their lives in an exact simulacrum of the world. As with the *Terminator* story line seemingly inspired by the letters of Samuel Butler, we see hints of the main premise of *The Matrix* film franchise in Natalija's nightmare. Indeed, contemporary representations in popular culture about all-powerful media and machines are never really all that new nor original.

The patient talked about her influencing machine as some kind of magic lantern or cinematograph, noted by Tausk as making her see one-dimensional images of reality. During his sessions with her, Tausk deduced that this machine—described as a jumble of wires, screws, batteries, metal plates, handles, knobs, and cranks—was a projection of the patient's own deep desires and frustrations. Before he could conclude his analysis, the patient broke off all contact, suspecting he had become under the influence of a machine as well, like everyone else in her life. Other analysts followed up on Tausk's work, suggesting that this case can be seen as a typical byproduct of the industrial age, a time understood by many as gradually reducing humans to automatons—powerless cogs in the machine of industry and ever-advancing technology. It can be argued that the recurring worries and fears about media in society are similar expressions about "influencing machines" that make us lose touch with reality and force and direct us, alienating us from our bodies, feelings, and selves, just like the device did in the case of Tausk's patient, the unfortunate Natalija.

Tausk's essay is a classic in psychiatric literature, its influence stretching far and wide—including in popular culture (such as in the 1962 novel and 1975 film *One Flew Over the Cuckoo's Nest*), and most notably in science fiction, providing the inspiration for many to suggest worlds where people live as or among body doubles, governed by an omnipresent technological society or universal machine (for more prominent examples, see appendix 1). Throughout literature, comic books, animation, films, television, and digital games, numerous cases can be found that feature parallel virtual worlds and social systems determined by powerful machines—involving behaviors and life-forms that blend humanity with media and machines and offering a technological illusion of reality that almost without fail ultimately becomes problematic, potentially endangering our survival.

In short, media are considered powerful because they are perceived as coming between us and who we really are, thereby corrupting the chance for real interaction, pure communication, and genuine understanding. It is possible to argue that what fuels our most fundamental interest in media and all things mediated—and what continues to inspire and shape theory and research in this area—is our continued longing for true communication, pure and unadulterated interaction, and the immediate experience of reality. When media and communication students and scholars ask questions about influence, impact, and effects, about how to improve the transmission, distribution, and reception of messages, about how people process and give meaning to content, and so on, they are giving expression to a deep-seated desire for perfect communication.

---

*In short, media are considered powerful because they are perceived as coming between us and who we really are, thereby corrupting the chance for real interaction, pure communication, and genuine understanding.*

---

Whenever we privilege dialogue over dissemination, direct social contact over mediated quasi-interaction online, or interpersonal communication over broadcast communication, we are projecting our anxiety about the lack (or loss) of true communication onto media as

our influencing machine. It is for this reason that questions about values and truth in the context of a life in media are significant, as the strategies we pursue to get closer (or back) to each other and to reality are codetermined by the place and meaning we assign to media in life. Whether we worry about the power of magic lanterns or the subtle steering of algorithms and artificial intelligence, in doing so we give shape and form to our longing for authentic individual expression and the possibility of shared narratives in society.

The three core strategies for our actions vis-à-vis the increasingly intimate entanglement of media, society, and everyday life are to wage war against the machines, to submit to a mediated existence, or to become media. Each of these strategies is discussed in terms of its historical analogy, resonance in popular culture and public debate, and scholarly response.

**The Options We Have: War against the Machines**

The 1909 short story "The Machine Stops" by the British author E. M. Forster (best known for *A Passage to India* from 1924) describes a society that is completely dependent on machines and technology. In this world, people live underground in hexagonal rooms that are fully equipped with a variety of networked devices and screens, supplying food, health care, and entertainment all controlled by a single powerful machine that stretches across the globe. Communication takes place via video and instant messaging—with which Forster seemed to presage such contemporary media as the internet, Skype, and WhatsApp, although the technologies necessary for telecommunications and videotelephony already existed in his time. Nobody gets outside anymore, and everyone seems satisfied with the connections and services that the machine provides. Everything is fake in this world—even the video link between a mother and her son is filtered by the machine so that they do not really have to see each other's true feelings and facial expressions—especially when they express any kind of negative emotions such as sadness, anger, or fear. Instead, they are presented with a sort of sanitized version of each other—a version that is only just sharp enough to distinguish (reminiscent of the "touch up my appearance" option or beauty filter in applications like Zoom and FaceApp). Through the machine, people make contact with many others—with hundreds, even thousands of others—even though people no longer meet in real life, and no one ever touches anyone anymore. The manual of the machine (and with it of all aspects of life) is sacred scripture for the world's population. Eventually the machine falls into disrepair, and when it finally stops, society falls apart completely. Most people die, but a few manage to escape, naked and vulnerable, to the world above.

Forster concludes that the same technology that freed people from their worries also made them utterly dependent, unable to truly express themselves or experience life and incapable of fending for themselves. "The Machine Stops" has been a major source of inspiration for all kinds of books and films about problematic relations between man and machine, as authors, artists, and directors have further explored its theme of a machine-controlled society. Examples of influential popular cinema in this vein include Fritz Lang's epic *Metropolis* from 1927, George Lucas's 1971 graduation film *THX 1138* (made six years before he rose to fame with *Star Wars*, extending Forster's story with an underground society controlled by androids), the *Matrix* franchise, Austrian director Michael Haneke's 2005 thriller *Caché* (focusing on the haunting experience of surveillance) and the Oscar-winning drama *Her* (2013).

Nearly a century before the worldwide breakthrough of the web and social media, Forster's story suggests that it is precisely these kinds of apparently well-intentioned machines, technologies, and media that are capable of destroying human society. The very fact that in his dystopian society people worship their devices as a kind of God is fascinating, indicating a relationship between people and media that is much more profound than pure convenience or pleasure. Apart from this interpretation, it is good to realize that in all his work Forster subtly criticized the repressive class society in England of his time. His glorification of the outdoors, natural life, and the liberating experience of the machine breaking down in this story is also a representation of his dream to escape that kind of (political, economic, social, and sexual) oppression. The theme of ubiquitous machines and media bringing about a rather mechanical and loveless culture is also reflected in many contemporary novels, such as throughout the work of the notorious French author Michel Houellebecq—especially his *Possibility of an Island* (2005), where people endlessly clone themselves to achieve immortality while living in electronic cocoons and having no feelings, free from both pleasure and pain.

Samuel Butler perhaps said it first: war against the machines can be a primal response to the prospect of ubiquitous and pervasive media that potentially produce an oppressive, heartless culture, rendering humans and humanity obsolete. Of course, he made his point quite literally, calling for the destruction of any and all machines. Contemporary variations of his call to arms can be found in *neo-Luddism*, which is a particular anti-technological point of view taking its name from a movement of disgruntled English textile workers in the early eighteenth century, who destroyed factory machinery to put pressure on employers to improve working conditions. Recent examples of neo-Luddite (or reform Luddite) movements tend to be less violent, instead advocating an overall slowing down in our media use, arguing for a disconnection of devices (described as a digital "detox" to counteract online media overuse) and deliberate nonuse of information and communication technologies. In this context people develop such tactics as taking a news vacation, leaving their smartphone at home when they go out, or picking hotels while on holiday that do not offer Wi-Fi access. Electronics manufacturers like Apple even market such products as smart watches as supposedly coming in handy when you do not want to be bothered by your smartphone while out with your friends. Such a "soft" war against the machines is not entirely unproblematic, as for many people being permanently available and online is a sheer necessity because of their work (think, for example, about freelance and so-called gig workers on short-term projects and assignments) or their particular social bonds—consider diasporic families with parents, children, and grandparents scattered around the world, trying to maintain kinship over distance, for whom (mobile) phones and internet connectivity are crucial technologies of love.

Although all-out war against machines is a well-worn narrative device in literature, television, film, and digital games, in public debates about technology in general and media in particular, concerns tend to be less alarmist, albeit still expressed quite forcefully. Well before the global coronavirus crisis of 2020, warlike wrangles about the perceived dangers of (and proposed limits to) media have been common across all levels of society, such as

- widespread parental unease about screen time for children and teenagers, especially since the introduction of game consoles (like the PlayStation, Xbox, GameBoy and Switch);

- surveys among media users in general (and young people in particular) consistently finding that many feel their media use has become excessive, experiencing a lack or loss of control and finding it difficult to withdraw;

- researchers from a variety of fields (especially since the late 1990s) focusing on the question of problematic media use, with "internet gaming disorder," for example, being added in 2013 to the fifth edition of the authoritative *Diagnostic and Statistical*

*Manual of Mental Disorders* (a primary classification system for psychiatric disorders) and intense scholarly engagement on such issues as compulsive internet use, pornography addiction, and smartphone dependency; and

- political apprehension about the role and impact of disinformation campaigns, the extraordinary reach and perceived power of internet platforms (such as Facebook), and the lack of privacy protections for citizens online, inspiring new laws around the world such as the 2016 General Data Protection Regulation (GDPR) and 2022 Digital Services Act of the European Union, a newly amended Act on the Protection of Personal Information in Japan (2017), Chile amending its constitution to include data privacy as a human right in 2018, and numerous governments around the world introducing similar legislation concerning the protection of personal information in the 2020s (including Brazil, India, South Africa, Australia, New Zealand, Switzerland, and China).

Generally, anxieties such as these have been common regarding all mass media throughout history, yet this was never followed up by much official response. In fact, it could be argued that the political and regulatory response to media (both as an industry and as a formidable force in society) has been generally lackluster, with state actors opting for nonintervention and deregulation rather than strict policy interventions (with the possible exception of copyright legislation). Similarly, in the scholarly and educational community, early approaches to media education primarily centered on promoting national media cultures, for example, teaching students about the rich history of local and regional film and journalism. With the rise of television as a mass medium and the emergence of video games (in the 1980s and beyond), this changed, as policy makers, educators, and researchers around the world proposed and pushed for more direct action against the media. Numerous media literacy initiatives were launched, culminating in a wide variety of workshops, courses, and even entire school curricula intended to train people to survive a context of media abundance, warning pupils against a variety of risks and harms associated with media use and generally using strictly skills and protocols-based approaches in teaching about media (more on media literacy in chapter 8).

With the introduction and rapid rise of the World Wide Web, mobile communication, and online gaming, critical political and pedagogical approaches further proliferated around the world, generally inspired by a protectionist philosophy—whereby digital literacy is considered a defense against (potentially) dangerous online content and experiences. As literacy initiatives have matured and become more evidence based, since the early twenty-first century calls have been made—such as by the United Nations Educational, Scientific and Cultural Organization (UNESCO) in consultation with teachers and researchers from all over the globe—for programs that are more geared toward empowerment, while still recognizing the potential perils associated with media. Within the overall war against machines frame, learning about the constructedness of media seems to be possible too, with programs focusing on the exploration and enjoyment of media, as well as helping children to make their own media.

Interestingly, a prominent source for a more hopeful and less fearful stance regarding media and technology, yet mindful of their powerful presence in society and everyday life, comes from the work of Ivan Illich, an ordained Austrian Catholic priest and philosopher, living and working for most of his life in Puerto Rico and Mexico, as well as in the United States, Germany, France, and Italy. Illich wrote and published the seminal book *Tools for Conviviality* in 1973 to worldwide acclaim, wherein he argues that technology should be a tool with which we must learn to live together to the benefit of all. Illich envisioned a relationship with machines as potentially freeing people from their dependency on political elites, corporations, and other top-down institutions (including schools). In earlier work, he advocated using computer networks to achieve a truly universal education system, by peer matching people to learn in cooperation. What makes his work relevant to considerations of a life in media is the influence he had on the people involved with designing and building the first personal computers in the 1970s and 1980s. These (mainly American) engineers and computer scientists considered Illich an inspiration for building machines that would enable each individual user to have their own terminal, connected to peers anywhere else, making them less dependent on the industry-owned and operated mainframe computers of the time. As today's media are, in some way, all versions of those very first personal computers and networks, the devices that make up our digital environment embody an empowering ideal, even though they carry potential danger as well, and now mostly are commodities produced and marketed by global commercial corporations. History suggests, however, that our digital environment is by design generative and potentially emancipatory—in that we can connect, create, and program in and on top of it.

In the practice and discipline of media studies, a warring approach to omnipresent media in society and everyday life continues to inspire a vast area of research focused largely on

- the (consequences of) concentration of media ownership and state (or corporate) control over the media,
- hidden messages in the content of the media (generally assumed to promote a consensual, mainstream and distinctly commercial view of the world), and
- the extent to which people understand (and act on) the media content they consume.

Fears about the real or perceived impact of media in this tradition of scholarly research frame people predominantly as powerless audiences addressed by companies and the state. Media tend to be seen as relatively uniform, promoting a corporate or dominant political view of the world, stifling diversity and dissent in media production and content. Raising critical awareness about this presumed role of media becomes a key strategy in the academic war against the machines. This is especially the case in more authoritarian and dictatorial regimes, where control of the media and regulations against their undue influence are part of the policy toolkit of ruling parties. In such contexts, whatever the dominant narrative is that gets produced and published tends to have real consequences for the people and communities involved, and struggles about reality and truth in media have profound political significance. Consider, for example, the way a war gets framed and promoted in the national media of an aggressor, versus those of victims and defenders. Such framing is not just a matter of representation, as it tends to have significant consequences related to people's willingness to accept armed conflict, or to pursue peace.

## The Options We Have: Surrender to a Mediated Reality

What if there really is no difference at all anymore between life and media, between the real and the virtual, between humans and machines? What if we, in other words, surrender to a comprehensively mediated reality? Although this may seem like a far-fetched scenario, some of the earliest formal considerations of such a likelihood reach far back into history. The fusion of human and nonhuman with direct reference to machines owes tribute to an emerging mechanistic worldview in the seventeenth century. During that time, intellectuals increasingly came to see the world less as divine creation and more as a vast machine subject to human manipulation—a worldview inspired by the many mechanical inventions at the time (such as the printing press, steam engine, and calculating machine). The English philosopher Thomas Hobbes, credited with developing the now common notion of the need for a social contract in society between a governing authority and its citizens to prevent warfare and chaos, in 1651 wrote (in his most famous work, *Leviathan*): "For seeing life is but a motion of limbs, the beginning whereof is in some principal part within, why may we not say that all automata (engines that move themselves by springs and wheels as doth a watch) have an artificial life? For what is the heart, but a spring; and the nerves, but so many strings; and the joints, but so many wheels, giving motion to the whole body, such as was intended by the artificer?"[3]

For Hobbes, seeing humans in terms of mechanical parts and understanding society through the metaphor of a machine offered the freedom to imagine and construct an ostensibly hierarchical, authoritarian system of civil government within which everyone would function as parts of the whole in orderly fashion. His contemporaries expounded upon these ideas, including the French physician Julien Offray de La Mettrie, who in his 1747 essay "L'homme machine" (translated into English as "Man a Machine") put forward the argument that human beings did not have souls, as all processes in our minds and bodies are related, and we should therefore think of ourselves as machines following the laws of nature (rather than those of, for example, dictates of the church): "I think that everything is the work of imagination, and that all the faculties of the soul can be

---

3. Thomas Hobbes, *Leviathan*, ed. Edwin Curley (Indianapolis: Hackett Publishing, 1994), 3–4.

correctly reduced to pure imagination in which they all consist. Thus judgment, reason, and memory are not absolute parts of the soul, but merely modifications of this kind of medullary screen upon which images of the objects painted in the eye are projected as by a magic lantern."[4]

La Mettrie's conceptual move beyond a mind and body distinction and beyond divine governance (while invoking the notion of reality as projected by a magic lantern similar to Victor Tausk's patient Natalija almost two centuries later) got him into trouble with the authorities, forcing him to flee his native France. However controversial, his ideas were just another expression of what the German thinker Gottfried Leibniz wrote about earlier, in 1714, as part of his comprehensive theory of everything. At the end of an extraordinarily prolific career, Leibniz tied all the strands of his work and thinking across a variety of fields (including mathematics, philosophy, science, religion, and politics) together in "La Monadologie," an essay putting forward the argument that everything in the universe—humanity, technology, nature, and the cosmos—is made up of monads: simple, invisible substances that are all interconnected, acting in preprogrammed unison and harmony without or beyond human intervention. He considered every organic body as a "divine machine or natural automaton," drawing no distinctions between people, animals, and machines. Unlike La Mettrie (and to a lesser extent Hobbes), his work did not run afoul of the church, as Leibniz was quick to add that the only necessary monad in his entire system was God and that therefore the world we live in is and always will be "the best of all possible worlds" (as quoted from his earlier book *Théodicée*, published in 1710).

The anecdotal examples of Hobbes, Leibniz, and La Mettrie are used here to show how our current distress about a mediated reality (where distinctions between true and fake, real and virtual, and between humans and their intimate technologies are imperceptibly disappearing) have powerful, far-reaching historical antecedents. This in turn is important to prevent a certain "nowism" in our analyses of life in media, avoiding the point of view that whatever we are observing or experiencing right now has no history, is unique and particular to this moment (marked by the technologies of the day), and can only be understood in its own temporal context. Making sense of media can fall victim to *shiny new toy syndrome* and *technomyopia*, meaning that we tend to be overly invested in the latest technological breakthrough, fancy new smartphone, hit TV series, or tentpole movie franchise, thereby forgetting the genealogy of such media and underestimating the long-term impact of any particular media artifact, activity, and social arrangement. At the same time, I have to caution against selectively quoting from venerable texts to sustain arguments about what is happening right now, as that runs the risk of obscuring historical nuance and context. For our consideration of a life in media, it is simply fascinating to note that the boundarylessness between humanity and technology has a deep history and

---

4. Julien Offray de La Mettrie, "L'homme machine" ("Man a Machine,"), ed. Mitch Abidor, Marxists.org, https://www.marxists.org/reference/archive/la-mettrie/1748/man-machine.htm.

cannot just be seen as an exemplary aspect of only our current experience of virtual reality games and documentaries, augmented reality mobile apps, and extended reality (XR) software—all contributing to an overarching metaverse, where social and mediated reality collapse into each other. Revisiting age-old debates that have relevance to our current predicament illuminates where our thinking comes from and possibly offers some inspiring ways of sense making while preventing earlier mistakes and pitfalls.

From the seventeenth century onward, machines became increasingly complex and gradually took over more aspects of human action. People's reliance on technology grew—and with it came accounts throughout literature, the arts, and intellectual life of human-machine blending and coming together, expressing deep ambivalence about our relations with technology. In 1731, during the time of Leibniz and La Mettrie, the Irish author, clergyman, and satirist Jonathan Swift—famous for his 1726 book *Gulliver's Travels*—composed the poem "A Beautiful Young Nymph Going to Bed," in which he describes in detail how a sex worker comes home at night and before going to bed strips off everything that makes her beautiful: a wig, a glass eye, mouse-hair eyebrows, and a steel corset. Without all these tools and devices, there is nothing left of her when she wakes up in the morning. A century later, this poem inspired the American writer Edgar Allen Poe—credited as the inventor of the detective fiction genre—for his short story "The Man Who Was Used Up" (published in 1839). He describes an encounter with a nearly perfect man—one John A. B. C. Smith—who, on closer inspection, turns out to be nothing more than a composite mechanism of cork legs and arms, a metal chest, wooden shoulders, dentures, and wig, and even his palate was specially crafted by a doctor. In both stories, technology conjures an image of pure perfection that ultimately serves to disguise (and solve the problem of) the fragile and messy nature of the human body. Another important influence on Poe was the work of the German author E. T. A. Hoffmann, especially his story *Der Sandmann* (appearing in 1816). In this novel, a young man (Nathanael) falls in love with a woman (Olimpia) who seems absolutely perfect to him, and they get married, only for him to discover to his horror that she is in fact an automaton. He runs away from her and tries to come to terms with himself again but eventually cannot tell the difference between humans and mechanical beings anymore.

The literary works of Swift, Hoffmann, and Poe fit into an artistic tradition and era in which humans begin to develop an uneasy alliance with machines – in part inspired by important works in the fields of mechanical engineering and technological design of (at times collaborating) Arab, Jewish and Christian scholars from the 11th century onward, and innovations coming from trade and conflict between European, Southeast Asian and Sub-Saharan peoples. On the one hand, technological applications offer all kinds of outcomes for people: a wig makes you look more youthful; glasses, binoculars, telescopes, and microscopes allow us to see better and further; electricity makes life easier; industrial production simplifies and speeds up the work process. On the other hand, people's dependence on all this technological ingenuity systematically advances—and with it our

apprehension accumulates. It is in the inextricable correlation between opportunity and dependence that we may lose ourselves to machines, technology, and media, in the process discovering different ways of making sense of the interdependence that ensues.

At the beginning of the twentieth century, we meet Sigmund Freud again, as the Austrian neurologist uses Hoffmann's story of the bizarre love of Nathanael for a robot to describe and diagnose what he considers a primary psychological process: the fundamental unease we all sometimes feel about who we are and how we stand in the world. In two famous essays from 1919 and 1930, Freud elaborates on his proposition that, although technological developments allow people to overcome all kinds of imperfections of body and mind, we are also increasingly further away from a direct, natural experience of reality. Being both what Freud calls "a feeble animal organism" and a "prosthetic God"[5]— divine because of our superior powers made possible by the technology, machines, and media—fills us with agonizing ambivalence and uncertainty. Freud suggests that even what seems comfortably familiar and thus known to us becomes (or can become) profoundly unfamiliar, unsatisfying, and even eerie when we look at it differently—like happened when Nathanael looked at Olimpia, his bride to be, through a spyglass, both falling deeply in love with a seemingly perfect companion and disgusted at the sight of a soulless automaton. In the process, we cannot tell real from fake anymore, causing confusion and anxiety, highlighting how the experience of losing touch with a settled notion of reality can be vexing, even horrific. Freud coined such fundamental feelings of ambivalence and internal conflict about our place in an increasingly technological world as a general sense of *uncanniness*. It is an uncanny experience that shines through all accounts, throughout history, of what happens when we gradually let go of holding on to strict frontiers between humans and machines and between media and (the reality of) life. The experience of uncanniness is quite unsettling, as all of a sudden the rules, protocols and rituals that structure everyday life become meaningless and even absurd. In the context of a life in media we consider how media amplify and accelerate feelings of the uncanny in ways that are both unnerving as well as encouraging new ways of observing, learning and understanding.

One of the best-known examples of our all-too-human feeling of uncanniness in relation to technology comes from the field of robotics. In 1970, the Japanese roboticist Masahiro Mori published an article in which he explores how far our tendency to anthropomorphize machines—also known as the *ELIZA effect* (after the 1966 chatbot ELIZA that mimicked a psychotherapist and, unexpectedly at the time, elicited strong emotional responses from its users)—stretches when robots are designed to be more human looking. In his work, Mori noticed that robots—such as a prosthetic hand or a mechanical puppet—were more appealing when they looked more human, up to a certain point. When robots appear very close but not entirely human, people tend to feel uncomfortable or even disgusted. This feeling of uncanniness disappears, Mori argued, when robots become truly lifelike. He

---

5. Sigmund Freud, *Civlizations and Its Discontents*, ed. J. Strachey (New York: Norton, 1989), 44.

called the gap between robots as somewhat realistic (yet making us feel apprehensive) and as perfect human replicas the "uncanny valley"[6]—a concept that researchers all over the world have explored and tested since, finding evidence particularly in the media industries. Examples include children's responses to early screenings of the character Princess Fiona in the 2001 animated film *Shrek* (produced by DreamWorks) as too real and a public outcry on social media upon the release of the 2019 film adaptation of *Cats* (directed by the British Australian filmmaker Tom Hooper) for its creepily realistic depiction of almost-human felines. The uncanny valley is vitally used in horror movies, giving non-human entities such as dolls human characteristics and using the unsettling ambiguity that follows for full dramatic effect.

In addition to being a source of inspiration for Freud's psychoanalysis, Hoffman's account of the intimate relationship between man and automaton formed the basis for the most famous and influential science fiction film of all time: Stanley Kubrick's *2001: A Space Odyssey* from 1968, based on a novel by the English futurist Arthur C. Clarke. The film mainly revolves around the relationship between an astronaut and the onboard computer of his spaceship. This computer, HAL (short for "*h*euristically programmed *al*gorithmic computer"), is an advanced artificial intelligence capable of having and expressing feelings such as pleasure, suspicion, jealousy, fear, and anger. Although the relationship between the astronaut and HAL is very warm at the beginning of their space journey, the computer turns out to be untrustworthy. An intense cat-and-mouse game ensues between the two, in which they try to kill each other. The so-called replicants in Ridley Scott's *Blade Runner* films from 1982 and 2017 (based on Philip K. Dick's 1968 novel *Do Androids Dream of Electric Sheep?*) are also modeled as artificial life forms after the automaton in Hoffmann's narration. The replicants resemble humans to such an extent that the difference can only be proven by means of a special psychological test. In fact, even the replicants themselves do not know whether they are robots or human (this enigma also applies to the role of one of the main characters, in both movies played by the American actor Harrison Ford).

The evolution of machines from Swift's sex worker to the replicants in *Blade Runner* in popular fiction is remarkable and fairly consistent. Each time, technological applications and innovations come a little closer, and they take over more and more functions and actions of humans, only to be exposed as hostile and dangerous. Even some more recent and nuanced popular film adaptations of increasing intimacy and boundary-blurring relations between man and machine, such as Spike Jonze's *Her* from 2013 and the low-budget production *Ex Machina* from 2015, do not escape a story line in which quite lovable media—a computer operating system and an intelligent humanoid robot, respectively—ultimately leave human beings behind.

---

6. Masahiro Mori, "The Uncanny Valley," trans. Karl F. MacDorman and Norri Kageki, *IEEE Robotics and Automation*, June 2012, https://spectrum.ieee.org/the-uncanny-valley.

Beyond the arts, the public and political response to the boundary-blurring capacities of life in media has been equally mixed. Policy makers as well as individual people around the world have passionately protested the camera-equipped cars from Google's Street View taking pictures in their neighborhoods, cities, and countries. After Street View's launch in 2007, lawsuits and formal complaints were filed in countries as varied as Brazil, Canada, India (where Street View has been rejected by the government), and Germany (where hundreds of thousands of households have opted out of the system). However, in most places, Google continued operations after the initial turmoil. Equally ambivalent has been the use of augmented reality applications on mobile phones, such as the *Pokémon Go* game developed by Niantic (originally a startup within Google) together with Nintendo. It uses mobile devices to locate, capture, train, and battle virtual creatures, called Pokémon, which appear as if they are in the player's real-world location. When the game launched in 2016, its global popularity regularly caused traffic jams and urban chaos around the world in places where rare virtual beings were said to be found. While the game was banned in some locations—such as historic and religious sites, election polling stations, and military bases—the developers were also called on to increase the number of Pokémon spawns and Pokéstops (places where players can get free items for the game) in minority neighborhoods and rural areas to enable a more diverse range of people to participate.

The developments in augmented and extended reality hardware and software development generally proceed without much political or public intervention, suggesting that the ongoing blending of the real and the virtual has become more mundane. The worldwide use of COVID-19 smartphone contact-tracing applications provides a case in point, with countries all over the world commissioning and advocating the use of such apps, embracing a bewildering variety of technological systems and protocols without a coherent regulatory framework in place—for example, to protect user privacy and prevent widespread surveillance. Some apps are supported and developed by local or regional governments (such as in North America and Russia), some government mandated (as in China, South Korea, and Singapore, as well as France, Norway, and Hungary), and some based on a joint interface by such corporations as Google and Apple (incorporated in their operating systems as of late 2020, used by Germany, the Netherlands, and the United Kingdom among others). These apps are based on various techniques that can provide users with infection-risk-level information specific to their communities and notify users who have been exposed to COVID-19. At the same time—and with occasional rapidly rising infection rates—people report feeling frustrated with living with a "pingdemic" as such apps continually notify users after close contact with someone who tested positive with the virus.

The combination of a near-global rollout of mobile software applications for digital contact tracing and vaccination passes with ongoing critical debates about privacy and surveillance is a good example of misgivings signaled throughout historical accounts of man-machine association, as is the general consensus that without such programs our options for fighting a pandemic would be much more limited. A similar instrumental position is taken up in many, if not most, early twenty-first-century approaches to digital and media literacy curricula for primary and secondary schools, which often tend to focus on what some critics would call "button pressing" on computers and mobile devices and pared-down skills, such as managing privacy settings in an online social network or installing and using a virtual private network (VPN) to prevent tracking and securely go online. While such competences are clearly important, the absence of critical uses of media (including deliberate nonuse) suggests that the war against media may be over.

The scholarly strategy in media studies within an overall surrendering position sees in the media less a controlling and corralling force, instead appreciating more complex and reciprocal understandings of the relations between people, technologies, and media. Researchers in this tradition look closely at how people appropriate, understand, and give meaning to media in recognition of the fact that none of this may align with the uses and interpretations as intended or preferred by media creators, companies, or corporations. Rather than consider mediated communication a one-way, linear process leading from a select few people that make and send messages—governments, media corporations, professional media makers, elites—to influence relatively indistinct masses and shape public opinion, media are seen as part of an entire *circuit of culture* indicative of the interdependent relations between people and media, featuring five interconnected elements:

- production, or the process of making media (from conception to creation, marketing and distribution);
- identity, or who the people, groups, and networks (and their norms and values) are involved in the various stages of making, distributing, regulating, and using media;
- representations, or the form, format, and genre of a media product or message;
- regulations, or the formal and informal rules and controls regarding media (including laws, policies, cultural norms, and expectations) and how these are enforced; and
- consumption, or all the different ways in which people engage with media.

Such multifaceted approaches to the study of media indicate a gradual laying down of arms against them. Instead, media are seen as part of a much broader system of meaning making in society, perhaps unavoidable yet certainly not all-powerful. A key assumption is that people have agency at various stages in the circuit of culture, even though there is still much apprehension about our increasingly intimate relations with media.

## The Options We Have: We Become Media

As shown before, prominent thinkers have argued as far back as the seventeenth century that humans can perhaps be seen as machines. The account of man-as-machine is part of even older narratives, dating back thousands of years. Consider, for example, the ancient Greek myth of Talos, a giant bronze man built by Hephaestus, the Greek god of invention and blacksmithing. What made Talos an early imaginary form of artificial life was a tube running from his head to one of his feet that carried a mysterious life source, enabling him to walk around the island of Crete and protect it from invaders. The same inventor designed several other mechanical beings—including Pandora, an artificial life form sent to earth to mingle with humans and unleash her box of evil (as punishment for the discovery of fire).

These kinds of stories are not particular to Western legends and religions. In fact, most creation myths around the world involve deities creating human-like life forms out of raw materials of the earth such as clay, sand, twigs, and water and charging them with some divine or otherwise vital spark. This includes the Chinese goddess Nüwa creating people from river mud as she felt lonely, while Sumerian mythology suggests the primeval mother Namma gave birth to humans by kneading clay and placing it in her womb. World religions such as Judaism, Christianity, and Islam all have similar origin stories where God breathes life into humans after creating them from dust and clay. The potent mix of earthy ingredients and a divine life force at the heart of all these narratives provides the groundwork for the third option we have in media life: to take to heart the intrinsic and inevitable interwovenness of humanity and nonhuman elements (including technology) and to become media.

What makes the part-human, part-machine characters in literature, film, and other works of fiction creepy is their lack of life force. The various mechanical creatures, automatons, robots, and cybernetic organisms populating popular culture (as mentioned throughout this chapter and book) may look like humans, move like humans, and talk like humans—but they all tend to miss something. However, this seems to be changing. Contemporary imaginary human-machine hybrids often are compassionate living entities, worthy of affinity and even admiration and capable of loving and being loved. This subtle shift in perspective—from waging war against the machines via some kind of uneasy coexistence to an empathic embrace of artificial life forms—coincides with the worldwide popularity of the World Wide Web, a global shift toward mobile broadband communication, and the ongoing computerization of all kinds of everyday devices (initially including phones, cameras, printers, and cars, while the tech industry intends to make just about everything "smart").

To some extent, the apex of benevolent human-machine convergence in modern fiction has been the aforementioned *Matrix* franchise (including motion pictures, animation series, comics, and digital games, debuting in 1999). The film tetralogy is part of an overall storyworld—a narrative universe consisting of (potentially) unlimited stories. The central theme of the *Matrix* storyworld is the technological fall of humankind, in which the creation of artificial intelligence led the way to a race of self-aware machines that imprisoned humankind in a virtual reality system—the Matrix—while their bodies are farmed as a power source. The storyworld approach taken for the *Matrix* is now considered a benchmark for *transmedia* storytelling efforts throughout the media industries (more about this in chapter 7 on making media), where different media are used for different parts of the story, with each medium deployed as it is most suitable for that particular part of the world.

The comic *Bits and Pieces of Information* (written by the directors of the *Matrix* films, Lana and Lily Wachowski, published in 1999) opened the overarching narrative with a backstory about self-aware robots, the start of a global civil rights movements for machines, and the beginnings of what would become an all-out machine war (which the robots initially lost). This story was retold in an animated short called *The Second Renaissance* as part of *The Animatrix*—a collection of nine such short films commissioned in Japan by the Wachowski's after the release and global success of the first *Matrix* motion picture. The other shorts served to add details to the overall narrative and bridged story lines between the three main films and several video games, told mainly from the machine point of view. The game *Enter the Matrix* (released in 2003, created by the American game studio Shiny Entertainment) connects the story of the *Animatrix* short *Final Flight of the Osiris* with the events of the second *Matrix* film *Reloaded* (released a week after the game), while the online video game *The Matrix Online* (2005) acted as a direct sequel to *Revolutions* (which was released later in 2003, as the second and third installments of the franchise were originally intended to be one film).

What made the *Matrix* storyworld more than just another rehash of Samuel Butler's call to wage war on the machines was its complex and deliberately confusing story line, where the fine line between the human resistance versus all-powerful machines was continually questioned and problematized throughout the various installments of the franchise. The Wachowskis' main source of inspiration for this was the work of Jean Baudrillard. The French philosopher's most famous book, *Simulacra and Simulation* (1981), can be seen in the opening scene of the first *Matrix* film, as Thomas Anderson (the generic human name of Keanu Reeves's character Neo) opens it to a chapter entitled "On Nihilism." In the movie, the book turns out to be hollow, serving as Neo's hiding place for black market software. In this influential work, Baudrillard develops an argument—which he would sustain throughout his career until his passing in 2007—that in a comprehensively mediated society, whatever we think "reality" is, is in fact a simulation consisting of symbols and signs we derive from the media. At the time of writing his book, Baudrillard primarily used magazines, television, and especially advertising as his frame of reference. He considered the real world as a lifeless desert, unable to grow and generate new versions of itself, instead generating endless copies of images and ideas taken from the media, which seem to have more vitality than their now hollow, emptied out, and meaningless originals.

In true intertextual fashion, Baudrillard was inspired for his argument by a conceit offered in a short story titled "On Exactitude in Science" (originally published in 1946), written by the Argentinian friends and coauthors Jorge Luis Borges and Adolfo Bioy Casares. In this narrative, they explore the devastation caused by an empire that commissioned a map of itself completely true to 1:1 scale. The map became so all-encompassing that the empire lost itself within it. Subsequent generations, lacking the skill and knowledge of the original cartographers, could not see any more that they were living in a perfect map of the empire rather than its now deserted and foregone original. Baudrillard comes to a similar conclusion in his take on the role of media in people's perception and experience of reality, suggesting that, ultimately, in a fully mediated existence any relation with an original place that would have existed before its rendering in media gets destroyed: "Abstraction today is no longer that of the map, the double, the mirror or the concept. Simulation is no longer that of a territory, a referential being or a substance. It is the generation by models of a real without origin or reality: a hyperreal. The territory no longer precedes the map, nor survives it. . . . It is the real, and not the map, whose vestiges subsist here and there, in the deserts which are no longer those of the Empire, but our own. The desert of the real itself."[7]

Baudrillard's dramatic reference to a "desert of the real" (an allusion to Borges and Casares) is used in the first *Matrix* film, where the character Morpheus introduces the hero Neo to the only part of the earth not yet taken over by machines with the words "Welcome to the desert of the real." What sets the Matrix storyworld apart from earlier approaches to the humanity-versus-technology conundrum is its deliberate choice to complicate the existence of these two factions. Is the real still there, or is it a figment of our imagination? Or perhaps it is just a simulation created by computers and machines? Even more so, could it be that the humans in the story are, themselves, machines—and vice versa? None of these questions gets effectively answered, leading Baudrillard, when asked what he thought of the franchise, to remark (in an interview with French magazine *L'Obs* from July 2004): "The Matrix is surely the kind of film about the matrix that the matrix would have been able to produce."[8]

The theme of being lost in an alternate reality produced by (or in) media and being unable to escape or return runs throughout twentieth- and early twenty-first-century popular fiction. In some instances, the heroes of such stories ultimately succeed in getting back to reality—which perpetuates earlier conceits about a supposed clear boundary between humanity and technology. Especially in speculative fiction that explores emotional aspects of human-machine collusion, another narrative can be found: that of the possibility of real connection, empathy, and even love. The aforementioned motion pictures *Her* (2013) and *Ex Machina* (2015), for example, deliberately create situations where we, as an audience, are meant to feel empathy and even affection for the artificially intelligent virtual assistant and intelligent humanoid robot featured in these films, thereby questioning our own humanness (as the main human characters in these films

---

7. Jean Baudrillard, *Selected Writings*, ed. Mark Poster (Stanford, CA: Stanford University Press, 1988). https://web.stanford.edu/class/history34q/readings/Baudrillard/Baudrillard_Simulacra.html.

8. Jean Baudrillard, "The Matrix Decoded: Le Nouvel Observateur Interview with Jean Baudrillard," trans. Gary Genosko and Adam Bryx, *International Journal of Baudrillard Studies*, July 2004, https://baudrillardstudies.ubishops.ca/the-matrix-decoded-le-nouvel-observateur-interview-with-jean-baudrillard/.

do). Similarly, the traditional science fiction franchises of the *Terminator*, *Star Trek*, and *Battlestar Galactica* series in their latter iterations feature cybernetic organisms as creatures capable of love and being loved (see chapter 5 for a more detailed discussion of media love).

Real-world telling examples of ongoing communion between media and life can be found throughout the strategy documents and business plans of tech companies. Speaking at the World Government Summit in Dubai in February 2018, Tesla, Twitter, and SpaceX chief executive Elon Musk told the audience that people need to become *cyborgs* to remain relevant in an artificial intelligence age, claiming the necessity of "a merger of biological intelligence and machine intelligence."[9] Musk's involvement with the neurotechnological company Neuralink (founded in 2016) speaks to his belief in electronic brain-computer interfaces as a future for all humanity, not just for paralyzed people—who are among the first beneficiaries of brain implants and eye trackers to be able to communicate directly with and through computers.

Google also works hard to integrate hardware and software in human operations. Since 2008, the company has funded the work of Ray Kurzweil, a technology forecaster and

---

9. Sarah Marsh, "Neurotechnology, Elon Musk and the Goal of Human Enhancement," *Guardian*, January 1, 2018, https://www.theguardian.com/technology/2018/jan/01/elon-musk-neurotechnology-human-enhancement-brain-computer-interfaces.

evangelist of artificial intelligence. Kurzweil got hired full-time by the company in 2012 to develop software able to independently learn, predict, and replicate human language. Before this, he is credited with inventing a computer scanner, the first speech computer, and, together with musician Stevie Wonder, a synthesizer that mimics the sound of a piano. The core of Kurzweil's inventions is always replacement: digital scans instead of books, a synthesizer for a piano, software that substitutes speech and writing. In addition to being an inventor and engineer, Kurzweil regularly publishes books in which he foresees a future of complete human-machine fusion. In *The Singularity Is Near* (2005) he predicts that artificial intelligence will develop so rapidly that humanity will no longer be able to keep up by 2045. The only option left to survive, Kurzweil suggests, is to merge completely with intelligent machines. He sees that moment—called the *singularity*—as a future full of promise, proclaiming that we can all live forever in (and as) media. This quest for immortality is as old as human history. One of the earliest known literary writings in the world, The Akkadian poem *The Epic of Gilgamesh* (written down on stone tablets around 2200 BC), tells the story of a Sumerian King in search of everlasting life out of grief over a lost friend. Throughout the world's literature, references can be found of people seeking out eternal life through special herbs, fountains, stones, and potions. The historical link between immortality and the nonhuman gets renewed articulation in our current anxiety about artificial intelligence extending or even replacing life.

Kurzweil's views on the fantastic implications of a universal artificial intelligence are not particular to Google (or its parent company, Alphabet). Beyond Silicon Valley and the tendency of its dominant companies to see human beings (and our problems) primarily in terms of technological and computing solutions, the expectation of an immortal future for humanity as fused with machines tends to be shared by others, including the Russian media mogul Dmitry Itskov. In 2011 Itskov founded the 2045 Initiative: a nonprofit organization with a mission "to create technologies enabling the transfer of an individual's personality to a more advanced non-biological carrier, and extending life, including to the point of immortality."[10] This "neo-humanity" (as Itskov calls it) would also necessitate a new religious and political order to cope with people living forever. In the same vein, the former Google engineer Anthony Levandowski, along with some other technology enthusiasts, in 2017 founded the religious company Way of the Future, whose main mission was to "develop and realize a God based on artificial intelligence and through understanding and worship of this God contribute to the betterment of society" (the church closed in 2021).

An equally ambitious proposition comes from the Japanese roboticist Hiroshi Ishiguro, director of the Intelligent Robotics Laboratory at Osaka University, whose Geminoid Project aims to build a robot that is as similar as possible to a live human being. Ishiguro believes it may be possible to build an android in the near future that is indistinguishable

---

10. Dmitry Itskov, "Frequently Asked Questions," "2045" Strategic Social Initiative, July 17, 2021, http://2045 .com/about/.

from a human. In his 2014 book *How Human Is Human* (published in English in 2020), Ishiguro documents his lifelong experiences of designing increasingly sophisticated humanoid robots (including a copy of himself), exploring the principle of body ownership transfer by downloading his identity, his memories, and his emotions into his android double—to live forever through his surrogate.

Perhaps all of this sounds a bit fantastic, figments of the lively imagination of privileged emperors and kings, Hollywood filmmakers, technology entrepreneurs, roboticists, and the superrich, well beyond the reach of the rest of us. Yet it can be argued that, in many ways, you and I have already become media. This point can be pursued by following the definition of media (as outlined in chapter 2)—as the artifacts we use, the activities we undertake with media, and how all of this fits into the daily routines of our lives—regarding how all of us become media.

***Becoming media: as artifacts*** We operate smartphones through touch and voice control, wave our arms around and dance in front of a motion sensor to operate game characters on a screen, sometimes shop at supermarkets and convenience stores using self-checkout systems enabled by electromagnetic tracking and scanning, and are increasingly confronted with digital biometric identifiers and authentication systems (including facial and voice recognition software, fingerprint scanners, automated analysis of keystroke patterns, and DNA matching databases). In each instance, some part of who we are permanently becomes digitized and part of a vast machine-to-machine communication infrastructure, which has ramifications for how things work in our immediate environment. Quite literally, our bodies become remote controls, joysticks, wallets, and passports.

Another way we are becoming media through the artifacts in our environment can be traced via fast-paced developments in the design and implementation of a global internet of things (IoT), the miniaturization of computer chips, and all kinds of wearable technologies. This includes activity trackers, heart rate monitors, smart watches, virtual reality headsets, augmented reality glasses, biocompatible electronics and digital tattoos, as well as e-textiles and smart fabrics in the fashion industry that combine clothing with electronic components. Granted, many of these devices and technologies are not available (yet) all over the world, but it can be argued that the effects of media are not just *instrumental*—as occurring when people use specific media—but also *environmental*, meaning that media contribute to shaping the kind of activities and arrangements that seem possible for each specific medium from the moment we imagine and talk about such devices (well before people's actual use of them). The worldwide push to make the objects in our environment increasingly smart is an example of the power of imagination, shaping how we envisage the implementation and use of future media. On a sidenote: this is why science fiction (in literature, film, and digital games) is such a fruitful area to study, as it provides a touchstone for the design and intentionality of emerging technologies from the perspective of today's expectations, hopes and fears.

Benchmarked by the ability to tag objects with information that can be read at a distance by other objects, the internet of things provides ways to identify, track, and coordinate flows of information, products, and people between different technological systems, across time and space. Human implantable microchips and remote biosensor systems—including the kind of enhancements intended by Musk's Neuralink—further conjure up uses for health care, defense, and education, as much as commerce and entertainment. The world of objects, devices, and technologies around us ceases to be one that is passive and inert. Instead, we are now moving—whether we want to or not—through a progressively animate environment, where artifacts are not just meaningful through our handling of *them* (i.e., switching something on or off, clicking, touching or swiping, voice activating) but that are also powerful as they handle *us*—by sharing data gathered in real time and enacting automated and algorithmic decision-making protocols on the basis of this exchange. In a context of global surveillance and an internet of things, maintaining a clear-cut distinction between "us and them" is perhaps less useful than careful consideration of how the entire human-machine environment acts in unison—as always not seamlessly but rather messily.

***Becoming media: as activities***  Quite often, when we use media, we forget not just what we are doing (as outlined in chapter 2), but we also unlearn that we are interacting with distinct hardware and software. We tend to respond to media in fundamentally social and natural ways, equating media with real persons or places. When a laptop freezes just when we want to share an important document, the television shows a blank screen at a crucial point of the game, or an automated voice assistant on the phone is particularly kind to us, we have an emotional response to the computer, TV set, and telephone that is comparable to what we would have if these were just other human beings: we become upset, angry, or friendly. The notion that media are not simply tools or channels for transmitting and receiving messages but function as actual and active participants in our social world and that we respond to them as such is at the heart of this kind of thinking about the role of media. This can be coupled with the crucial awareness that media also perform a variety of actions affecting us without any human intervention—such as when computers are taken over by a virus or when an algorithm determines that you pose a risk as a traveler or taxpayer.

We tend to take our media intensely personally. People spend so much time using and being exposed to multiple media that those media become their life, in effect treating media as their home. This results in typical household idioms and rituals being translated into media contexts, for example, by repeatedly decorating devices (putting stickers on a laptop, dressing up a smartphone with a cool cover), adding colors, emoji and other personal touches to social media profile pages, dressing up and customizing avatars, and rearranging and tidying up the files and folders on one's personal computer or laptop. Just as the home gets infused with media and reshaped in the process—for instance, with a dedicated home cinema or media room, the addition of specific furniture to store and use media hardware, and complex cabling running everywhere—our media become domesticated spaces typified by mass personalization: customized ringtones, individualized screensavers and wallpapers, fine-tuned arrangements of favorite websites (as bookmarks) and television stations, all kinds of emotes specific to your character in a game or online environment (such as dance moves in Fortnite, dedicated emoticons on Twitch, and unique filters for Instagram), and so on. When doing all of this, we extend ourselves—our personalities and identities, norms and values, tastes and affinities—to our media. In this boundary-blurring process, the mediated experience becomes more meaningful to us.

Another excellent example of how intimate the merger of humans and their media can become in terms of our activities is the case of virtual personal assistants, available for a multitude of media devices—from televisions to mobile phones, wristwatches, and personal computers. A genealogy of such assistants includes

- the earlier mentioned ELIZA chatbot in the 1960s,
- innovations in voice recognition technology in the 1970s and 1980s,
- the introduction of text-based digital companions in the early 2000s,
- the inclusion of Siri on the Apple iPhone in 2011,
- and the growing popularity of smart speakers in the home, such as those powered by Alexa (Amazon's voice-controlled intelligent personal assistant), launched in 2014.

Just like ARPANET—the 1960s computer network that provided the backbone of today's internet—Siri was originally financed and developed by the U.S. Department of Defense as part of a large artificial intelligence research project. The software program that came out of this enterprise, the Cognitive Assistant that Learns and Organizes (CALO), was sold as a separate company to Apple, laying the groundwork for Siri. According to Apple, its digital assistant fulfills billions of requests each month, each time letting the program learn from its users in order to better predict and recommend what we would want next. After some initial hesitation, Siri (and other programs like it) became popular all over the world, which was an achievement largely accomplished by making its voice more human. Today, to many people, such a computer companion is just nice to listen and talk to. In the process, people's interactions with Siri and her counterparts have become quite chummy, additionally giving users the illusion of control while at the same time outsourcing more operations and agency to the computing systems and algorithms working underneath its interface.

People often use multiple media simultaneously—mostly unaware of the concurrent exposure involved. This multiplication of mediated experiences not only contributes to a lack of awareness of media in our lives; it also amplifies and accelerates an ongoing fusion of all domains of life (such as home, work, school, love, and play) with and in media. Such intense and immersive media use—while also an ordinary aspect of everyday life—can be seen as turning us into helpless addicts, slaves to machines: zombies. Perhaps we are zombies in that we mindlessly succumb to the drive of our devices; we are zombies because we use media in ways that erase our distinctiveness as individuals, our preferences and predilections increasingly determined and automated by algorithms. As much, if not most, of our media use and exposure is anything but deliberate, the zombie metaphor seems somewhat appropriate. In Germany, the winner of the annual Youth Word of the Year contest in 2015 was *smombie*: a merger of *smartphone* and *zombie* to indicate how people can at times completely tune out the world around them when they are on their mobile devices. When questioned, many of us would say that a lot of our media use is a way to resurrect dead time—for instance, listening to music on headphones as we traverse the city, browsing magazines or endlessly scrolling through social media updates and smartphone apps while waiting for a train or a friend to arrive. The zombie metaphor can be deployed widely to appreciate a life in media, as we are potentially less aware of our surroundings, less tuned in to our senses, and thus living more like lifeless automatons. Yet at the same time, living in media extends our senses and enhances our ability to connect with others, to witness and experience different cultures and ideas, and helps us to manage the growing social complexity of our world. If our media activities turn us into beings neither living nor dead, entities beyond conceptual capture in terms of the aliveness of humanity or deadness of technology, the zombie is perhaps a fitting abstraction to appreciate the increasing integration of our lives (and bodies) with technology and media, inevitably producing ambivalent and incongruous outcomes, experiences, and feelings.

***Becoming media: as arrangements***   Throughout the history of media, there have been periods in which they were first heralded as platforms for innovation and change, connecting people to the wider world and vitalizing anything from the economy to democracy, culture, and education. Soon, however, such lofty expectations were countered with equally over-the-top dystopian assessments—media came to be seen as dangerous, as the enemy, as threatening the very foundations upon which society is built. Such moral panics about media often coincide with a certain device, text, or practice reaching mass status—as when the popularity of a rockstar or game franchise, particular smartphone application or high-definition screen, quickly reaches widespread uptake among the population in any given country. Most recently, this role is reserved for the internet. It was initially heralded as a force for good, connecting the world and bringing more people into the global conversation. We now find governments, research institutes, international nongovernmental organizations, and a whole host of other groups and people sincerely worried about the consequences of unfettered access to what was once touted as the "information superhighway" (way back in the 1970s in policy documents, statements of futurists, and artists' manifestos around the world). Various stakeholders in these debates have joined forces in the Forum on Information and Democracy (FID), a network created in 2019 by eleven organizations, including Reporters without Borders, the Digital Rights Foundation, Free Press Unlimited, and Research ICT Africa, endorsed by 43 countries around the world at the United Nations, exemplifying the global unease about the digital environment we find ourselves in.

Current anxious assessments about the state of media generally and that of the internet in particular tend to fall in two related categories. First is an ongoing concern about the spread of false or misleading information online about the global coronavirus crisis—labeled an infodemic by the World Health Organization (WHO) early in 2020. The FID is just one of a profusion of organizations, networks, policy briefs, statements, and declarations worldwide by governments, academics, and other stakeholders alike, all critically discussing and lamenting the proliferation of disinformation, invariably labeling the way misinformation travels online as a crisis. The core components of the infodemic include misleading rumors, conspiracy theories, racism, fearmongering, and various instances of fake news. The WHO additionally pushed for the development of a scientific discipline: infodemiology—a term coined earlier, in 2002, to describe the study of the determinants and distribution of health information to identify possible gaps between best evidence (what experts know) and practice (what people do or believe). As stated at the outset of this book, the conflation of a pandemic with an infodemic and considering these just as dangerous to people's lives all over the world is a clear indication that our lifeworld has become mediatized (see chapter 5 for detailed discussion of mediatization).

The term *infodemic* is more a framing device than an accurate, evidence-based description of what happens in people's lives as they come to terms with the coronavirus. Following the WHO announcements in February 2020, news media all over the world uncritically adopted the term, using it to frame their coverage of the world's response to the virus. Scholarly reports in fields as varied as data science, education, human geography, immunology, and cardiovascular medicine similarly refer to infodemic as a concept—generally without making much effort in defining and operationalizing it. In media and communication studies, this tends to be met with some concern, as there is little evidence that information—misleading or otherwise—spreads the same way a virus does, and the underlying assumption in the use of epidemiological language that individuals can somehow be "infected" by information and unthinkingly carry it over to others implies a rather problematic assumption about how people ordinarily use media and handle information. In fact, research on how people use media consistently documents that

- people's exposure to deliberately misleading information (about COVID-19 or any other issue) is generally quite limited;

- when confronted with an overabundance of information, people typically process all of this in a variety of active, idiographic, and at times quite critical ways and are overall less likely to take things at face value if the topic concerns something they have direct personal experience with;

- the source of most misinformation tends to be political leaders and other authorities rather than social (or any other particular) media; and

- throughout periods of mass upheaval (such as the global coronavirus crisis), people tend to turn to legacy news media and trust science more than they otherwise do.

None of this means that we should not worry about disinformation campaigns, the deliberate manipulation of facts for political gain, the role and influence of those peddling fake news, and so on. During the coronavirus crisis, there are significant segments of the population around the world that are hesitant about getting a vaccination, in part because of false or misleading information spread on social media. However, resistance to immunization has a long history, going at least as far back as the early 1800s when the first smallpox vaccine was developed in England, somewhat inspired by the practice of variolation common in Turkey, China, and across Africa. Throughout history and around the world, significant parts of the population in any given country have questioned or outright opposed vaccination. The role that online social networks and platforms play in all of this today is the role that all media play: that of amplifier and accelerator. It is questionable that the role of media in people's lives is that of an independently operating outside agitator making us think and do things we would otherwise never consider. What is much more likely—and what the available evidence bears out—is that most people generally are quite comfortable living in media and making up their own mind, somehow coping with high-choice media environments, negotiating and navigating an endless variety of channels, systems, genres, and mediated experiences in generally nondeliberate but nonetheless active and distinct ways.

*It is questionable that the role of media in people's lives is that of an independently operating outside agitator making us think and do things we would otherwise never consider.*

Beyond the problematic frame of a dangerous infodemic affecting us all indiscriminately and in ways unprecedented in history, a second category of concern about media today is at the heart of all discussions in this chapter: How do we know whether something—anything—is true or real? Who or what can we trust? Considering our lives in media in terms of how we make media fit into our daily routines and ways of making sense of the world around us, the extent to which a degree of unreality has creeped into everyday life seems disconcerting. Take, for instance, the various ways in which people continually manage and maintain all their profiles, avatars, and accounts in online environments as varied as digital games, social media, platforms, and e-commerce contexts. Such intense digital curation is a skill crucial for (surviving, let alone enjoying) a life in media yet entails some degree of editing, which in turn raises the question to what extent the myriad digital versions of ourselves are accurate representations of us and to what extent we are an approximate version of them.

We are so used to images we are confronted with in media being manipulated that most of us would not hesitate to expect any picture to be photoshopped—verbing the name of a computer program for editing photographs originally released in 1980. Image adjusting, retouching, editing, and manipulation has been possible (if prohibitively expensive and

complicated) since the late nineteenth century, remaining the provenance of professionals until the introduction of digital cameras and editing software in the 1990s. Digitization made image adjustment simple, giving rise to much debate among both scholars and professionals about the prospects for photographic truth. The introduction (in 2010) of dedicated filters for images taken with (generally substandard) smartphones by Instagram and the video-sharing platform TikTok (owned by China's ByteDance, introducing video editing features such as camera face filters, animations, stickers, music add-ons, and themes when it launched in 2016) made image doctoring even more straightforward—and fun! Social media influencers often use their own proprietary filters to give all their messages the same look and feel, offering followers the opportunity to purchase their presets—adding yet another level of subtle deception. Sharing anything on social media with #nofilter (or #nocrop, #nozoom, #original, #allnatural, and other similar claims expressed via hashtags) represents a fascinating collective claim to authenticity in a context where realness is almost impossible to check and verify and where the veracity of any truth claim can be contested by the fact that anything captured on camera inherently represents only a partial and incomplete account of reality.

It is people's daily navigation of unreality—and our apparent comfort with it—that gets eagerly exploited by those wanting to push certain narratives to challenge what is actually happening and further their own agenda. Examples are politicians and other state actors distributing disinformation to rally support for their cause, as in the case of Russia flooding the media both at home and abroad with numerous alternate theories and stories during its ongoing war with Ukraine since the Russian annexation of Crimea in 2014 and particularly after its invasion in 2022 to confuse and diffuse public opinion. Within Russia, media are only allowed to refer to the war as a "special military operation" (or risk fines and prison time), and Russian institutions abroad—such as embassies, missions, and consulates—and their followers share press releases and social media updates attacking foreign news coverage of the war or submit alternate versions of events on the ground. Russian state-owned media outlets publishing in English, such as Russia Today (RT) and Sputnik, also contribute to such propaganda efforts, as RT editor in chief Margarita Simonyan explained during an April 2022 broadcast on Russian television, defending censorship and blaming freedom of the press for the fall of the Soviet Union.

In the early years of the World Wide Web, Usenet newsgroups, and so-called MUDs (short for "multi-user dungeons," which are text-based online communities), experts predicted people would play around with their identities a lot, especially swapping one's gender (also known as *gender-bending*), age, or ethnicity. As it turns out, such behavior did not become as widespread as predicted. It happens primarily in massively multiplayer online games (such as *World of Warcraft*, developed by Blizzard Entertainment in California, with a global player base of roughly six million people) and seems to be done primarily by male gamers for a variety of reasons, including the experience that female avatars are treated better and more likely to receive gifts and help from other people in the game. This is an extension of the "what is beautiful is good" stereotype, also known as the *halo effect*. Beyond the all too human tendency of people wanting to present a socially desirable, positive self-view to others, the halo effect plays a part in how people present and perform themselves in media, generally making themselves look slightly better than is the case AFK (i.e., "away from keyboard"). Such exaggerated self-presentations contribute to concerns about the degree of unreality in our digital environment. On the other hand, studies show that all these self-expressions, while somewhat idealized and fanciful, tend to not veer too far off one's real persona and experience, as the social nature of internet platforms customarily acts as an accountability mechanism. In other words, as with online dating, on the whole it does not pay off to completely misrepresent yourself, especially if you also plan to meet people offline. A similar argument goes for the propaganda efforts of politicians, diplomats, and other state actors: it is hard to maintain alternate truths when the same media you use to do so can be effectively used to gather and show evidence to the contrary.

Every individual always balances delicately between self-idealization and authentic self-expression, both online and offline. A certain degree of misrepresentation is therefore to be expected in any kind of communication—especially in the context of mass self-communication. The role of media in this unavoidable experience of unreality makes it more likely that we will share somewhat inauthentic information about ourselves through the interaction between the circuit of artifacts, activities, and social arrangements of online social networks, including:

- platform interface design that encourages the sharing of beautiful, funny, and ostensibly happy slices of life;
- built-in software properties that enliven photographs and videos to fit the characteristics of visual communication people are accustomed to in commercial media production;
- positive, visually rich, and emotive updates tend to become more visible (as pushed by platform algorithms that determine what kind of posts people get to see first) and get more likes and comments, which inspires people to contribute more of the same kind of updates;
- the same reasons we can have for participating in a global surveillance society (see the conclusion of chapter 3) compel us to sharing the best parts of our lives online: smart technologies making sharing straightforward and easy, sharing stirring news tends to get more likes and comments and keeps us connected to people near and far, making us feel seen and heard (in ways that we can somewhat control), all of which in turn can make us feel good.

While we may not be surprised if celebrities and stars (such as musicians, actors, social media influencers, and models) are not exactly as they seem to be in media, there is still some debate over the representation of politicians and other authorities. Pundits, professors, and media professionals alike regularly express their fears about a "post-truth" and "fact-free" political culture particular in the twenty-first century and extend their alarm to an emerging "post-reality" altogether. In 2016 Oxford Dictionaries announced *post-truth* as their word of the year, defining the adjective as "relating to or denoting circumstances in which objective facts are less influential in shaping public opinion than appeals to emotion and personal belief." Ten years earlier, Merriam-Webster made "truthiness" their word for the year 2006. American TV comedian Stephen Colbert first introduced the term in October 2005, meaning, in his words, "truth that comes from the gut, not books."[11]

The conflation of truth with what feels true rather than (or next to) what can be factually verified as truthful tends to be attributed to our digital environment, with its easy opportunities for editing and manipulation, the fast-paced networked way in which rumor and misinformation travels online, and a superabundance of alternate narratives and versions of the truth. Beyond the editing, curation, and manipulation of texts and images, the representation of the real in video is the subject of recent concern, coinciding with the rise of moving images autogenerated by artificial intelligence software. Such examples of

---

11. "The Word of the Year: 'Truthiness,'" CBS News, December 9, 2006, https://www.cbsnews.com/news/the-word-of-the-year-truthiness/.

"deepfake" (a combination of *deep learning* and *fake*) or *synthetic media* (referring to any kind of media that is produced through automated means) already have a rich history in pornography, political parody, acting in film, and of course in art. With the introduction of free next-generation video-editing software such as Unreal Engine 5 (released in the spring of 2022 by Epic Games, the company behind the popular free-to-play *Fortnite* game), and image-generating artificial intelligence model DALL-E (made available from 2021 onward by the non-profit OpenAI research laboratory), anyone is potentially able to construct still and moving images and even entire virtual worlds that are almost indiscernible from reality.

Before the particulars of our digital environment, similar claims about the impossibility of concepts like truth, authenticity, and originality in and through media were made by prominent theorists in the late twentieth century (including the aforementioned Jean Baudrillard). Suggestions have, for example, been made that some kind of *mediology* should replace ontology as the primary source of how and what we can know about the world. Media historians hasten to add that complaints about fake news and less-than-truthful information and journalism go back much further, reaching as far back as the start of the first newspapers in the sixteenth century. Throughout this chapter, we have seen that worries about the media as coming between us and reality run throughout the course of human history. While a certain degree of unreality may be an ordinary feature of human beings and their media worlds, the rise of mass media in the twentieth century and our current digital environment have supercharged the ambivalence and confusion about what constitutes reality and truth—an uncanniness exploited to full effect by propaganda peddlers, populist politicians, and authoritarian regimes around the world.

The media as an industry contributes to this process of reality-unreality boundary blurring, exemplified by the professionalization of media production—in particular journalism and advertising. Over the last century, both professions experienced an uneasy transition from a propensity to produce reality-based narratives to increasingly including more tendentious and emotional ways of telling stories and relating to audiences, thereby contributing to making the "authenticity puzzle"—the delicate negotiation between media producers and consumers about what is fake or real—just a little more complicated than it already is.

Journalism, starting out with distinctly partisan and unabashedly biased reporting, came to embrace the norm of objectivity during the second half of the twentieth century, preferring the self-image of a discipline of factual verification. Although this development meant different things in different cultures of journalism, in general journalists all over the world self-identify with similar core values such as telling the truth and getting news to the public quickly. As the century progressed, prominent voices in the profession proclaimed the necessity of additional approaches to the news, for example, suggesting that journalism needed to embrace more engaged, inclusive, constructive, and attached forms of reporting—whether that would be in an effort to nurture indigenous and native forms of journalism (e.g., throughout Africa, Australia, and Latin America) or to counter detached ways of covering the news as a way to reconnect with increasingly disenfranchised and distrustful audiences (in much of Europe and North America). This apparent tension is paramount in contemporary news media, and it is quite common to find—within a single news broadcast or newspaper online and in print—strictly fact-based coverage of high-profile events right next to highly opinionated pieces and emotionally charged reports, as well as branded content sponsored by advertisers (see also chapter 7 on making media). While the differences between these genres may be clear to the reporters and editors involved, most people visiting a news site or reading a magazine do not make such distinctions when assessing content.

Advertising is, much like journalism, a profession that matured in the latter half of the twentieth century after having been around for quite some time. Initially, most advertising was focused on promoting specific products based on evidence-based arguments on why they were supposedly better or just good for you. Advertisements would often include all kinds of facts and data about the product, using scientific-looking persona to make the case (men in white lab coats are to this day a particularly prominent feature in campaigns for a variety of products). This traditional model of advertising got upended quite radically in the 1990s. Of key importance is what the industry calls its "Marlboro moment" in 1993, when Philip Morris, the parent company of cigarette manufacturer Marlboro, slashed the price of a packet of cigarettes by 20 percent in an effort to keep up with cheaper brands (in the United States, its key market). All of a sudden, it became clear to everyone that advertising on its own simply did not compel people to keep buying a certain product. Despite having one of the longest-running and most recognizable advertising campaigns in history—the famous "Marlboro Man" ads originally created by the American advertising executive Leo Burnett—the brand apparently could only continue to appeal to consumers by price. The stunning move produced a cascade effect among other globally operating consumer-products manufacturers (including Heinz and Coca-Cola), with companies losing billions in the process. This paved the way for a new type of branding to become dominant throughout the global media industries, focusing on a mix of emotive messages related to lifestyle, identity, and engagement as the key elements of contemporary advertising campaigns—whether to promote a particular product (such as a soft drink or car) or a person (like a politician or influencer).

The role of media in election campaigns provides a special case, where the duplicity of media and society comes together with profound consequences for the way in which people live. Arguably one of the most referenced examples of the role media play in choosing a political leader is the first-ever televised presidential debate in the United States, between Senator John F. Kennedy and Vice President Richard Nixon, broadcast on September 26, 1960. As few people in the country had a television at home, many prospective voters listened to the debate on the radio. The results from the debate were mixed. The majority of the people that listened to the debates on the radio tended to think that the candidates were evenly matched, while TV viewers leaned more toward John F. Kennedy. Nixon looked pale and weak—in part because he refused makeup—while Kennedy was in great shape. It changed the dynamics of the election. Ever since, campaigning for election to positions of national significance has become as much about how a candidate looks and feels like to voters as it is about the substance of their political message. This fragile orchestration seems to have tipped over into post-real territory in the 2000s, as illustrated by a much-quoted statement attributed to a White House official by American journalist Ron Suskind in a 2004 piece on the Bush presidency: "The aide said that guys like me were 'in what we call the reality-based community,' which he defined as people who 'believe that solutions emerge from your judicious study of discernible reality.' . . . 'That's not the way the world really works anymore,' he continued. 'We're an empire now, and when we act, we create our own reality."[12]

To many, claiming to belong in a "reality-based community" is a badge of honor, a statement of pride. On the other hand, it is hard to ignore the phenomenon that national and international politics around the world has become as much about how it is presented in the media as how it represents substantial governance.

In media studies, the dissolution of media and life boundaries takes shape in a wave of theoretical and empirical work emerging in the late 1980s, while gaining traction in the early 2000s, critiquing typical taxonomies and traditional distinctions upheld between media production and audience reception, between sending and receiving mediated messages, between professionals and amateurs making and circulating media, and between different kinds, genres, and channels of media fighting for people's attention. Instead, there is now a wealth of research detailing how such categories and definitions of media lack explanatory power when we try to make sense of our digital environment. Questions are asked about media production as an element of consumption, and vice versa—for example, when investigating the role of fans in the viability of film or game franchises and television shows. The usefulness of drawing distinctions between television, radio, film, games, newspapers, magazines, mobile media, and the internet is similarly called into question, with scholars increasingly opting to consider all media in concert, as an ensemble or environment that people move in and through on a daily basis. Furthermore, problems of accurate or biased representation—while certainly still highly relevant (see chapter 8)—are perhaps less pronounced today than questions about the embodied, emotional, and affective nature of people's relations with media and via media with the world. Unique methodological and theoretical challenges emerge from such work, suggesting a shared reckoning with both an age-old and brand-new mediated reality.

To consider all the different ways in which media, as a social arrangement, have nestled themselves in even the most fundamental processes of society and across the intimacies of everyday life (as we explore in the next chapter on love), we are confronted with the very real possibility that trying to maintain a line between us and our media may take

---

12. Ron Suskind, "Faith, Certainty, and the Presidency of George W. Bush," *New York Times Magazine*, October 17, 2004, https://www.nytimes.com/2004/10/17/magazine/faith-certainty-and-the-presidency-of-george-w-bush.html.

away a significant option we have when faced with a digital environment of pervasive and ubiquitous media: to (re)connect with reality in and through media.

## Real Life

All the developments outlined in this chapter—relating to what media supposedly do to people and how media (as technologies, the way we use them, and how media as industries operate) possibly contribute to a post-real experience of the world—underscore how a position outside of media may be impossible for us. This inability for reality to remain real may be a deeply frustrating one, but it is not caused or uniquely determined by our current digital environment—it is a phenomenon that runs throughout human history. The key strategies we have for engaging with the dilemma of the real in a comprehensively mediated context—waging war on our machines, finding some kind of precarious coexistence, or becoming media—each contain much promise.

Keeping alive the struggle against omnipresent and pervasive media can be done through structural investments in digital and media literacy, both for primary and secondary education as well as through programs of adult instruction. Media literacy, long left to the sidelines in discussions about the principal components of lifelong learning, in some parts of the world has become a panacea for many societal problems—especially those related to a real or perceived infodemic in the context of the global coronavirus crisis, political polarization, and concerns about information overload (see chapter 8 for more discussion on digital and media literacies around the world).

Surrendering to the significant role of media in society and everyday life comes with its own set of potential schemes to counter their influence in democratic societies aiming to empower people to stay one step ahead of the media in their life—or in the very least learning how to maintain some sense of control. Understanding basic computer skills, knowing how to bypass or even hack the settings and requirements of smart devices and software programs, and learning how media as industries operate are dedicated tactics within this strategy.

Letting go of clear or necessary boundaries between media, the world, and us involves a significant rethink of man-machine relations. This third perspective acknowledges a certain indivisibility between media and life and would require acceptance of the uncanniness that inevitably follows. While all three strategies are valid and useful, it is important to note that they are not mutually exclusive. Certain situations and contexts clearly call for different approaches. In some circumstances, questioning reality and truthfulness is a luxury effectively exploited by propagandists, while in other cases willfully embracing a more fluid reading of information can contribute to better understanding between people. Much depends on the extent to which we are willing to take responsibility for our choices in media.

Despite media's entanglement with life and a certain degree of unreality that ensues, it still matters what kind of claims someone makes and what kind of world is presented to us. The difference with earlier strategies can be found in letting go of either reality or one's own feelings as the benchmark to which any kind of mediated version of people and events is tested against. Instead, the focus shifts toward the possibility of creating, sharing, and cultivating shared narratives in media, to understanding media through making your own media in collaboration with others and learning how to listen—how to be a witness rather than just a consumer—in a mediated context.

In December 2021, Keanu Reeves and Carrie-Ann Moss embarked on a public relations tour to promote both the release of the fourth Matrix film (titled *Resurrections*) and the Unreal Engine 5 tech demo *The Matrix Awakens.* The film continues the plotline of questioning boundaries all-too-easily drawn between humans and machines, and between reality and a computer simulation thereof. The demo features a minigame where you play as Reeves and Moss, fighting off agents while driving through a vast photorealistic city. In one of the interviews—with American technology news website The Verge[13]—the two Canadian actors discuss the uncanny experience of seeing themselves as avatars, pontificating about a possible future where they can just "stay home" while their avatars star in all kinds of projects. The conversation quickly moves on to philosophizing about the digital world "that is becoming more and more real," for example, addressing the already well-established practice of archiving one's digital self for future reference (and commercial exploitation). When asked how they feel about this near-complete blending of the virtual and the real, Keanu Reeves offered an anecdote about having dinner at a friends' house and trying to explain the premise of the Matrix franchise to a teenager. As he talked about the struggles of his character—Thomas Anderson or Neo—figuring out his life in a virtual world and a real world, his friends' daughter asks him why this is important: "Who cares if it's real?" Reeves considered her indifference to whether a digital life is more or less real than an analogue, embodied and physical one, "awesome," whereas Moss remained quizzical. In a nutshell, the girl's trivialization of any meaningful distinction between the real and the virtual in the context of digital life, Keanu Reeves's awe of how much such a perspective opens up opportunities for expression and experience, as well as Carrie-Anne Moss's incredulity all capture the uncanny experience of life in media. People find their way in media through an at times unsettling mix of device dependency and the dizzying freedom information and communication technologies provide, juggling attention, boredom, and distraction in an always-on digital environment, trying to strike a balance between authenticity and visibility. It is, above all, an emotional roller-coaster, while also providing plenty of challenges regarding media and information literacies.

What real life in media is, ultimately, requires a realization that people feel quite strongly about (their) media. In fact, it could be argued that our relation with media is,

---

13. Alex Heath, Vjeran Pavic, and Phil Esposito, "Keanu Reeves and Carrie-Anne Moss on making The Matrix Awakens with Epic Games," The Verge, December 9, 2021, https://www.theverge.com/22825102/keanu-reeves-carrie-anne-moss-interview-matrix-awakens-epic-games.

first and foremost, affective. In chapter 5, this heartfelt aspect of life in media gets explored in detail, as it gets us closer to ourselves—via media. This visceral relation we have with ourselves, each other, and the world in and through media furthermore fuels the way people engage with the world—especially if there is something about it that they want to change. This is the subject of chapter 6. Finally, as we have seen for the media professions of journalism and advertising, strong feelings are the primary currency for the media industries and determine the way media are made—which we explore in chapter 7.

# 5

## Love Life

> People really love media. What is more, our love lives increasingly involve media in all kinds of intimate ways. This affective entanglement of media, life, and love raises fascinating questions about our interdependence with media and technologies, issues that are tackled in this chapter by focusing on the ways in which people use media in everyday life to find and maintain love, exploring the circulation of information and ideas about love via media, and looking at how we express and understand our love for media.

- I heart my media. I really do. Who doesn't anyway? I love the way they make me feel, the way they make my life easier, even the way they look.
- I heart media because it is the most human part of me. I want to be part of media like it is part of me.

These quotes, taken from my website WhyIHeartMyMedia[1]—where media students from all over the world leave comments and stories documenting their feelings about media—express a widely shared sentiment: we love media. Couples, sports teams (and their supporters), and even entire revolutionary social movements have their favorite song signifying their passion and commitment; teenagers experience symptoms of depression and anxiety when going offline for a while or when they lose their phone; parents use the threat of limiting screen time as a way to exert parental authority; fans fawn over their favorite character in a book or television series, motion picture, or digital game; many households have storage boxes somewhere full of "ancient" media technologies that are not discarded because they have some emotional significance. Users of Apple and Microsoft computers argue passionately about the merits of their respective device and software ecosystems, gamers wax poetically about their console of choice, and in the context of the global coronavirus crisis, animated debates rage about the pros and cons of various video-calling programs. It is clear that our media are intimate and intensely personal. We genuinely care

---

1. https://whyiheartmymedia.com/.

about media and about the experiences we have in and through them: the friendships we make and maintain, our triumphs and failures and those of the people we follow, what we have learned, what we have achieved, and what we can escape from—even if it is for just a moment. And let's not forget it is not just the users of media that are so deeply enamored—the people who make media are generally in it for the love, too. "I can't believe I'm getting paid to do this!" expresses a common sentiment among those who get to make media for a living (see chapter 7). Media love, in all its manifestations, is everywhere.

As discussed in the previous chapter, our already strong feelings about and for media get supercharged by our desire for true, authentic communication. We wish to be seen and understood; we are frustrated by the forever imperfect nature of making ourselves heard in a complex and globalizing world; we want to belong to communities of like-minded people. In all of this, media play both an enabling and corrupting role: it is through media we can build and sustain relations at a distance, yet it is those same media that amplify noise, misunderstanding, and *polysemy* (as content always contains multiple meanings, dependent on context). Our media love comes—like all instances of love—with a variety of intense emotions, including desire, longing, amazement, and joy but also annoyance, anger, boredom, and disgust.

In all of this, it is important to remind ourselves how much fun and enjoyment we get from media. Sure, we need to be critical and reflective about the way media are made and used, but it would be equally, if not more, important to recognize the profound pleasure that media bring, as devices as much as content, as what we do with them and how they fit into the way society and everyday life works. In our endeavor to think deeply about media, to dissect and scrutinize the media, and to build sophisticated theoretical and methodological models and approaches to study media, it is easy to forget our love for media. All this love clearly warrants careful consideration. In this chapter, we dive deeply into the role love plays in how we use, talk about, and give meaning to the media in our life. First, this chapter covers how the "erotics" of media—how media amuse and entertain us, make us laugh and cry, and connect to (and reflect) our deepest desires—has historically been looked at in scholarship and research and what love's role is in our current digital environment. Secondly, we consider all the different ways in which we can think about media love based on the three distinct building blocks of thinking about media, society, and everyday life used throughout media studies:

1.  Understanding media as *practice* related to love, allowing us to focus on the various ways in which people use media in everyday life to find and maintain love, to make love, and what happens when their love for media gets out of control

2.  Exploring the *mediation* of love, an approach that deals with the circulation and appropriation of information and ideas via media (as institutions) about love—such as popular depictions in literature and film of romantic love in general and of love involving humans and their media in particular

3.  Appreciating the *mediatization* of love, a speculative way of looking at the central and historical role media play for our understanding and expression of love—specifically our love for media

The conceptual taxonomy of media as practice, mediation, and mediatization allows us to determine specific areas of investigation, of research focus. To consider media as practice, we ask a simple question: What are people doing in relation to media in all the different contexts of everyday life? Rather than assuming that watching television or scrolling through social media means the same thing for everyone, a practice perspective suggests that everyone does things differently and makes sense of what they are doing in different ways—and that these differences matter. Such an approach additionally reminds us that media are not the center of people's existence, even though our lives are awash with media. When studying the mediation of something, we are most interested in what media, as institutions, tell us about the world we live in, helping us to understand our role in it, and how these stories reflect the (changing) customs, conventions, and culture of a particular community (or society as a whole) with respect to the topic or issue we are interested in. Mediatization offers an even broader point of view, as it assumes that media play a formative role in whatever it is we are researching and that this role has deep historical roots. It furthermore reminds us that media are not just the artifacts, activities, and social arrangements of our environment. Media, as social institutions, also play a powerful role in the way society functions and need to be considered as such—in tandem with how people use and make sense of media. Within each of these areas of investigation, this chapter offers examples to help us understand and explain our evolving and deeply affectionate relationship with media and through media with our innermost feelings.

In conclusion, I return to the overarching theme of this chapter: the role and meaning of love in everyday life. Following the insights of love studies as an emerging discipline (with roots in feminist scholarship, social psychology and philosophy, anthropology, and historical studies), the point is made that taking love seriously involves an appreciation of its transformative power. Love makes us take action, often doing things we did not contemplate before and allowing us to take risks, to step outside our comfort zone in an attempt to make something happen. Love also inspires and encourages hope—for a better way of living, a better world even. In other words, given the notion that people love media, it follows that media (can) play both *aspirational* and *transformative* roles in society and everyday life. All of this forms the prelude to the next chapter, where we explore the various ways in which people engage with their situation in and through media to change it, focusing on media activism around the world.

## Media Love

The erotics of media are significant for any consideration of the role of media in society and everyday life. The pleasure (and sometimes pain) we get from our media is the product of any relationship between the makers of media, the content of the media, and the audiences of media. All the associated feelings of the erotic—sensation, desire, pleasure, excitement, and so on—tend to be notably absent from formal scholarly discussions of media. As our experiences with media inevitably contain elements that arouse us, that bring us pleasure, that inspire and seduce, it seems clear that we should make this explicit. However, even as consumers and users we are often apologetic about our love for media—people tend to talk about binge-watching television or playing a mobile game as a "guilty" pleasure, respond to researchers who study their media behavior with socially desirable answers (by either downplaying their media time or overreporting the hours spent each day), and often state that they "should" follow the news and watch the latest documentaries instead of seeing a rerun of a sitcom or the latest superhero movie. It can be hard to acknowledge that we love media—even though we all clearly do.

When we study people and their media devices, the way people respond and give meaning to mediated messages, how people make media professionally, we are studying forms and aspects of love. Historically, researchers either problematized such love—passionate media professionals are clearly blind to their own exploitation, fans are suckers for corporate franchising practices, and device love gets classified as addiction and disorder—or glossed over all the love we have with a throwaway line about people's care or passion for a favorite character, show, or game. Such resistance to, reluctance about, or subtle redirection of the topic of love in media studies can be explained by a variety of reasons:

- Media and communication researchers until quite recently did not really consider nor study the human body and its role in how we use and make sense of our media. As all emotions are inevitably embodied and involve some kind of expression (in other words, we feel something and talk about it), studying people's feelings about media necessarily involves taking the body into consideration.

- Love traditionally has been associated with the realm of women, the home, the private, and the apolitical, and all its related emotions and experiences therefore were considered less relevant or important compared to the privileged place rational discourse, politics, and the economy were afforded in scholarship and public debate.

- Until well into the twentieth century, love was generally not considered to be such a fundamental feeling—its presence in academic discourse, in literature, and in music still quite limited and its role as a cornerstone of, for example, committed romantic relationships seen as much less fundamental than it is today.

- Love as a distinct field of study—love studies—is still a nascent field, emerging from a growing awareness in numerous scientific disciplines that our feelings and bodies play a profound role in how we make sense of the world around us and that such an awareness matters in that it opens up new lines of inquiry and new ways of appreciating how the experience of life (in media) is different and unique to every individual.

---

*When we study people and their media devices, the way people respond and give meaning to mediated messages, how people make media professionally, we are studying forms and aspects of love.*

---

Generally speaking, media and (mass) communication theory and research throughout much of the previous century tended to take love either as a given, studying it only in "extreme" manifestations (e.g., in studies on fandom), or otherwise saw it as troublesome. Some of the dominant theories and concepts from twentieth-century scholarship about media have a direct bearing on this problematization of people's love for media. People were seen as somehow dependent on media—for information, entertainment, and understanding the world. Our limited capacity to process all the media around us was seen as a problem to be solved, and the enormous amount of time people spent with their preferred media—especially since the rise of television as a mass medium—led to much speculation about problematic media use as a possible addiction or disorder. With the rapidly growing popularity of the World Wide Web at the end of the 1990s—and perceiving people's passion to express themselves online (in chat and discussion groups and on weblogs)—came academic debates about whether we feel more or less now that our lives had become digital, including the question of whether our feelings as shown in media (*OMG! LOL!*) are genuine or just performed and perhaps always over-the-top.

Today, media studies as a field tends to tackle love head on. Love is not necessarily seen as a problem anymore—although there is most certainly a lot of fretting about the problem of misinformation, fake news, and disinformation campaigns online as corrupting people's access to (and participation in) well-informed public discourse. One way of interpreting the persistence of misinformation and the fact that it tends to spread faster than factual information verified by society's experts is through the lens of love. Why would people willingly and voluntarily ignore the truth, even when confronted with irrefutable evidence as provided by the falsified work of scientists, the reasoned voice of institutional authorities, and diligent fact-checking efforts by journalists? One answer could be that we prefer the people and ideas we already know and love—including our self-love fueling the feeling of being "right"—over the supposedly rational discourse of strangers.

Similarly, the widespread concern noted throughout the popular as well as scientific literature about filter bubbles, echo chambers, "telecocoons," and other forms of social polarization and fragmentation online that are seen as either determined or at the very least amplified by technological affordances involves love, in that people are considered as either loving themselves and people like them too much (and therefore only bonding with like-minded individuals and networks), or as not having enough love for those that look and act differently. On the other hand, the available evidence tends not to support a

diagnosis of widespread polarization online, so perhaps we underestimate the complexity of our love in and through media—love seems to allow for more kaleidoscopic understandings of the world than a simple notion of people directly swayed and manipulated by a handful of targeted messages on Facebook or TikTok.

As discussed in the previous chapter, beneath the largely latent and oblique relation with love runs another current throughout all the research and theorizing about media: our deeply felt desire for perfect union with one another, a true fusion of souls—whether between romantic partners, between citizens and society, or among all of us together in humanity vis-à-vis the vast emptiness of the cosmos. Desperately wanting perfect communication to be possible—in our shared all-too-human quest to be heard, seen, understood, and recognized—we turn to the rigorous study of media and communication to control, fix, or restore the process. The origins and foundation of the field of media and (mass) communication studies can be characterized as a quest to solve the profound problem of communication: its imminent imperfection. Hence the more or less exclusive direction of early twentieth-century studies toward media effects, impact, and influence, focusing either on achieving the effects that were preferred or on preventing those that were considered to be problematic.

Love is not just something media scholars articulate, however implicitly, in their work on popular culture, fandom, or problematic media making and use; nor is it solely a smoldering fire inspiring so many questions, concepts, and themes in media and (mass) communication research. Love clearly inspires all of us in how we use media. Especially during the global coronavirus crisis of 2020 onward, variations of "media love" have been in full swing during the various lockdowns and social distancing protocols, policies, and experiences around the world. Consider, for example, the exponential growth of video calling, of (old and new) social network applications on smartphones and tablets, and of streaming services (for games, film, and television) facilitating a range of shared experiences. Another example would be the rapid rise of telemedicine and digital health applications, such as smartphone contact-tracing apps, as well as social media hashtag campaigns and activism—including trending topics #StayHomeStaySafe and #YoMeQuedoEnCasa—connecting people all over the world in the fight against the virus. Even the rise of online communities around conspiracy theories and other so-called alternative truths can be seen as an expression of love—an intense love for only a particular side of the story that is the global coronavirus crisis. It is all indicative of our desire to connect and communicate with ourselves and each other, and of our inexorable interdependence with information and communication technologies.

It is one thing to acknowledge the role love plays in why we use, make, and study media. It is another to appreciate how this has always been the case. Throughout history, people have relied on media of all kinds to inform themselves, to establish rules and rituals governing community life, to establish a sense of belonging, and simply to be entertained, including but not limited to

- cave paintings in Spain and Indonesia depicting hunting routines and shamanic practices;
- clay tablets documenting trade, royal protocols, and divine commandments;
- papyrus scrolls containing political and trade agreements for the library archive;
- hand-painted advertisements for sword fights, wrestling matches, and theater plays on the walls of ancient cities (across the Roman Empire, the Indian subcontinent, and China);
- individually made books about important political figures, religious traditions, and the myths and legends of a particular society; and
- all the mechanical and electronic media since the invention of the printing press, truly opening up media making and use to almost anyone.

In all these media, we observe a common theme: media are primarily used for telling stories that serve the purpose of letting us know who we are, where we belong, what we are supposed to do. In doing so, we use media to express love—for ourselves, each other, and the culture and community we belong to. An important difference between the walls on which we painted in the past and today's Facebook wall is the local specificity of our media. Where media in the past were particular and often unique to a locale and the traditions of a group or community, platforms like Facebook offer a singular (and intensely commodified) experience for all their billions of users worldwide. Given that universality of experience, it is no surprise that so many of us try our best to adorn the platform experience with localizing details—such as national flags and regional symbols, using locally specific hashtags and filters, expressions in local languages added to profile pictures, and so on. Media can be seen as crucial for the formation and survival of communities, both local, regional, national, and even global. Our love for media is, to some extent, also a love for communion, for connection and contact with others (like us). And it is through media that we not just establish social bonds and build bridges between different people and groups—we also find, maintain, and sometimes break up romance and relationships.

**The Practice of Media Love**

People use media in a variety of ways to look for love. Especially at a time of globalization and urbanization (as over half the world's population lives in cities), meeting someone new, falling in love and taking the time to explore the possibility of a relationship, is perhaps not as straightforward as it was less than a century ago, when most people would get together with someone from the same neighborhood or coupling was arranged through family ties and other networks based on proximity and kinship. Today, hundreds of millions of people around the world turn to online dating services and apps to facilitate their search for (some kind of) love. In doing so, the practice of using media for the purpose of love also creates a specific way of arranging and being in love. This does not mean that love found online is necessarily different than one explored elsewhere (at school, work, parties, or in the neighborhood)—it just suggests that the role of media in love arranges and shapes the ensuing possibilities of romance. Considering media as practice is a way to be mindful of this world-making capacity of using certain media for specific purposes in everyday life.

Finding love in and through media, whether through dedicated services or via all the other ways in which people come into contact with each other, has become a common, normalized aspect of romantic life—and certainly is not particular to the online environment as love letters, talk radio programs, and dating reality television programs predate the current *platformization* of love through a wide variety of sites and apps online. Online dating, as an industry, tends to get little love from scholarly observers, maligned for its muddying of romantic waters with cold-hearted mass consumption, endless choice, efficiency, and standardization. On the other hand, the ultimate goal of people using any of these services is still to meet someone new, a stranger—which necessitates some kind of honest, authentic self-disclosure, exposing our vulnerabilities even though we may be looking for a one-night stand.

An extension of our quest for love in media is the practice of enacting, sublimating, and automating sex and sexuality in media—from consensual sexting and consuming (and producing) pornography to using *teledildonics* (using technology to mimic and extend sexual interaction) to be in touch despite distance, all the way to the 2018 launch of the world's first commercially available sex robot called Harmony. As with all media, there is a dark side to such playful acts of sexual exploration. Sexting without mutual trust or consent can have devastating consequences, there are reports of exploitation of performers throughout the porn industry, and the Campaign against Sex Robots (starting in 2015) makes critical points about the objectification and commodification of the human body. In all these activities, media play a formative role, and in the process people's ideas about love and sex both change—yet also stay very much the same, requiring careful attention to detail and respect for all possible positions.

Smart sex toys (such as voice-controlled oral sex machines and long-distance vibrating devices) and humanoid sex robots may seem at the cutting edge of innovation when it comes to relationships meshing humanity and technology but, at the same time, signal a continuation of rather traditional practices and routines—for example, reinforcing the "coital imperative" in the design and marketing of devices for virtual sex. Sexting similarly on the one hand can be a liberating practice, taking control over your own body and sharing intimacy with a significant other in media, while on the other hand it can come with social pressures that reinforce gender stereotypes—for example, as girls tend to be judged and criticized much more than boys, both if they practice sexting and if they do not want to. When it comes to porn, media scholars tend to highlight the emergence of new, often playful yet also sometimes abusive rituals around romance and sexual activities. All of this serves to underscore both the significant place media and technology has in our most intimate encounters and the ambivalent, complex nature of such technologies and experiences.

Beyond dating apps, sexting, pornography, and smart sex toys, media play all kinds of roles in love life, such as through music—with couples identifying their love through particular tunes that played when they met, when they kissed for the first time, and when they got married or when romantic partners draw on various media to express their adoration (e.g., when one says to another "You are the Mork to my Mindy," "the Kirk to my Spock," or "the Bella to my Edward"). Media love also plays a role in the ways in which couples enact and give meaning to their togetherness, for example, making a ritual out of deleting a dating app from each other's smartphones to signal their commitment. In family life, parental monitoring and control of media plays a profound role in shaping and expressing relationships, hierarchies, and connections. Parents threaten their children with device restrictions as a way to maintain discipline, while teenagers circumvent parental supervision by flocking to sites, networks, and services that give them more control over who can access their personal information. At the same time, families try to establish joint rituals—often involving media, such as watching a TV series together (or its exact opposite: all family members agree to switch off their smartphones at the dinner table).

Beyond such literal instances of media love, we can and should also ask questions about what can happen because of our love for media—especially when such attachment leads to problematic media use. Excessive mediatime has often been viewed as harmful and unhealthy, leading to addiction, dissociation from reality, reduced social contacts, diversion from education, and displacement of more worthwhile activities. Television has traditionally been the usual suspect, and before that, films and comics were regarded similarly—even radio was once considered harmful (especially to children), and book reading before that. Today, digital games, the internet, and social media have become the latest perpetrators. Specific genres—especially those related to sex—tend to be singled out for concerns about media.

Given people's love for their media, recognizing the fact that we spend the majority of our time concurrently exposed to media and considering how these devices converge and become increasingly mobile and personal in an always online capacity, substantial attention and scholarship have been dedicated in recent decades to the question of media addiction, and discussions abound on such issues as problematic internet use and smartphone dependence. Internet gaming and online gambling disorder was, for example, included in the 2020 version of the International Classification of Diseases (ICD-11) of the World Health Organization. Harmful media use tends to be defined either as a psychiatric disorder or as part of a broader set of behavioral conditions involving excessive human-machine interactions. Such disorders differ in definition across the literature but generally contain two main components: compulsivity (the inability to control a certain type of media use) and impairment (how such media use harms or interferes with a person's life).

Overall, there is much important scholarly debate about the difference between addiction, harmful use, dependency, and "high engagement" with media, about the appropriateness of addiction criteria when making sense of time spent with media, the neglect and significance of context, and a general lack of expert consensus on how to approach and measure disorders and addictions regarding media. What is uncontested, however, is that people's relationship with media is invariably intimate, which at times can lead to problematic or harmful media use. While media scholars would hesitate to contribute to the medicalization of media use, it would behoove our field to stay mindful of people's intense, intimate, and indeed loving relations with media in all their various forms and manifestations.

## The Mediation of Love

The study of mediation most certainly includes what people do with media yet can be applied more broadly to account for the various ways in which media contributes to (and subsequently shapes) public debates and sense-making practices in society. From a mediation perspective, meanings are formed and social, cultural and technological forces operate freely according to various logics, with no predictable outcome—other than the fact that media never play a neutral role. The process of mediation inevitably influences or changes the meaning received, and there is a documented tendency for "reality" to be adapted to the demands and criteria of media presentation rather than vice versa. In other words, the way things look like in media affects how we see things outside of media, which in turn prompts people to record and present experiences in terms of media. This, for example, explains that people's photographs of their holiday travels to popular cities like Amsterdam, Sydney or Rio de Janeiro all look remarkably like the typical images of these places in travel guides and as the backdrop of films and television shows. Mediation also works the other way around, as governments around the world invest heavily in convincing studios to turn their countries into filming locations for major franchises—such as Star Wars, Harry Potter and Game of Thrones—in order to attract tourists.

In the context of this chapter, the concept of mediation inspires us to look at the different (re)presentations of media love in, for example, literature and cinema as an indicator of how our sensemaking of media and love has changed and evolved over time. When referencing and discussing specific films, television shows, and digital games, I use specific stories and scenes to illustrate my point. In doing so, I am analyzing media as texts—a methodological approach that is foundational to the academic discipline of media studies. Anything can be a "text": an interview, a blog post or online social network update, a photograph, a book, a film or television program (or even just one particular scene), a game, and so on. The analysis thereof generally involves a *close reading* of all aspects of the text, documenting the process in detail to find out what it is trying to say. Given the media industry's convergence and its creative and commercial quest to integrate characters and story lines across a variety of media (see chapter 6), textual analysis increasingly involves investigating relations between different texts as well as between the makers and audiences of a text to find out what its intended meaning is. Such scholarship of *intertextuality* looks for and builds bridges between, for example, a game, its advertisements and promotional materials, related toys, fan communities, the studio, other entities involved with its production and design, and so on (see chapter 2 for examples featuring Taylor Swift and BTS). A third approach to textual analysis looks even more broadly at the text as part of how specific communities or even society at large makes sense of itself. The assumption here is that media, as texts, represent something more than just the creativity and intentions of their maker(s), the strategies of the media as an industry, or the interpretations and uses of its audience. Texts, in whatever shape or form, are considered significant as a source of meaning for (and about) a particular culture or community. This *cultural analysis* of media supplements intertextual scholarship and close reading as the three key approaches to study texts in media studies.

Examples of love and sex in media are countless—it is safe to say that it is the dominant theme in the world's literature, music, film, and television. That said, it is remarkable that the two most popular media genres featuring love around the world—popular romantic fiction novels and soap operas on television—take up much less space in the academic literature than, for example, work on political and economic structures (such as democracy and capitalism), despite valiant and rigorous scholarly efforts regarding romance books and their authors, soap operas and women's magazines and their audiences, pornography, and digital games. As researchers in these fields would note, formerly distinct style conventions, design elements, and narrative structures from all these popular genres can now be found across all other storytelling forms, increasingly complicating all-too-easy categorizations of media. Indeed, a crucial aspect of the history of media studies is its recognition of popular culture as worthy of scholarly research.

To reconstruct a complete picture of the stories we tell and are told about love and how we in the process construct an idea of what love is regrettably falls beyond the scope of this book. Here I focus on the specific issue of love for and with media as a theme in popular culture. Consider, for example, narratives about the human love for artificial beings, such as in Nobel Prize–winner Kazuo Ishiguro's 2021 novel, *Klara and the Sun*, telling the story of the love of an "Android Friend" for children from the perspective of the machine. Such human–nonhuman relationships date as far back as E. T. A. Hoffmann's influential short story "The Sandman" from 1819 (as discussed in detail in chapter 4), wherein the protagonist falls in love with an automaton. Whereas the robot in Hoffmann's story was decidedly eerie, the android in Ishiguro's novel is in fact the protagonist, demanding our empathy. However, in the end, even she gets discarded to a robot scrapyard.

A theme of more or less fundamental incompatibility of humans and machines recurs across literature and the arts more generally, at least until the 2000s, when more generous readings of our intimacy with technology emerge. Consider for example the role of cybernetic organisms such as the Cylons (short for "cybernetic lifeform nodes") in the *Battlestar Galactica* TV series and movies (in 1978, 1980, and 2004), the Terminators in the *Terminator* franchise of motion pictures and TV series (from 1984 to 2009), and the Borg collective as part of the *Star Trek* universe (originally appearing in 1989 in the second season of *Star Trek: The Next Generation*). Cyborgs were originally depicted as cold, heartless villains, murderers, and assassins, reflecting the kind of deep-seated anxiety about human-machine mixing also found in Hoffmann's work, whereas in later iterations of these respected popular franchises, more personal, even intimate relationships between humans and cyborgs evolve, featuring Cylons, Terminators, and even the Borg as creatures capable of love and being loved.

The history of human sexuality as interfaced in all kinds of fascinating ways with technology finds glorious expression in the genre of speculative fiction, as it gives artists considerable freedom to imagine alternate universes, different societies, and spectacularly atypical realities. A prominent theme running throughout late twentieth-century and early twenty-first-century popular culture is that of technologies providing people with sexual pleasure, such as

- the "orgasmatron" in the 1964 French science fiction comic book created by Jean-Claude Forest and its motion picture adaptation *Barbarella* (1964, directed by Roger Vadim and starring Jane Fonda);

- an electromechanical cube also called the "orgasmatron" in Woody Allen's film *Sleeper* (1973, starring Allen and Diane Keaton); and
- high-tech headgear intended to substitute sexual intercourse in *Demolition Man* (1993, directed by Marco Brambilla, starring Sandra Bullock, Wesley Snipes, and Sylvester Stallone).

In all instances, people achieve sexual pleasure by outsourcing their orgasms to machines—yet these media texts, from different eras, show a subtle evolution of the role of technology in sublimating sexual pleasure. In *Barbarella* the orgasmatron was designed for torture but conquered by Barbarella for her own pleasure, whereas the orgasmatron in *Sleeper* and the simulation headgear in *Demolition Man* are featured as rather mundane aspects of people's sex lives in a near-distant future. Despite this normalization of human-machine relations, there still seems a slight unease at work—as exemplified by Sandra Bullock's character in *Demolition Man*, who appears appalled (as are all other people in her time) by the idea of sexual intercourse without media:

John Spartan [Sylvester Stallone]: Look, Huxley, why don't we just do it the old-fashioned way?

Lenina Huxley [Sandra Bullock]: Eeewww, disgusting! You mean . . . fluid transfer?

Another fascinating example of media entering into our innermost feelings and experiences related to love and sex is the Oscar-winning 2013 film *Her*, where the main character (Theodore Twombly, played by Joaquin Phoenix) develops a romantic relationship with Samantha (voiced by Scarlett Johansson), an artificially intelligent virtual assistant living in the operating system of his (and everyone else's) computer. It is both an unusually warm and intimate story about human-machine relationships, as it is a stark reminder of technology's supposed otherness in that Samantha in the end abandons Theodore, leaving him desperate and alone.

In these examples of American popular cinema, the mediation of love (and sex) in media suggests an evolution of media love from scary, unsettling, and even hostile to ambivalent, sometimes reciprocal, and possibly benign. This is not a linear progression, of course—I am reminded of the Oscar-winning 2014 thriller *Ex Machina* (written and directed by Alex Garland, starring Domhnall Gleeson, Alicia Vikander, and Oscar Isaac), wherein a computer programmer falls for the intelligent humanoid robot Ava, who ultimately manipulates and betrays him, in the process making him question his own humanness. In a fascinating marketing twist, a Tinder profile of Ava (using the image of Alicia Vikander) appeared upon the premiere of the film in the United States (at a festival in Austin, Texas), leading her matches to the Instagram account promoting the film.

Our love for media may have matured somewhat, and people perhaps recognize and appreciate their omnipresence and sway, but this does not mean we have truly come to

terms with media unambiguously. Moral panics and public freakouts about media endure, across all cultures, regardless of more complex, complimentary constructions of human–machine relations in literature, the arts—such as in the work of new media artists like Nam Jun Paik (from South Korea) and Linda Dounia (from Senegal)—and science fiction. Despite their familiarity and intimacy, media to some extent remain a somewhat alien other, at times credited with great power.

### Mediatization of Love

In recent years, it has become clear to many, if not most, scholars that media and mass communication are not just influencing established processes in society but are also creating routines within and across society's institutions on their own. It is, for example, not too far-fetched to consider how contemporary politics seems to be oriented toward getting and maintaining media attention, as much as conducting good governance. The same can be said about managing a business: without a significant media presence, it can be hard to survive for businesses large and small. To grasp the far-reaching consequences of this *double articulation* of media and society, the concept of mediatization helps us to appreciate the profound and ubiquitous role of media next to other metaprocesses shaping societies all over the world—such as globalization, urbanization, and individualization.

The mediatization of love works in two ways. First, as a process in which all society's institutions—governments, political parties, companies and businesses, religious and nonprofit organizations, communities, and neighborhoods—have to adapt to the rules, aims, production logics, and constraints of the media to function. In all this, institutions find themselves facing a digital environment that requires a continuous mediated presence, including

- operating and regularly updating a website,
- being responsive to phone calls and email queries,
- developing a social media strategy,
- looking to earn publicity through actions and campaigns that warrant the attention of media, and, in general,
- always being available to the spotlight of media attention.

It is not just institutions that over time find themselves increasingly oriented toward media to secure their survival—a similar observation can be made about all of us as individuals acting in and interacting with the world. As mentioned in chapter 2, a popular online expression such as "hashtag or it didn't happen" serves to remind us of the necessity of a mediated element to almost any kind of social event. Managing our public persona in media has become an ordinary aspect of life in media, and curating ourselves online can be done to ever-increasing levels of sophistication. Given the abundance of digital tools, filters, and effects, it is clear that we care greatly about all this—or at the very least are inspired to care by the media around us. As always, this process is not particular to the contemporary digital context, as an orientation to media life can be said to be emerging ever since those who featured in media were not just people of nobility, authority, and celebrity but started to include everyone. As the cameras of media turned on us in the course of the twentieth century, we subtly and inevitably changed our behavior. As we have seen in the discussion about our role in a global surveillance society (in chapter 3), participation in an all-seeing system of observation comes at a price. The good news is that all this surveillance has not made us all the same; nor have we uniformly become more disciplined and controllable because of it. On the other hand, it would be naive to assume that media simply offer us a window to the world and a mirror to ourselves. As explored in the previous chapter and considering the perspective of mediation, media intervene in all kinds of ways and are anything but neutral, so whatever is seen by media transforms in the process.

The presence of media in all aspects of society influences and changes the way decisions are made, what kind of products and services are developed and introduced, and how any kind of information travels and arrives at its destination, up to and including how we use and make sense of sounds, images, and words in the identity-building practices of everyday life. In the realm of love, mediatization would suggest that media affect how we find, organize, and take care of romance, what kind of relationships evolve in and through media, and how we give meaning to such dalliances. Online dating, for example, to some extent changes the geography and socioeconomic status of relationships, as people are more likely to connect and hook up with others outside of where they work and live and beyond their usual social circles. On the other hand, the prevalence of increasingly specific and boundaried dating services—exclusive for farmers, particular to certain income or education levels, only for Muslims—negates such diversification of amorous engagements. Similarly, it can be argued that the relative ease with which to find potential paramours makes romantic encounters much flimsier and more ephemeral. On the other hand, the primary allure of online dating is not so much finding a soulmate—as there is no evidence that the matching algorithms of websites and applications are effective in predicting love—but can be better understood as an effort of people trying to assume control over their love lives in a digital age. In all this, we are once more reminded of how profound the role of media is in all matters that matter and how ambivalent this influence is.

Beyond such an institutionalist perspective on the mediatization of love, a second approach involves a careful consideration of the role media play in the arrangement, experience, and expression of our most intimate feelings—in other words, to acknowledge the love people feel for their media. Clearly, we all have strong feelings for and about our media. These are feelings of frustration, anger, fear, and hate as well as very warm, affectionate, pleasurable, and even passionate emotions. Capturing such feelings, exploring what they tell us about our relationship with media, and considering how this contributes to our understanding of life in media are doubtlessly of great significance.

Since 1998 I have been teaching an undergraduate course at various universities around the world about our life in media. These courses generally have hundreds of students, with on average at least one-third of students coming from countries outside the one I happen to be teaching in. As part of my pedagogical approach, I regularly conduct an exercise inviting the students to talk about why they love (their) media so much. They do this by posting their personal answers on the group weblog "Why I Heart My Media" (whyiheartmymedia.com), which started in 2011. Students are free to share all their feelings about media, including more critical or otherwise fearful ones—the exercise is all about unearthing the full bandwidth of emotions about media. After perusing the responses of these young people, the affective motivations they have for (their) media can be divided into four thematic categories: self-expression, identity, belonging, and passion. Together, these feelings help us to understand love for media, as well as appreciate how fundamental this love is to our humanity.

First of all, the students that contribute to the group blog love media because these devices allow them to express themselves. This can be done by sending or uploading something yourself—such as making a video and putting it on YouTube or TikTok or maintaining a photo gallery on Instagram—or by simply enjoying a nice movie or cool game. As one student puts it, "I like media because it allows me to escape from everyday problems." One could argue that media are primarily so seductive because they offer us the opportunity to self-express and thus be ourselves (in whatever form or version of ourselves), and media companies and professionals tend to make good use of that temptation. Examples thereof are incentives to share what we are consuming with others online (via social media or directly through a synchronized "watch party" on a streaming service) and to interact with content in different ways (by leaving comments, sharing stories with friends, liking and favoriting posts, voting for candidates on reality television shows, calling into radio broadcasts, submitting letters to the editor, etc.). Throughout the media industries, there has been a marked shift in thinking about audiences from reaching as many people as possible to getting a significant group of people really engaged with a specific product or service. With the decline of mass audiences—expected to congregate around the same news headlines, television programs or blockbuster films—an emerging business strategy is that of engagement, trying to get us involved with a product using as many media as possible. The key to all this is to see the audience as more than just consumers, focusing on people's preference to interact, express themselves, and be recognized when doing so. In this way, media industries tap into one of the most powerful emotions we have that explains our attachment to our various devices, channels, and platforms.

In addition to self-expression, the ability to discover who we are explains a large part of our strong feelings about media. "Media is my life," writes one of the students. "I wouldn't know what I would do without media." Another student adds, "I don't exist without media," referring to maintaining a profile on various online social networks. Others note that media benchmark their existence, for example, by tagging, recording, and archiving the places they have been, including the people they were with at the time. We furthermore associate media with who we are—our identity—mainly because we can use and shape media in all kinds of ways as we please: "Media is practical, it is entertainment, media is really whatever you want. Maybe that's why I love media: the way media can adapt to any lifestyle, including the kind of life you'd like to lead." Here one of the students points out something significant: in the game that we play with our identity in media, we can give free rein to our hopes and ambitions. From a critical perspective, we might raise questions about how we in this way run the risk of creating a fantasy world for ourselves, losing all sense of reality in the process. A more optimistic look at this phenomenon recognizes that media provide a space for people to be themselves in a way that is perhaps not safe elsewhere (especially at home) because of issues related to identity, such as their sexual orientation or (lack of) religious beliefs.

Looking around to see where you belong and how you fit in is a natural desire for humans as the social animals we are. A third aspect of our love for media concerns all the feelings that accompany this notion of belonging. "I love my media because it keeps me in touch with my family and friends," says one student after another. One of them further explains: "Last night when I was in my room my phone stopped working. I felt lost and cut off from everything and everyone. It's sad but my phone is my connection to the world. That's why I love media." For many people, their significant others—loved ones, friends, and relatives—are scattered everywhere, across the country, region, and world. For them (and especially in pandemic times for so many), media are indispensable.

Beyond self-expression, identity play, and developing a sense of belonging, a fourth love for media can be distinguished from the various accounts on the WhyIHeartMyMedia website: the ability to have, express, and give meaning to what could be considered more extreme emotions. The chance to express strong emotions, to be passionate in whatever shape or form for which there is or seems to be no space elsewhere, makes media powerfully attractive. Some contributors to the group blog describe how they found old social media posts of themselves expressing heartache after a breakup or total elation seeing an amazing live performance. We probably all remember the first time hearing a certain song or seeing a particular movie scene that changed our life in some way, of ducking behind a friend's shoulder during an especially scary moment of a film or laughing until you cried and regularly reliving that moment by looking up the clip of that segment on YouTube. Sometimes, as students and scholars, it is easy to forget such magical moments in and of media, so deeply involved as we are in our critical analyses, sophisticated research methods, and rigorous documentation of our findings. The various posts of students on the WhyIHeartMyMedia blog remind me of how amazing media can be and how important it is to be mindful of that. At the same time, the relative freedom to express extreme emotions in media also comes at a price: some people—predominantly men—clearly feel quite

uninhibited to use online social networks to disseminate hate and find pleasure in attacking others (especially women and any kind of minorities). Such *dark participation* (as also discussed in chapter 4) is a regrettable byproduct of the love we have for media.

This is by no means a complete or comprehensive study, but I hope this account of the mediatization of love—in institutionalist terms, as the ways in which media scholars come to talk about and conceptualize our affective digital environment, and in constructionist terms, how we, in everyday life, give expression to our love for media—may be useful to further explore the concept, make it explicit, and contribute to our overall understanding of life in media.

## Media Love

This chapter has been a deep dive into a variety of ways to consider, investigate, and understand our love for (and with) media. The importance of this topic cannot be underestimated: people care about their media, our devices and what we do with them are personal to us, and all the feelings associated with this color and shape the way we interact with and give meaning to the media in our lives. Taking this intimacy seriously is a way of taking responsibility for the fact that we study media in the first place. We love media.

The exploration of media love followed a fundamental taxonomy of approaches to media, society, and everyday life, understanding media first as practice (focusing on the various ways in which people use media, second to look at the role of media as a form of mediation), dealing with the circulation and appropriation of information and ideas via media, and finally considering the mediatization of love—involving an appreciation of the central position and influence of media as one of society's key institutions and of the ways in which we all make sense of our love for media. The principal material for this analysis were selected media texts—novels, television shows, films, online services, and mobile applications—that were subjected to a close reading, connected to each other through intertextuality, and articulated with broader cultural transformations. In doing so, the key theoretical and methodological tenets of media studies as a scholarly discipline were applied to achieve a slightly better understanding and appreciation of people's love for media.

If anything, what we have learned from this exploration of media love is, first and foremost, how our love for communication fuels all media studies. Second, the issues discussed in this chapter inspire a renewed appreciation of the porous boundaries between media and life, opening our eyes to the intense feelings we have for and about media and how intimate our relations with and through media have become. When operationalizing media love as practice, we can take this literally and explore the use of media to find romantic love and to enable relations at a distance. A practical perspective on media additionally helps us to ask more fundamental questions about what happens because of our love for media—for example, when all this infatuation leads to problematic media use. In all these activities, we find much ambivalence and vulnerability, as well as adventure and fun. When exploring the mediation of love (and sex) in media, we can document an evolution of love between humans and media from scary, unsettling, and even hostile in the past to mostly equivocal, sometimes reciprocal, and possibly benign in contemporary literature, film, and television. Finally, when considering the mediatization of love, we get to appreciate the affective role media play in people's lives through the various reasons people have for loving (their) media.

Love is not just important as a topic because it is an essential element of all our motivations for using media—it is pivotal as a primary force inspiring and moving us. Considering love carefully involves an appreciation of its transformative power. Love makes us take action, often doing things we would not normally do in an attempt to make something happen, in the process changing us and our environment. Love inspires and encourages hope—for a better way of living, a better world even. The potent mix of hope and action at the heart of love offers inspiration little else can aspire to. The various examples of media activism (as documented in chapter 6) all serve as reminders of the affective and aspirational qualities of media, and how people in and through media not just make worlds, but also inject their love, hopes, and dreams into those worlds.

---

*Love is not just important as a topic because it is an essential element of all our motivations for using media—it is pivotal as a primary force inspiring and moving us.*

---

In the next chapter, we look more closely at how media love can also translate into media activism, as throughout history people have taken to media to change their world.

# 6

## Change Life

> People do not just use media as consumers nor more or less exclusively as producers. For many of us, we use media to change something in our lives and to change something in the world. It is in these instances that the boundary-erasing properties of a life in media become apparent, and this offers us a tremendous opportunity to both study and experience what living in media is really like.

In 2013, the American freelance programmer Ivan Pardo created the smartphone application Buycott to promote conscious consumerism. The premise of the app is that shoppers should know whether the products they buy meet ethical standards. With the app, you can scan the barcode of any product and find out information about the company that produces it. Buycott also allows users to support a wide variety of campaigns aimed at spreading awareness of unethical business practices.

In 2009, the British artist Duncan Speakman organized the first "subtlemob": an offline assembly of people, all listening to prerecorded soundtracks on their personal media devices—an integration of social and physical space intended to make the participants look more closely and carefully at their surroundings. In 2019, the project celebrated its tenth anniversary, having hosted such gatherings around the world.

During 2014, many people in Hong Kong peacefully demonstrated for direct democracy, protecting themselves with umbrellas against police pepper spray and the use of tear gas (in turn inspiring a movement of "umbrella art" throughout the city). The protesters used Facebook to organize the event, Twitter to keep the world updated as it was happening, and YouTube to show the events afterward—using the hashtag #UmbrellaRevolution to track all posts, uploads, and comments across platforms and generating much support and inspiring solidarity rallies all over the world (including in mainland China).

From 2014 onward, the world has been shocked by the footage of executions and other horrific propaganda materials distributed via social media by the militant group IS (Islamic State; Daesh in Arabic) operating in Iraq and Syria, with affiliate networks in

Afghanistan and elsewhere. The posts, songs, and videos by IS were intended to terrify as well as serving as a recruitment tool, yet they also galvanized efforts throughout the Arab world to remix, parody, and make fun of their militant interpretation of Sharia (Islamic law).

A 2017 post on Twitter by the Hollywood actress Alyssa Milano using the hashtag #MeToo started a global trend of sharing experiences online of sexual abuse and harassment. The phrase was used originally in 2006 on the MySpace online social network (between 2005 and 2008 the largest of its kind), and similar postings about widespread sexual harassment spread virally across social media before 2017 (such as in Ukraine in 2016 with the hashtag #янебоюсясказати, meaning #IAmNotAfraidToSpeak) and afterward in different languages.

In all these instances, people used media to affect social change—engaging with the world through media in ways that obliterate the boundaries between online and offline life. The examples mentioned here indicate a variety of contemporary forms of media activism (each with different levels of societal relevance and impacts):

- using media for fun (or simply artistically and aesthetically) to create awareness, instigate ethical debates, and deliberation about one's mediated presence in public spaces;

- using media to inspire conscious shopping and other acts of participating more critically-reflexively in the daily rituals of everyday life;

- using media to organize, coordinate, publicize and promote, archive and document, and galvanize support for political protests;

- using media to cause terror and to radicalize supporters for militant causes;

- using media to take part in a variety of activist campaigns (about issues profoundly affecting people's private lives) without necessarily being physically co-present.

This is just the tip of the iceberg of media activism, of course—more recent forms include data activism, involving practices by people and organizations using big data for political purposes or simply advocating more critical awareness about the collection of large amounts of data on citizens. Also, deliberately not using or even disabling media has become a form of media activism. In such cases of "disconnection activism," media are used to disrupt or corrupt digital systems—for example, through coordinated denial-of-service attacks on certain websites or by installing software anonymizing one's identity while using online devices and services. More subtle examples of disconnection activism are lifestyle choices, such as taking a news vacation, quitting online social networks and encouraging others to do so (the online service Web 2.0 Suicide Machine, for example, automates the "killing" of your accounts on Facebook, Twitter, and LinkedIn), going on a digital "detox" for a while, strictly curtailing screen time (for everyone in the household), and adopting company policies against sending work e-mails after office hours. Finally, as media come to matter more to people's lives, there is much media activism about the media—for example, when fans take to the internet to create their own versions of existing franchises (i.e., fan fiction) or when avid viewers use social media to campaign for additional seasons of their favorite show or to protest the continuation of the contract for a particular musician, anchor, or actor based on their actions offstage or off-screen. Such types of media activism suggest a worldwide vigilant media audience, scrutinizing what people (especially public figures) do in media and calling and actively campaigning for either their continuation or cancelation.

Media play a crucial role in processes of social change, and it certainly seems that in our current digital context people all over the world discover and appropriate media in a variety of ways to further their goals—whether these be personal or explicitly political. Following our definition of media, their involvement in societal transformation is never neutral:

- as distinct artifacts, media enable and shape activist or revolutionary networks, groups, and movements in specific ways;

- using media for activist purposes inevitably includes and excludes particular people (based on having the ability, access, and motivation to use particular technologies); and

- different kinds of media become a principal part of all the ways in which individuals as much as entire social movements organize and express themselves.

In this chapter, we explore the history and role of *media activism*, looking at various uses of media for social and political transformation while distinguishing between different types of media activism and considering what is new and significant about the current context of digital (often singularly social media–based) activism—such as #MeToo and #FridaysForFuture globally, #BLM (BlackLivesMatter) in the United States, #YoSoy132 in Mexico, #EleNão in Brazil, #FeesMustFall in South Africa, and so on. It is also crucial to note some problematic and critical aspects of media activism and engage with the debate about the kind of impact and success—if any—such actions and campaigns have in the "real" world of politics and social justice. Furthermore, we have to recognize the dangers of digital activism, as governments around the world tend to prosecute activists as criminals or even terrorists—often to protect the corporate interests of technology and telecommunications firms.

The chapter concludes with an appreciation of the widespread appropriation of popular culture—such as the *Avatar*, *Harry Potter*, and *Star Wars* franchises, as well as rock and pop music—for civic and protest purposes, in true intertextual fashion (see chapter 2 as well). For better or worse, our media seem to have an aspirational quality, throughout history contributing to our desire to improve our lives and that of others. While industries seek to capitalize on our affective engagement with media, people and communities are definitely not without tactics or strategies to counteract, resulting in a fascinating and enduring struggle for (communication and symbolic) power. This is a truly global phenomenon—which is a further indicator of the pervasiveness and ubiquity of media in people's lives.

---

*For better or worse, our media seem to have an aspirational quality, throughout history contributing to our desire to improve our lives and that of others.*

---

### On the Inseparability of Media and Social Change

In the context of activism, media are a medium for activists to communicate and interact, a space to raise awareness and provide visibility for social causes and revolutionary aspirations, and a battlefield where different people and groups try hard to get their voice heard and their cause recognized—a fight over impact and meaning in media that takes place in many localities, as well as on a global stage. Given the proliferation of media activism all over the globe, it may seem as if this engagement is a distinct sign of our times, a phenomenon engendered by the current omnipresence of (digital) media. As with every aspect of our life in media, this is not the case. Social movements, revolutionary groups and associations, people banding together in protest: the origins of all this run deep in the histories of societies around the world and can be seen as part of the engine that make contemporary democracies (as much as dictatorships) work. In all these histories, media play a formative role.

From the middle of the nineteenth century, women in various countries—including the United States, England, Germany, Australia, New Zealand, Finland, Norway, and Denmark—started to organize themselves around their right to vote in elections and around the issue of equal rights more generally. This, for example, led to the foundation of the International Woman Suffrage Alliance in 1904 in Berlin (Germany), an organization that continues to this day as the International Alliance of Women, a nongovernmental organization with representatives in major international councils, such as the United Nations, the Arab League, and the Council of Europe. The history of the international fight for women's suffrage is, to a significant extent, a media history. Key to the success of the campaign after many decades of struggle was its effective use of media in myriad forms, including various printings, posters, postcards, editorial cartoons, advertising campaigns, and news releases to invite coverage of its parades, pageants, mass meetings, protests, and other *mediagenic* gatherings. During the twentieth century, another important role of media was the editorial support the movement enjoyed from some popular mainstream newspapers and magazines, next to the continuing professionalization of women's own print publications. The significance of creating news outlets of their own cannot be underestimated, as such defiant efforts can be credited with helping to redefine what it meant to be a woman in male-dominated contexts and cultures. The movement also made good use of pamphleteering: distributing inexpensive printed statements everywhere to be read aloud in churches, in taverns and pubs, and at meetings. This practice had its origins in sixteenth-century England at a time of great debates around religious controversies and political upheaval. Next to newspapers that were still very much a medium for a small literate elite, pamphlet warfare proved to be a relatively quick and easy way to reach many people, even those without the skills to grasp the intricacies of the written word. Interestingly, despite most newspapers initially being openly hostile toward the suffragettes, they reported extensively on their meetings and events and published their letters—in part because the women's movement made for sensational headlines helping to sell a lot of papers.

The campaign for women's rights around the world is but one among many examples of how activist campaigns cannot be understood separately from the role various media play in it. Its history also underscores the ways in which social movements can tactically co-opt the commercial mechanisms of corporate media to further their own goals—for example, by staging spectacular protests to generate attention-grabbing headlines—next to strategically developing autonomous media. The same conclusions can be drawn about more contemporary struggles, such as the various so-called indignation movements emerging in the 2010s. This includes the Arab Spring (throughout the Middle East and North Africa), the 15M movement and Los Indignados in Spain, the Occupy Wall Street initiative in the United States of September 2011 (in subsequent months spiraling into a global Occupy movement), the Come to the Street Movement (also known as the Brazilian Spring) in 2013 and its more conservative counterpart Movimento Brasil Livre (Free Brazil movement) from 2014 onward, and less high-profile yet equally significant forms of widespread resistance and political mobilizations involving the frenetic use of many different media, such as favela media activism in Brazil, township media activism in South Africa, and countless examples of community and Indigenous activist initiatives, projects, and movements throughout Africa, Latin America, and Australia (and elsewhere) involving community radio, online social networks, poster and flyer production, mobile messaging campaigns, staged events for the news media, and crucially the production of independent media. Common themes in many of these social movements are collective action for housing and other basic human rights, protests against police and state violence, and critical reactions to discrimination and racism—all in the context of intense and autonomous media use for activist purposes.

Historically, social movements tended to emerge tied to a specific physical place, involving people operating in a local network and context, addressing issues close at hand: farmers rebelling against their landowners, citizens rallying against discriminatory practices in their neighborhoods, people protesting a social injustice experienced by someone in their community. Today, social movements are all that and much more in part because of media—sometimes even operating almost exclusively online. A seminal example thereof was the Kony2012 campaign, propelled into the global spotlight by a short documentary published online on March 5, 2012, about the indicted war criminal Joseph Kony, a Ugandan cult and militia leader. The professionally produced film went viral on YouTube—it was the first video on the platform to reach over one million likes (and well over one hundred million views within the first six days of its release). The Kony2012 video was part of an awareness campaign about the atrocities committed by Kony and his followers in the hope of getting him hunted down and arrested. In the months to follow, the organization behind the documentary—Invisible Children—utilized a variety of media to get the public actively involved to continue the campaign, for example, by asking people to put up posters and wear T-shirts with a Kony2012 logo and making action kits available that included campaign buttons, posters, bracelets, and stickers. Various (American) celebrities endorsed the campaign as well, generating much publicity. Online, the response was massive—while offline participation remained rather limited. The phenomenon is an example of what is sometimes mockingly called *slacktivism*, as the campaign inspired people to participate through little more than sharing the video (and its core message) via media. The use of "slacking" in this context is interesting, as it suggests that those who choose to pursue their activism in media lack ambition, do not take responsibility, and are not really engaged with the issue at hand—simply because they solely (or mainly) use media to express their support. As a side note, the Kony documentary was the culmination of many years of lobbying and charity work by Invisible Children and its founder, Jason Russell—often via quirky media schemes, including the production (in 2006) of a silly dance video staged at a high school, featuring the message that children in Uganda were "in bad times" because of the horrors inflicted upon them by Kony's army.

Grassroots and citizen activism, in short, occurs throughout history and can be plotted on a continuum from small-scale, street-level organizing to global and largely virtual forms of participation and engagement. While the role of media in processes of social change needs to be taken seriously, it would be a mistake to consider media as the causes or even the main drivers of protest, organization, cooperation, and coalition building by people in an attempt to change the world. The specific (social, cultural, economic, and political) context of a situation primarily affects both the ability of people to gain access to different kinds of media, as it shapes their motivation to take to the streets.

## Addendum: The Role of Media in War and Conflict

When states and governments opt for war, this can be understood as a horrific way to enact change in the world as well. The role of media in war and conflict has always been close, especially because of the major news value of warfare. War sells, and news coverage of combat and military action can contribute to the escalation, as well as de-escalation, of hostilities. The presence of journalists, cameras and microphones in hand, can prompt soldiers to start shooting. At the same time, ongoing news coverage of fighting and bloodshed has been known to negatively influence public support for war. Conflict journalism is complex, difficult, and often traumatic for the reporters involved. An often-used quote—attributed to several prominent politicians around the time of the First World War—is that in war, truth is the first casualty. While it can be true that journalists often struggle to report in times of war, a deliberate response can be to invest in conflict resolution journalism or peace journalism, whereby reporters and editors make choices of what to report and how to report it focusing on nonviolent opportunities to resolve conflict, uncovering the causes behind a war, and making sure to humanize all victims.

In the professional practice of peace journalism, there is specific attention for the various ways in which news and information gets manipulated by the warring parties. This highlights the role of governments and other state actors in times of war, including the military, when it comes to media. During (the run up to) the war, states tend to expect compliance and support of their national media, often introducing new laws in service of the war effort. During the Russo-Ukrainian war (from 2014 onward), for example, the Russian government made it illegal for their news media to refer to the Russian invasion of 2022 other than as a "special military operation," while the Ukrainian government issued a decree that combined nationwide TV channels into one platform—arguing that "in a state of war, the implementation of a unified information policy is a priority issue of national security." Especially since the rise of mass media in the early twentieth century, nations at war increasingly rely on media as part of the military effort. This includes the use of propaganda during the First World War (1914–1918), which led to much handwringing afterward about the perceived power of such persuasive media. Before that, colonial powers such as France and the British Empire enacted deliberate policies to suppress rebellious movements and influence subjugated populations, whereas the United States in its wars in Vietnam and Iraq committed to a strategy of "winning the hearts and minds" of the local people. These influencing tactics always included media, such as producing news in local languages to promote efforts (by occupying forces) to help Indigenous people, dropping leaflets from planes calling on people to support their efforts, broadcasting positive news on local and community radio in native languages, and so on.

What makes the contemporary context for the role of media in war and conflict of such importance for our considerations of life in media is the military notion of *hybrid warfare*, whereby a nation's government and military combine information warfare (specifically, propaganda and disinformation campaigns) and cyber warfare (attacks to weaken or destroy an opponents' information and communication infrastructure) with conventional warfare (which would involve an invasion with troops, tanks, and other military equipment). The concept is not necessarily new as it can be traced back as far as to the wars between Athens and Sparta in ancient Greece, and it has been taken up in military discourse around the world (including, for example, in China and among terrorist networks like IS). There is a specifically Russian equivalent to hybrid warfare called *gibridnaya voyna* that defines the country's approach to war, the primary purpose of which is to

subvert, manipulate, and confuse the enemy—in Russia's case, the West. The governing idea is that media are used in all kinds of different ways to undermine whatever the enemy is doing—including meddling in the elections of other countries, fueling social and political polarization (e.g., through coordinated campaigns on social media), engaging in hacking attempts of foreign computer systems, and especially by producing an endless variety of stories, alternate theories, and counterfactual statements aimed to distort and pervert any kind of consensual message or narrative coming from the enemy. The notion of hybrid war seems directly inspired by the digital environment we all find ourselves in, turning warfare into an unlimited (and possibly never-ending) enterprise.

Within the current digital environment and in the context of hybrid warfare, media play a profound role in armed conflict. Beyond the parties involved trying to control media and steer the narrative, it must be noted that in the twenty-first century everyone directly and indirectly involved in war produces media on the conflict, making any kind of real control over the message an almost impossible task. Keeping the Russo-Ukrainian war example as a case in point, it is truly striking to see all kinds of parties engaged in telling stories and providing versions of the war, including (but not limited to)

- the respective presidents and their governments;
- the armies involved, both officially sanctioned communiqués and soldiers on the ground uploading smartphone and drone footage of what they are doing;
- thousands of Ukrainian and foreign journalists (and their local helpers, often called fixers, stringers, or producers);
- all kinds of other professional media makers in Ukraine, such as famous film directors, musicians and bands, game developers, and advertising and marketing creatives; and
- people both locally and all over the world using platforms, chat applications, and social media, such as TikTok, Facebook, YouTube, Telegram, and Twitter.

Each actor on the media battlefield produces, distributes, and shares a complex variety of media, utilizing an equally bewildering variety of platforms, services, and technologies. A direct consequence of this collective construction of the war is how the conflict is made in media differently, depending on where and who you are (in the world).

### Revolution, Rebellion, and Activism in Media

While protest movements, rebellions, and revolutions have always been a crucial aspect of human history, activism is a relatively new term, introduced in the mid-1970s to refer to the ability and intention of people to act to bring about social or political change—without necessarily resorting to violent uprisings or otherwise radical action. The term gained prominence largely in response to a rapid expansion of protest movements and identity politics around the world, as Indigenous populations—from Basque nationalists in Spain to Native Americans in the United States, including a Māori renaissance in New Zealand and anti-apartheid student uprisings in the townships of South Africa—as well as groups such as gay men, lesbians, and women organized to oppose discriminatory laws and pursue government support for their interests. While some of these struggles certainly involved violent action and suppression, it did not lead to the kind of civil wars, mass rebellions, or nationwide armed conflicts as with, for example, the American, European, Russian, or Chinese revolutions (occurring between the late eighteenth to the mid-twentieth centuries).

The relatively recent switch from rebellion and revolution to activism in the way people band together to change the(ir) world does make one wonder what made it such a global force from the second half of the twentieth century onward. Beyond the particulars of local conditions, place-bound protests and grievances, and struggles particular to a certain community, country, or region, some general trends can be discerned. First and foremost, we have to consider the political context of widespread protest. Scholarship suggests that the larger the gap between what people want from their society and a government's ability to adjust to the demands of the people (further framed by the level of democracy in a country), the more likely people are to take matters into their own hands. Economic indicators also play a prominent role, as precarity proves to be a powerful motivator for activism. Beyond such general indicators predicting the potential for activism, people's personal values and ideals cannot be ignored. Since 1981, the World Values Survey has been documenting social, political, economic, religious, and cultural values of people through regular surveys in more than 120 societies all over the world. The project's goal is to assess how people's values change over time. Researchers involved with the project note that rising levels of education combined with the spread of self-expression predict the motivation and tendency of people to demand (and gain) more freedom of choice in how to live their own lives. In other words, in the modern era, a high level of existential security encourages openness to change, diversity, and new ideas—as much as poverty, oppression, and misery propel people into action. Regardless of important differences in culture, location, and context, the breeding ground for activism seems to have deepened and broadened in the last couple of decades, and cannot be disentangled from people's access to—and comfort with—media.

In a context of generally rising standards of living around the world (especially for those born after the Second World War), the role of media comes into sharper focus. Regarding the mass media of print newspapers and magazines, radio, television, and film, throughout the twentieth century, people increasingly got exposed to ways of living, value systems, and information from sources and places unseen and unheard of before. Media are a universal comparison machine, showing us how "the other half" lives. With the introduction of (personal) computers, the internet, and later on the World Wide Web and mobile communication, we are drawn into a world that we can participate in—with the push of a button, the click of a mouse, the movement of a joystick, a screen tap or a swipe, and the rattling of keys on a keyboard. Despite the efforts of corporate publishers and technology companies to close down and control the ways in which we use our devices—for example, through closed app stores (instead of maintaining an open marketplace for all transactions), cloud-based storage (rather than using your own hard drive), and subscription software (contrasted with one-off sales or open-source software that would free up programs for us to modify and use as we please)—people all over the world find supporters for their cause, self-organize, express their opinions and protest, and make a difference beyond the screen.

In our digital environment, we tend to use the same platforms and networks that activists use to publish and consume all kinds of content, making it much more likely that we stumble across critical debates, political messages, and calls to action that we would otherwise miss. Despite efforts by such platforms as Facebook or TikTok to be seen as neutral facilitators, they end up being quite political in how they are used as vehicles for both funny memes and campaigns for social change. The Russo-Ukrainian conflict shows that even in times of war digital media and online social networks play crucial roles for mobilization, either to rally people to support and defend a country under attack or to influence and steer people toward propaganda and disinformation intended to promote an aggressor. Interestingly, when platform providers step in to, for example, ban certain practices—or when governments threaten to shut down popular sites—people can all of a sudden become activists simply because their favorite network or service to share fun facts with friends stops working. Sharing, connecting, and participating are all activities that are designed into our digital environment—whether we, companies and corporations, security forces, dictators, or governments like it or not. Other than completely pulling the plug, there is always a possibility that people get involved with some kind of activist cause or campaign who otherwise would not consider themselves to be politically engaged at all.

From a life in media perspective, it could be argued that the potential for (and global expansion of) participation and activism is, at least in part, also due to a subtle shift throughout the twentieth and twenty-first centuries from a "sit-back-and-be-told" media culture (particular to mass media) to today's abundant "making-and-doing" media (based on personal computing). This corresponds with a historical evolution from "read-only" culture—where most people could only read, listen to, or watch prepackaged and professionally produced media—to todays' ubiquity of "read and write" technologies, inviting more interactive uses of media. This development got its start with audio cassette recording (in the 1960s), home video taping (starting in the 1970s), programmable personal computers (appearing in households in the 1980s), and the introduction of rewritable CDs (in the late 1990s). In each instance, more people were drawn into an increasingly customizable, interactive, and participatory media environment. The global shift online further led to participation in internet forums, blogs and vlogs, podcasts, and all of today's social media. Of course, this history is neither a neat progression nor does it work similarly (or seamlessly) around the world. However, it can be argued that the material context of our digital environment prefers us to take action and to engage rather than to sit back, relax, and consume.

As shown in the case of the worldwide women's rights movement (and all other social movements), activism is inseparable from both its representation in news media and its ability to utilize and self-produce media. Throughout history, different types of media have been appropriated for activist purposes in medium-specific ways. Print media have been creatively used to copy and distribute leaflets (such as neighborhood newsletters), with people, for example, cutting and pasting clippings from magazines and newspapers to assemble new narratives. Radio from its early days also spawned a lively field of hobbyists as well as activists producing free radio, sometimes labeled as "pirate" radio—referring to unlicensed or otherwise unauthorized transmissions through the airwaves. On television, the movement toward activist and do-it-yourself media has been perhaps less widespread, even though groups of people set up independent stations and signal transmitters around the world, in places as varied as Jamaica, Saudi Arabia, Ireland, Israel, Canada, Italy, Greece, and Spain. Although quite a few of these pirate media were distinctly commercial in structure and operation, the philosophy of the movement tended to be much more activist in character—with the operators and facilitators of such stations pointing out that the airwaves belong to the people, emphasizing their allegiance to amplifying the stories and opinions of individuals and communities not able to express themselves in media otherwise. The power of scheduling and influencing public opinion is not the only characteristic of mass media institutions. Indeed, the history of media as an industry is also very much a history of Indigenous, community, alternative, and diasporic networks claiming and securing voices of their own vis-à-vis state or corporate media offerings. In a digital context, the production of independent, or indie, media has proliferated, despite (as well as in outright opposition to) an ever-increasing commercial presence online.

A case in point is the rise of Independent Media Centers (better known as Indymedia, or Centro de Medios Independientes in Spanish-speaking countries) from 1999 onward. With protests against what many saw as unbridled capitalist expansion gearing up around the world during the 1990s—embodied in intergovernmental financial-economic networks, such as the World Trade Organization (WTO), the World Economic Forum (WEF), and the International Monetary Fund (IMF), as the targets of such criticism—plans were made by activists to network, coordinate, and publish the various actions, demonstrations, and initiatives on a single platform. The first Independent Media Center (IMC), made possible by donated computers and running on open publishing software, came online to cover the intense protests against a WTO meeting in Seattle in November 1999, followed by similar IMCs operating in Melbourne around protests against the WEF, and in the Czech Republic vis-à-vis a meeting of the IMF in Prague (in 2000). The activists involved coupled the end-to-end infrastructure of the internet—connecting computers from all over the world to each other without much control, oversight, or hierarchy—with emerging networks of resistance and protest worldwide. With the rallying cry "make media, make trouble," volunteers used the IMC website to mobilize support and organize protests, publish news and report in real time on events and demonstrations, write short posts, produce audio and video, and host active discussions. A print publication (called *The Blind Spot*) was also published to accommodate those without internet connection. Many more Indymedia sprang up all over the world, forming loose coalitions of a variety of existing and new protest groups and social movements. From a life in media point of view, it is fascinating to note that observers and participants at the time likened the fluid organizational structures of IMCs to the infrastructure of the internet itself.

At the time, I was working with a colleague in Amsterdam, Sara Platon, who was involved with setting up Indymedia in Sweden and the Netherlands, witnessing up close the complexities involved. After much enthusiasm and energy in the first years, Indymedia faced some serious problems. In several countries, its volunteers were persecuted and arrested, web servers seized, and some centers ordered closed by law. Within the movement—a loosely organized, globally networked constellation of various groups without clear leadership—tensions emerged, for example, between those involved with maintaining the technological infrastructure and the activists pursuing various political agendas. There were also fierce debates online about (potential) sources of funding for the various Indymedia, illustrating the difficulty of reaching consensus across a global network. Similar struggles took place during the global Occupy protests of 2011 (and beyond).

While people in local IMCs organized face-to-face, many IMC projects had international involvement, and discussion about them happened primarily through email lists and, secondarily, through chat channels, breaking down the boundaries between local and global information flows and protests. The independence of Indymedia was also not strict, as the software code and content of the various websites were made and maintained by people that were, in one way or another, affiliated with many different groups and movements providing their own content, representing a wide variety of perspectives and goals. Independent in the context of Indymedia rather meant being free from commercial and corporate interests, as the main premise of Indymedia as a form of media activism was a vital critique of commercial, corporate mainstream media organizations, next to acting as a publishing and organizational hub for various protest initiatives. During the 1999 protests, the IMC site in Seattle had an average of two and a half million visitors every two hours. This high figure doubled during intense protests around the meeting of the so-called Group of Eight (G8) major economies in Genoa in 2001. At its height, there were well over 150 Indymedia worldwide. Today, only a handful remain.

Independent Media Centers and the various social movements and groups of protesters that powered them were rooted in a surge of what can be called *cyberactivism* starting in the mid-1990s as activists around the world embraced online bulletin boards and discussion forums, internet chat channels, email lists, and fledgling websites to get organized and spread the word about their causes and actions. A benchmark example thereof is the creative use of the internet to garner support for their cause by the Zapatista Army of National Liberation (from 1994 onward). The Zapatistas—a grassroots resistance movement based in southern Mexico—seek Indigenous control over land and resources, using guerilla radio, bulletin board systems, email lists, websites, and other media strategies to set up a network of Indigenous peoples in the region, across the Americas, and elsewhere. One of their rallying cries is to "become the media" as a way to develop networks of alternative communication rather than relying on established media forms and organizations. Starting in 1996, the Zapatistas organized annual Intercontinental Indigenous Encounters, at times hosting thousands of activists and their supporters from forty-two countries on five continents to discuss their struggles. In their proposal, the movement outlined its vision for alternate networks for communication online to provide a platform for sharing experiences, discussion of strategies, and exchange of a wide variety of self-produced media: "Let's make a network of communication among all our struggles and resistances . . . This intercontinental network of alternative communication is not an organizing structure, nor has [it] a central head or decision maker, nor does it have a central command or hierarchies. We are the network, all of us who speak and listen."[1]

This combination of local, place-based activism with the global reach of digital media provoked new awareness among academics and activists alike of the much broader (and older) phenomenon of Indigenous revival and struggle. It also got specific attention from media scholars due to the innovative and autonomous ways in which the movement appropriated (and claimed to be) media.

From the late 1990s onward, digital activism proliferated around the world, often linked to widespread concerns about corporate globalization, protest against the war in Iraq (in the context of the US-led "war on terror" following the attacks of September 11, 2001), calls for more democratic participation and addressing social inequalities in major economies (such as Brazil, Russia, India, and China), growing global awareness about government spying and state surveillance (especially following the revelations in 2013 by Edward Snowden about the overreach in global surveillance; see chapter 3), and common concerns about the accelerating pace of climate change.

The early Zapatista and Indymedia examples as much as contemporary forms of digital activism should remind us of how specific characteristics of media—in this case, the combination of the internet, mobile communication, and broadband connectivity—influence (as well as reflect) the particulars of different kinds of activism and social movements using such media, which in turn has consequences for the ways in which we can understand and appreciate media activism as a global phenomenon. A few features stand out in these and many other forms of networked media activism around the world:

- online activism generally extends from offline actions, whereas offline engagement at times proceeds from online exchanges;

---

1. Zapatista Army of National Liberation, "First Declaration of La Realidad for Humanity and against Neoliberalism," quoted in DeeDee Halleck, *Hand-held Visions: The Impossible Possibilities of Community Media* (New York: Fordham University Press, 2002), 415.

- the distinctions between local events, groups, and participants versus global initiatives and campaigns become less clear-cut as everything gets mixed up in the digital environment;
- different kinds of communities are co-present in an online context—interpersonal networks consisting of individuals and working groups, organizational networks consisting of local groups and affiliate organizations, and a global (virtual) community of participants and supporters—all interacting and mutually shaping the activism involved;
- in a networked context, the involvement of a diverse range of people and groups ebbs and flows, divides and fuses, and can be really intense while sometimes it altogether disappears;
- among all the participants, both near and far, there is a less-than-clear expression or experience of rank and hierarchy, and at different moments, different people take the lead or otherwise become influential.

All of this suggests both a strength and a weakness of contemporary media activism: on the one hand, these types of networked engagements for social change tend to be flexible and able to quickly adapt to rapid-paced developments on the ground and in the world at large, while on the other hand their loose organizational forms and fluid, often temporary structures make for a precarious existence, always teetering on the brink of falling apart. What tends to keep such new and networked social movements together are the distinct personalities and personal motivations of individual participants rather than (appointed or elected) leaders, bureaucratic organizational structures, the characteristics of technologies, or any specific political ideology. Even the Indymedia and Zapatistas, despite having much in common as grassroots struggles, are hard to pin down politically, as are their supporters around the world. Under broad headings (such as social justice, human rights, and anti-globalization protest), people from different walks of life and different parts of the planet come together in and through media to voice as well as act on their multifaceted concerns. Like with the aforementioned shift from rebellion to activism since the 1970s, the 1990s saw a shift from ideology to identity as providing the primary fuel for the formation of activist networks. In both shifts, the media of that era—respectively mass media and online media—played a formative role.

### Defining and Understanding Media Activism

When a significant event takes place—an election, a natural disaster or terrorist attack, a pandemic, or some form of mass protest—people turn to a variety of media to find out what is going on. This has historically been the domain and role of legacy news media. However, professional journalism tends to privilege state and institutional authorities and academic experts in their coverage of such phenomena, generally marginalizing or downplaying the voices of ordinary people—especially those who are protesting or otherwise agitating against the status quo or established social order. In today's context of social media and a ubiquitous interactive media culture—as well as an increasingly digitally literate population—news about what is happening anywhere in the world gets produced, disseminated, and shared on a real-time basis, where the voices of journalists and politicians have little more status than those of social media influencers, vloggers, the occasional eyewitness or bystander, and the networks of people (and bots) that like, favorite, share, and forward their posts. The exponential growth of self-produced media (in the context of our *mass self-communication*) can in part be attributed to the reluctance of conventional media to include a greater variety of voices. This hesitation can partly be explained by the pressure on mainstream media to address and attract a mass audience. In the context of a digital environment that encourages engagement and self-expression, it is perhaps unsurprising that many people and groups feel their concerns and perspectives are not represented in the media.

Activism is action for social, cultural, political, and national (including local) change. *Media activism* is a form of activism that either has the media as the object to be reformed or uses a variety of media to further its goals. All in all, media are essential to all forms of activism, and the way people use media for activist purposes in turn shapes such activism: how people express themselves, how they interact, and how things get organized in the context of the ongoing campaign. In each protest, during each campaign, different media tend to have different functions. Even a single medium, such as an online platform, offers diverse opportunities. For example, Twitter comes in handy to spread and update information in real time and has been used during demonstrations to quickly arrange alternate meetups or redirect the movements of a group of protesters based on the actions of security forces. Facebook is a practical tool to promote events or facilitate debates, while video hosting services like YouTube, TikTok, and Vimeo can at times assist in spreading compelling videos or even short documentaries about what is going on. Instant messaging applications like WhatsApp, Signal, and Telegram furthermore facilitate instantaneous communication with the additional benefit of different levels of security and encryption.

Prior to platforms and social media, media such as mobile phones, handheld video cameras, and personal computers changed the way social movements self-organized and documented their activities, and before that (community, Indigenous, and alternative) radio played an important role in activism, with print media—such as magazines, flyers, and pamphlets—also influential in various struggles around the world. Although the narrative of this history suggests a somewhat linear trajectory from the activist use of analog media to today's engagement with the digital environment, it is important to note that media of all shapes and sizes—from print to electronic, online as well as offline—are part of the activist's toolkit today. The notion of *tactical media* is useful here, as this concept does not privilege either digital or analog media to be deployed to further one's cause—the only requirement is that do-it-yourself media of all sorts can be used by groups and individuals who feel their voice or cause is not heard, recognized, or included. The use of tactical media also underscores how any kind of medium can be exploited for activist purposes, depending on the requirements, available resources, and particular context of the community or cause involved. At the same time, media are not a neutral playground of possibilities for activists and social movements to explore, as their uses are to a certain degree determined and prescribed by the *affordances* of specific technologies, interfaces, and policies (e.g., those set by platforms), and anything people do with media becomes visible—not just to supporters and stakeholders but also to the very institutions and authorities people are rebelling against.

In each instance, the role various media play in activism is that of

- a *tool* for planning, organization, and coordination, including fundraising, networking, and coalition building;
- a *vehicle* for participation and engagement, as well as mobilization and recruitment;
- an *amplifier* for messages, actions, events, and news involving the movement and its cause; and
- an *archive* for preserving the histories of social movements and protest networks.

Although activism, throughout history, cannot be considered separate from the media that shape and sustain it, it would be an overreach to give media a role beyond that of a (at times powerful) facilitator of people's political and social engagement. A significant increase in the use of media is much more likely to follow a significant amount of activist activity than to precede it. The new media environment seems to provide protesters and activists around the world powerful, easy-to-use and relatively low-cost tools, yet studies show that the greater the level of internet and social media penetration (which generally corresponds to people living in wealthier environments enjoying uncensored access to various media), the lower the level of protest in any given community. It is precisely those populations that have the greatest need to mobilize that tend to find it most difficult to exploit the digital environment due to lack of access, freedoms, skills, and resources.

In the specific context of today's digital environment, once we become aware and engaged to do something—using media for social or political transformation—our activism can be characterized along four types of activities:

- various forms of *slacktivism* or *clicktivism* and *metavoicing* (reacting to other people's online presence and posts), where we are mostly spectators or cheerleaders of activist campaigns;

- supporting activities, such as contributing to *crowdfunding* campaigns, signing up for activist newsletters (and other online registrations of interest), and signing digital petitions;

- engaging in direct action, for example, participating in gathering and analyzing data, helping to hack online sites and services (known as *hacktivism*), organizing and publicizing campaigns and events, coordinating actions online, and making media (including memes, short videos, and so on); and

- all types of actions where participants digitally organize yet lack a clear cause, getting involved in media without necessarily being all that committed to the specific group or objective, but still feel engaged, which can also be labeled as a kind of affective or somatic solidarity, often motivated simply by a desire to be connected and belong.

It is clear that all activities people (can) engage in may overlap and lack hierarchy—present-day media activism generally needs activities on all four levels to have any chance of succeeding. It could be argued that for many contemporary movements, the power of digital media is precisely its potential to raise awareness so that people engage offline—to inspire "unwired" interventions in streets, parks, and offices. There is an important relationship here between *avatars* and *bodies*: for any type of activist movement to propel people into action, it needs "bodies in the streets" (including in such operational roles as making phone calls, keeping social media updated, running all kinds of errands, and scouting locations for protest), whereas people's actions and involvement can quite effectively be triggered and mobilized through engaging their avatars (in other words, their virtual presence and profiles on countless networks, platforms, sites, apps, and services online). The online and offline aspects of activism therefore mutually reinforce each other, signaling an inevitable blurring of the boundaries between media and life when it comes to changing the world. While some would argue that the risks taken by bodies—for example, by physically facing police—are quite distinct from those by avatars, we cannot ignore the fact that online political engagement can be monitored and tracked by security forces, in some circumstances having severe consequences for the people involved. Political organizations and social movements today combine a variety of ways of organizing and mobilizing, mixing online and offline efforts, connecting local issues with global struggles and vice versa, participating in virtual networks as much as engaging with people on the ground, spreading as well as sharing risk, and engaging their bodies as well as their avatars to further the cause of political transformation or social change.

## Key Issues when Changing the World in and through Media

At any moment, there are numerous worldwide shareholder and investor activism campaigns underway to force corporations into greening their operations, hiring more female executives, and improving working conditions (an example of environmental, social, and governance oriented activism), making strategic use of media channels to publicize their demands and prompt greater pressure from other shareholders. Every year, there are more cases of such business-oriented activism around the world than the year before.

Francisco Vera is an enterprising twelve-year-old Colombian who makes headlines around the world for his brave environmental campaigning, specifically gathering support through Instagram (using the handle franciscoactivista).

The American film industry faced a massive strike in 2021 by members of the International Alliance of Theatrical Stage Employees, Moving Picture Technicians, Artists and Allied Crafts (IATSE), one of the most powerful unions representing film crews in North America, generating support with the hashtag #IASolidarity while fighting exploitative working practices of streaming services like Netflix.

Patients around the world who continue to suffer from the symptoms of COVID-19 months after they have been infected rally online on private Facebook community pages, using Twitter hashtags like #longCOVID, sharing videos on YouTube, and starting the web-based support group Body Politic—all of which got the attention of the World Health Organization, helping to get long COVID formally recognized as a medical diagnosis.

Employee activism in Australia and elsewhere has grown as a consequence of remote and hybrid working arrangements during the coronavirus crisis, which many workers want to keep (at the very least as a flexible option regarding caring duties).

Global brands such as Carlsberg, Oreo, Nike, and others partner with activist campaigns on a variety of issues, such as to advocate for the LGBTQ+ community and people with disabilities in the screen industry or to champion local businesses—using their mighty media marketing budgets to simultaneously promote themselves and the subjects they support.

Throughout 2021, students across all twenty-six public universities in South Africa protested—using such hashtags as #FreeEducationNow and #FeesMustFall2021—against high registration fees and for the provision of adequate funding and resources (in a repeat of similar nationwide protests in 2015 and 2016).

Overall, the Global Protest Tracker (maintained by the Carnegie Endowment for International Peace) counts well over 230 significant antigovernment protests erupting in 110 countries between 2017 and 2022 (many of which specifically related to the coronavirus pandemic).

Considering this global snapshot of people taking action in different ways (yet all involving media), it certainly seems that something is afoot. Activism, in all shapes and sizes, amplified by voices, channels, platforms, and services in the media, seems to be spreading and accelerating worldwide. So much so that US television network CBS *greenlit* a new reality television show in 2021 called *The Activist*, featuring a competition between six activists based on missions, media stunts, digital campaigns, and community events to promote health, educational, and environmental causes. The winner—based on online engagement, social metrics, and input from the show's hosts—would get the chance to lobby world leaders at a G20 Summit. The announcement resulted in a worldwide outcry against the show's rather cynical equation of social media success with achieving activists' goals and for having activists competing against each other. Soon thereafter the international advocacy group Global Citizen, coproducer of the show, released a statement saying that "global activism centers on collaboration and cooperation, not competition. We apologize to the activists, hosts, and the larger activist community—we got it wrong," whereafter the network postponed the broadcast date and changed the format to a one-time documentary special.

Our participation in all kinds of activism in, about, and through media exposes a critical tension of life in media: whereas our digital environment certainly invites participation, and there is some emancipatory, playful, and cocreative potential built into the technological infrastructure of the internet (and all its attendant technologies), at the same time these affordances are circumscribed by a distinctly corporate and overarching controlling context. Overall, the production and circulation of technologies and knowledge are predominantly guided by economic interests and profits. The technology, media, and telecommunications industries tend to be managed and controlled by a relatively small subsection of people and companies from less than a handful of countries—mainly the United States, China, and Japan—that fiercely self-defend their interests and privileges against any kind of democratic governance or public oversight.

The *inclusivity* of digitally mediated activism is exemplified by providing a relatively affordable and convenient arrangement for participation, while its *exclusivity* is amplified by different kinds of digital inequalities, divides, and forms of digital marginalization. Of particular concern here is the popular and sometimes necessary use of corporate social

media for purposes of crucial civic engagement. Such commercial online social networks both flatten as well as create hierarchies by a combination of their technological affordances, the ways in which they are used, and what kind of actions, expressions, and organizations they propagate. All of this greatly benefits larger and very active networks, favoring people with clear awareness of platform logics, and prefers any kind of noncontroversial discourse. Smaller groups with people coming from a diverse range of backgrounds, skill sets, and perspectives tend to find popular platforms like Twitter, Facebook, and YouTube less user-friendly. At the same time, these networks remain popular tools for activism as they tend to be stable and durable, are available almost everywhere, and remain difficult to censor or control completely.

A second and related insight about the history of media activism, particularly regarding the rise of digital activism in recent years, is the danger of universality. Digital media in relation to activism are often considered universal in the ways in which activists use them for their purposes. Given the omnipresence of digital media around the world, an assumption could be that building a website; maintaining a blog, vlog, or podcast; setting up a wiki or email list, coining a hashtag; and so on work and feel the same regardless of local, political, economic, or cultural context, that accessing platforms like Twitter, TikTok, YouTube, and Facebook provides the same affordances and challenges globally, and that all these digital services offer similar opportunities, thereby greatly democratizing people's civic engagement. Time and again, examples and cases of media activism turn out to be deeply situated and context dependent much more so than universally comparable. Another related expectation would be that, because of the appropriation of the digital environment, contemporary protest movements are necessarily spontaneous, horizontally organized, inclusive, and leaderless. Again, the evidence suggests otherwise. Hierarchies within social movements often emerge (or preexist) especially regarding the very digital tools and technologies that are supposed to enable participation for all—if anything because digital media are complex and require significant maintenance and upkeep.

Although (digital) media clearly play a profound role in all kinds of activism, the part various media play is difficult to generalize and seems more particular to the identity of a specific social movement (or even of the unique individuals involved) than whatever a platform or application claims or promises to offer. In fact, this insight highlights the unique tension between the universal and the particular that a life in media perspective offers regarding any phenomenon under investigation. The role of media in activism is both a feature of the technologies, platforms, and services used as well as distinctly rooted in local struggles and contexts, having some universal features alongside many quite particular characteristics. All this reminds us not to overly romanticize nor idealize the emancipatory potential of media activism.

A subsequent third insight is that digital activism, on its own, cannot be said to be very successful in a traditional sense. We continue the delicate discussion of "success" further on in this chapter; for now, the historical record generally suggests that media activism has a much better chance of achieving at least some of its goals if

- it is combined with offline, physical, and proximate engagement, such as street work, social organizing, and community and grassroots activities;
- the informal and flexible networks typical of mediated connections are supported by some kind of formal organization; and
- the strategies and tactics of the people and social movement(s) involved are politically aligned.

Such a realization of the necessary inseparability of online and offline actions fits the overall thesis in this book of a crucial coupling of life and media as the baseline for understanding media (and for contemporary media studies). The conjunction of events, networking, and community organizing happening on the streets and in neighborhoods with people connecting, sharing, and participating in media is more than just a recognition of mutual reinforcement—it necessitates an appreciation of interdependence. The identity of a movement and its technological infrastructure mutually constitute each other. What happens in media definitely does not stay in media—it shapes and influences circumstances, conventions, and conversations outside of media. The various ways in which people engage with civic issues and social transformations in turn direct and give form to particular expressions, actions, and networks in media. Although this insight governs all the arguments developed in this book, it is striking to see it at work so specifically in both the history and the contemporary practice of revolution, rebellion, and activism.

## On the Properties of Media Activism Today

Beyond their long histories, all the different kinds of currently proliferating forms of media activism have properties that give rise to unique types of civic engagement. As noted earlier, simply the fact that digital media are omnipresent—which makes them obvious gear in the activist toolkit—does not mean the groups, communities, or movements involved are necessarily egalitarian or otherwise horizontally organized; nor are the ways in which people use (digital) media for social change intrinsically empowering and spontaneous. That said, it is clear that popular, everyday digital media devices, platforms, and services enable many more people to engage, contribute, and establish new tactics for mobilization and resistance, generally lowering the (physical, financial, and to some extent even emotional) cost of political participation. Being part of protest or struggle becomes much more common and seems to be part of the *media repertoires* we all have when going online—even if it is just liking a post, using a hashtag for an update, and changing one's profile picture to show solidarity with a particular event or cause.

In all this, digital activism is not that different compared with offline social movements, although it looks and feels quite different from traditional political participation. Online activism can develop and blossom in an instant, dissipating almost as quickly as it emerged. The transient nature of online participation sometimes is mirrored in intense yet often brief offline engagement, such as in the case of a strike, protest march, or mass event. A distinct example of new opportunities for innovative action are the rise of what have been called *microrebellions* all over the world: acts of protest and resistance by one person (or only a few individuals), documented and shared on popular social media, spreading a message far and wide. Much of the activism of a network such as the protest group Femen (founded in 2008 by a group of Ukrainian feminists) gets organized as a series of microrebellions, with women around the world using their naked bodies to raise awareness and seek publicity about a wide variety of issues—including a lack of public toilets in Ukraine, the oppression of women in Morocco, the imprisonment of influencers in Iran, and the lack of prison time for sex offenders in South Korea. From its modest beginnings, Femen now has branches and affiliates around the world. This connectedness can have an energy effect on a local action or microrebellion, potentially turning it into an international event, as in the online context individual voices get amplified and can make a difference—sometimes even on a global scale. The same process can happen the other way around, as local groups or individuals co-opt a struggle from far away. Examples include the Black Lives Matter (BLM) movement, which found fertile ground in long-standing movements for racial justice, such as the Coalizão Negra Por Direitos (Black Coalition for Rights) in Brazil, and inspired marches in Japan and across Europe, while in Australia BLM got linked to the local issue of Aboriginal deaths in custody. At the same time, the networked inflammation of people's passions can result in rapid energy depletion, as people's passionate engagement is hard to sustain, especially if it has no corresponding local structure to back all the media activism up.

In the digital context, there occurs an interplay of individual, local, and otherwise distinct stories with much larger cross-regional or even global narratives and collectivities. This dynamic can be quite powerful for the chances of activists involved with underresourced and unpopular social movements, such as LGBTQ+ movements and support networks—especially in regions where questioning people and their supporters are few and far between and where it is risky to come out at such. For example, in 2011 the open discussion platform Ahwaa was launched for Arab LGBT individuals, using anonymity (through the use of avatars and pseudonyms) and game mechanics (helpful contributions to the site unlock special features) to protect and engage its community. Ahwaa is an initiative of Majal, a regional not-for-profit organization (founded in 2006) focused on "amplifying voices of dissent" throughout the Middle East and North Africa (MENA) via digital media. The founder of Majal, Bahraini civil rights activist Esra'a Al Shafei, also started Mideast Tunes (in 2010), a platform for underground musicians throughout the MENA region. Ahwaa, Majal, and Mideast Tunes are supported by a variety of awards, grants, and funding agencies, including the Omidyar Network (a self-proclaimed philanthropic investment firm established by eBay founder Pierre Omidyar) and the Arab Fund for Arts and Culture (registered in Lebanon and Jordan). Through a combination of international sources of funding and support with distinctly local activities, mixing both anonymity and public exposure and thus being able to both mobilize constituents as well as evade unwanted public attention, Ahwaa embodies the many fascinating contradictions of contemporary media activism.

Through digital activism new issues are continuously put into public contention. As shown before, at the moment there does not seem to be a shortage of options to protest and take action. All this engagement is enormously varied and highly unpredictable. People participate in activities and activisms with varying levels of commitment and belief, flexible engagement that is greatly facilitated by our life in media. Some people may get radicalized because of this—immersing themselves in endless YouTube playlists, intense hashtag-driven debates, and echo-chamber-like forums and channels online. Connection is convenient, it is easy to join in, algorithms tend to provide more of the same all the time, and sometimes all you have to do is to watch and cheer on real-life activities unfolding online. However, the same dynamic can be reversed, given the abundance and variety of viewpoints and sources of information available online, while disconnection is uncomplicated and generally without consequence. This precarious aspect of digital activism does not deter the formation and proliferation of transnational advocacy networks (TANs) around ethical issues faced by people around the world. TANs are composed of a variety of actors, including (individuals within) nongovernmental organizations, civic groups, international organizations, academia, media companies, and governments that come together around such issues as the banning of cyanide-based mining in central and eastern Europe, the rights and protections of children in the context of armed conflict, and support for the movement against female genital mutilation in such African countries as Guinea, Kenya, and Somalia. Some of these TANs are completely virtual, such as hacker groups like Anonymous, emerging in 2003 on the image board 4chan, regularly enacting loosely coordinated series of protests, pranks, parodies, and outright hacks against organizations like the Church of Scientology, numerous corporations, and prominent individuals (including captains of industry, government officials, and political leaders). Anonymous even declared formal "cyberwar" against the Russian government after its invasion of Ukraine in February 2022. The endurance of TANs is but one example that belies the supposed fragility of connective activism.

Overall, digital activism is both an extension and amplification of different ways in which media activism played out before, as it is a new kind of connective action, using the infrastructures of information and communication technologies to change both the content, activities, and arrangement of traditional ways in which people have used media to change the world. As digital media have become a constituent element of any kind of contemporary activism, social and digital media themselves have become political agents—whether they like it or not. No matter how much platforms and online social networks would like to claim their role as neutral or simply facilitative, their cultural and political footprint all over the world is unmistakable. Platforms like Facebook, YouTube, and Twitter to some extent curate their networks, removing posts and suspending accounts for hate speech, spreading harmful conspiracy theories, or inciting political violence—such as during the coronavirus crisis of 2020 onward. To a large extent, platforms take action well after the harm was done, after facing accusations of facilitating the hate speech and propaganda of autocratic regimes in Myanmar, the Philippines, Brazil, Poland, and Russia—often with dire consequences for the people involved. Facebook for example admitted in 2018 that it had not done enough to prevent the incitement of violence and hate speech against the Rohingya, the Muslim minority in Myanmar. An independent report commissioned by the company found that "Facebook has become a means for those seeking to spread hate and cause harm, and posts have been linked to offline violence."[2] As it is facing legal repercussions for this (and other scandals), the company is among many in the tech sector now investing heavily in both automated and in-person curation, albeit still without any kind of transparency or formal oversight. Chinese-owned TikTok is another case in point, in March 2022 stopping its users in Russia from live-streaming and uploading new content while blocking access to most foreign accounts in response to increasingly restrictive media regulations. In doing so, it effectively created a separate online environment for the country. Such actions of platforms on the national level contribute to the growth of the so-called *splinternet*, where platforms increasingly regulate content based on individual national laws. In doing so, these corporations add another unpredictable element to the already complex mix of properties that make up the contemporary global media activism landscape.

## Problems of Changing Life in Media

The potency of digital activism comes with potential drawbacks. To wage a successful campaign online, activists often make use of popular corporate platforms, which in turn privilege and promote specific ways of organizing and expressing yourself. To get noticed and raise awareness, those involved in all kinds of often precarious and dangerous struggles have to promote themselves and their causes just like you or I may want to highlight going on a fun vacation or attending a cool concert. Difficult, complex, and sensitive issues generally do not sit well with the way the algorithms of social media work—that tend to prefer quick, direct, and visually appealing updates (that users can like, favorite, or ♡) over lengthy texts and nuanced explication. Furthermore, by using popular online networks, digital activism runs the risk of platform dependence, which is problematic

---

2. Alex Warofka, "An Independent Assessment of the Human Rights Impact of Facebook in Myanmar," Meta, November 5, 2018, https://about.fb.com/news/2018/11/myanmar-hria/.

given that such companies have the power to evict, censor, or otherwise obscure users and groups on a whim (or in response to suddenly changing national regulations).

As mentioned earlier, the use of digital tools and platforms privileges those who have the necessary access, skills, and know-how—and are furthermore motivated to use these media accordingly. These different levels of digital exclusion have a compounded effect, reproducing and magnifying already existing offline inequalities. Beyond celebrating the participatory potential of new media for the purpose of changing the world, a critical awareness of various types of digital exclusion is necessary. A crucial question is how social media and digital platforms can be appropriated in alternative ways by individuals, groups, and movements beyond or outside the mainstream of society, ideally including the widest variety of voices and truly delivering on the empowering promise of participation for everyone. At the same time, such a participatory ideal can also be quite problematic when nonregulated platforms and services become the staging ground for communities deeply invested in fear-based conspiracy theories, for a variety of (now increasingly interconnected) hate groups, and for others championing less-than-peaceful rhetoric and ideals.

Beyond looking at how to tactically use or manipulate commercial platforms, it is vital to explore the possibilities offered by a wide range of nonprofit, cooperative, and commons-based initiatives online. A promising field of study, for example, focuses on the cooperative digital economy, looking at initiatives that provide communal and sustainable alternatives to commercial services like Facebook, Uber, Spotify, and AirBnB. Some examples of such *platform cooperatives* are

- the German eBay alternative global marketplace Fairmondo (launched in 2013);
- the collectively owned ride-sharing app Eva (started in 2017 in Canada);
- the UK-based music streaming service Resonate (operating since 2015), which is jointly owned by artists, listeners, and volunteers;
- several initiatives to challenge or provide an alternative to online social networks like Facebook and Twitter, including MeWe (since 2015), Mastodon (since 2016), Counter-Social (since 2017), and BeReal (since 2020), generally offering more control over privacy; and
- the Dutch coalition of organizations PublicSpaces, "committed to providing an alternative software ecosystem that serves the common interest and does not seek profit" (quoting from its manifesto, published in 2018).

---

*A crucial question is how social media and digital platforms can be appropriated in alternative ways by individuals, groups, and movements beyond or outside the mainstream of society, ideally including the widest variety of voices and truly delivering on the empowering promise of participation for everyone.*

---

The central idea behind such initiatives is the notion that our digital environment should not necessarily be more or less exclusively based on either economic profit or political control. The notion of platform cooperatives furthermore pushes people to imagine a different future rather than one dependent on corporate platforms and state actors. In the field of governance and political organization in the context of media, a similar set of initiatives under the header of *digital commons* can be found around the world. The impetus here is to come up with new, more inclusive, and democratic ways to envision public debates and the sharing of resources and decision-making processes less driven and determined by either market-based approaches (primarily oriented toward prices and profit) or states (that generally produce difficult-to-navigate bureaucracies and do not lean toward easy collaboration with other nations). Driven by a global ecological crisis and widespread concerns about unbridled commodification (and corresponding growing social inequalities) and enabled by the networked structure of the internet, digital commons projects pop up all over the world—with the online encyclopedia Wikipedia and internet browser Firefox being among the most well-known and successful examples. Free (and open-source) software, open standards, and open access policies are similar instances of digital-commons-inspired developments pursued by companies like IBM (through its open-source subsidiary Red Hat) and Google, such governmental organizations as the European Union, and universities all over the world in an effort to make science available for everyone.

A third problem facing digital activism around the world is based on recognizing that media are complex and difficult—not just how to operate them but also (and very much so) how to understand mediated messages. Mediated communication is never unambiguous and always involves *polysemy* (as words, signs, and symbols have multiple meanings), and messages come across differently across different media. Providing a platform for protests does not always mean everyone understands the content of a message or knows what action to take upon seeing or hearing it. Activists working in African and Latin American countries in particular sometimes note with frustration how their messages and campaigns are picked up by Western media and audiences—admittedly with great enthusiasm but often without much appreciation for (or understanding of) local contexts, which can hinder or even silence work on the grassroots level. Furthermore, the spread of inaccurate information (which is an inevitable byproduct of mass participation) about events, demonstrations, issues, and movements can certainly hurt the cause, adding another level of complexity to the work of social and political transformation.

A fourth potential problem that occurs as media activism goes online is that of a possible reversal of democratic gains, as state actors, security forces, and corporate entities are in the process of heightening their efforts to surveil, police, and sanction activist activity on publicly available platforms. Despotic regimes are notorious for spying and censoring, clamping down on the supposedly liberating space of social media, and stifling digital dissent with impunity. Yet (as seen in chapter 4), surveillance is a global phenomenon,

affecting anyone going online, and also works the other way around—with the actions of oppressors coming into full view online. Related to this is the additional risk women, minorities, and members from otherwise marginalized groups take when they speak out in media, often resulting in personal attacks. Women suffer from censorship and state surveillance throughout the MENA region and in the Global North are the primary victims of cyberbullying and online harassment by men.

A fifth and final problematic issue regarding digital activism is the question of whether all this engagement online and in media generally really challenges the powers that be in meaningful ways, or does it do little more than reinforcing the status quo, not necessarily drawing new or more people into participating in conventional techniques of protesting? Accounts of success and failure differ in their assessment of what constitutes social transformation, and the needs of different movements (operating in different parts of the world) vary wildly. Many researchers who study media activism and social movements participate in such actions, strongly believing in their purpose and value and taking responsibility for the inevitably hopeful nature of doing scholarly work. There are certainly ample opportunities for resistance and counterhegemonic struggle everywhere.

## On the Successes and Failures of Media Activism

A fundamental issue about any and all kinds of media activism is whether or not it is successful: Does the use of media for the purpose of changing the world—however modest or ambitious one's goals—have any real effect? Historically, it is clear that the use of media has contributed to processes of profound social and political transformation—as illustrated, for example, regarding the struggle of women's rights and indeed human rights generally. Similarly, continuing environmental activist campaigns have consistently built support, mobilized people, and spurred policy makers into action (albeit slowly) throughout history and all over the world. Environmental media campaigns stem from early publications about the need to protect nature in the eighteenth century, including an eloquent argument for animal rights from Jeremy Bentham, who also proposed the design of the Panopticon prison (discussed in chapter 3) which in turn inspired contemporary concerns about all-encompassing surveillance. More recently there have been prominent actions from such organizations as Greenpeace, Ecology Action, the Environmental Foundation for Africa, and the Delegación Joven (formed in 2021, representing Latin America youth at UNICEF to participate in debates about how to tackle the climate crisis) to raise awareness about the consequences of climate change. All this got supercharged by the lone voice and actions of then sixteen-year-old Swedish activist Greta Thunberg in September 2018 when she started her strike for climate action, inspiring the #FridaysForFuture hashtag and activist movements involving untold numbers of youths worldwide.

When it comes to digital activism, there is much debate about its effect and effectiveness. The use of (and possibly reliance on) popular social media, the brittleness of people's participation online, the relative quality of clicktivism, the obstinacy of offline power structures versus the affective outrage on social media—all these can be put forward as a forceful argument against digital activists' conceivable success. On the other hand, one could point at the role of Los Indignados in Spain, paving the way for the foundation of Podemos, a left-wing populist party that later became part of the country's first coalition government. Similarly, the so-called Tea Party, a fiscally conservative political movement in the United States (operating from 2009 onward, later becoming part of the so-called alt-right movement in American society and politics), contributed to the platform for Donald Trump's 2016 election as president of the United States. The #MeToo movement keeps leading to high-profile exposures (as well as cancelations and court cases) of sexual abusers and predators across the various media, cultural, and creative industries.

Beyond specific examples of media activism that quite clearly makes a difference, sustained social and political change seems to be somewhat lacking for many, if not most, digital campaigns. Much depends on how we define and attribute success in this context. Is getting people engaged, involved, and participating in itself a measure of success? Is structural change in a traditional sense—for instance, referring to political transformation, the introduction of new laws, and the better enforcement thereof—the best way to indicate success or failure? Several frameworks have been developed as guidelines for social movement success, mostly looking beyond the direct role of media. Indicators in such models include the clarity of the cause, the professionalization of methods and tools used, the lasting image of the activists in the public's eye, the extent to which any movement or campaign succeeds in articulating a clear and unified message, how many people are involved, and how much effort everyone puts into it. Other elements of consideration are the ability of activists to mobilize people and to keep people involved and the extent to which a movement can build coalitions that would also help sustain itself over time. It can be argued that, for all these aspects, media are indispensable, especially in the context of our lives as lived in media.

## Change Life

Considering media activism in terms of the kinds of media used for activist purposes should also include using the kinds of stories produced for and by such campaigns. This is what happens when activists adopt, adapt, remix, or parody elements of news and entertainment to further their cause and to enhance their narrative. It is here we find all kinds of playful, ironic, imaginative, and performative as well as serious and engaged ways of using media to change life.

Consider for example the inhabitants of Bil'in, a Palestinian village located west of the city of Ramallah in the central West Bank. Over the years, Bil'in has become the center of resistance and protest for the Palestinian community against the building of a wall by the Israeli Defense Force (since 2002) to separate Palestinians and their neighboring Israeli settlements. Israel calls this a "security barrier," while Palestinians refer to it as the "apartheid wall" or "segregation wall"—not in the least because it at times cuts off communities from family members and farmland. Since January 2005, the village community has organized weekly protests against the barrier. These protests take the form of marches from the village to the site of the barrier with the aim of halting construction and climbing sections already constructed. From a life in media point of view, the otherwise tragic plight of the Palestinian people is fascinating because of the way the community orients itself toward media to get support for its cause. Villagers used to operate a professional multilingual and multimedia website, and the protests themselves are often staged in mediagenic ways—dressing in costumes of national soccer teams around the time of a World Cup or as characters of famous movies. In 2010, for example, protesters painted themselves blue and wore bright blue clothes inspired by the Na'vi people in James Cameron's highly successful film *Avatar* (2009). The villagers' choice to dress like the Na'vi people in the film was explained by pointing out a parallelism in the narrative, suggesting that Palestinians also fight imperialism. This demonstration in particular garnered worldwide attention and outrage, as pictures and video of blue protestors sprayed with tear gas canisters spread across the globe—both by the participants' efforts to upload material to YouTube and through international news coverage. Cameron's films have been co-opted by numerous activist and community initiatives around the world, such as by the Dongria Kondh tribe in the eastern state of Orissa in India to fight against plans of the British mining company Vedanta Resources to dig up mineral resources around a mountain the tribe considers its homeland. The Kayapo Indians, living along the Xingu River in the Amazon, also called on Cameron's support to oppose the construction of a dam sponsored by the Brazilian government that threatened the livelihoods of the Indigenous peoples along the Xingu.

The *avatar activism* of Indigenous communities in Palestine, India, and Brazil is just one of the more high-profile cases of people appropriating media in terms of their *texts* as well as their *technologies*. Especially among youths around the world, a combination of using digital media with fan activism proves to be a popular way of engaging with political projects. The international nonprofit organization Fan Forward—initially founded in 2005 as the Harry Potter Alliance, renaming itself in 2021—has become a significant force, with chapters in thirty-five countries across six continents. The organization uses characters, elements, and story lines from popular media franchises, such as the *Harry Potter* books, films, and games to raise awareness and gather support for campaigns on such issues as literacy, gay rights, sexism, mental health, and climate change. The movement, initially started by the American comedian Andrew Slack and the wizard rock band Harry and the Potters, rallies around the idea that fans can use their passion and creativity to make the world a more loving, equitable place—an appealing notion embraced by hundreds of thousands of fan activists. In the early years, Warner Brothers sometimes would try to shut down the group regarding copyright issues, which ended up getting more fans involved and raising people's awareness about intellectual property and free speech issues.

Since 2005 the organization sponsors, collaborates with, and initiates campaigns around the world, while also providing coaching and training services for activists. The organization links people's stories and personal pleasures related to media with social justice issues, which turns out to be a powerful motivator to get like-minded people involved.

Next to activism related to popular media franchises, all kinds of products and brands today use media activism as a way to both differentiate themselves and to support a range of causes. Of particular relevance to the discussion in this chapter is the role of influencers on social media, such as Instagram, TikTok, and YouTube. On the one hand, part of their popularity (and often a significant source of their income) is the flaunting of a luxury lifestyle, promoting products, and showcasing their commercial enterprise. On the other hand, calling on people to get vaccinated against COVID-19 and pontificating about viral social justice issues, such as #BlackLivesMatter and #MeToo, have become very much part and parcel of gaining attention and being considered authentic as a public figure online.

Activism has gone mainstream, in the process getting co-opted by commercial forces as much as it has become a way for people from all walks of life and backgrounds to engage and express themselves about issues affecting their lives. It is not an inherently democratizing movement; nor is it necessarily constrained and exclusive given its reliance on digital media and use of corporate platforms. It can be profoundly pleasurable to participate as much as it is precarious and dangerous for many, regardless of whether they become active with their avatar or their body. All in all, media activism is a significant phenomenon particular to life in media, as much as it is a way of living life infused with social or even radical hope: the promise of collectively (and collaboratively) making a better world by taking action in it—despite misgivings as to whether any of this really can make a difference.

# 7

## Make Life

> To explore the inner workings and structures of the media as a global industry, we can look at film, television, games, journalism, music, and advertising from the perspective of the people at work in these worlds. Media professionals create stories that inform and entertain us, while media users increasingly engage in productive activities too. The result is an industry dynamic, full of contradictions, complex to navigate even by the most skillful of practitioners—and changing rapidly to accommodate our lives in media.

It is one thing to recognize the various ways in which media entangle with all elements of everyday life and how media shape (and are shaped by) the way society works. It is another to appreciate the historical nature of such mutual shaping, considering how who we are as people—as humanity—cannot be extricated from media, technology, and machines, as much as we are linked to nature. Throughout it all, there is a third factor to ponder, one that gives meaning and form to a life in media: the growing significance—in terms of both economic impact and cultural influence—of media as industries. From journalism to advertising, from the cinema to television, from book publishing, adult and social media entertainment, and digital games to music and recording, all uniquely benchmarked by today's digital environment dominated by platforms, media industries play a profound role in the kind of stories we tell, what our shared narratives are, how we connect and feel connected, and how we experience and cocreate our life in media.

Beyond their significance as profitable industries and producers of culture, the media (and the professionals at work in them) are crucial to consider because of the ways in which the words, sounds, and images of media not just reflect but also direct the world in which we live. Given the ubiquity of (mass) media in society, the history of making media is also a history of media practitioners producing a certain way of looking at the world. This does not mean we all see the world in the same way because we have all watched the same movies, binged the same (or similar) television series, or rocked out on the dance floor to the same tunes. It does suggest, however, that we have all grown accustomed to a stunningly similar way of experiencing the world and expressing ourselves in media. The particular

genre conventions, production techniques, storytelling formats, formulas, and scripts of media industries such as journalism, film, and advertising have become mundane to us. Tactically framing a photograph, angling the camera to get the best possible light, composing a message in a compelling way, using various songs to collate a soundtrack for our life, adding a filter or digitally touching up our appearance before posting something online—none of this seems strange or unusual. Even more so, so many of us seem to just know what to do in such circumstances—often without any formal training. We seem to sense what to do with media because we have been living with (and in) media for all our lives.

How media are made is important as the media produce, frame, legitimize, and circulate ideas, attitudes, and information in and about society. As we all participate in this process through our mass self-communication, we are inevitably caught up in this process of "making life" in media. It therefore makes sense to look more closely at those who make media professionally—such as the reporters and editors in journalism, the creatives and planners in advertising, the production crews and cast members in film and television, the developers and designers in digital games, vloggers, influencers, and so on—to see how they make it work, how they navigate the digital environment and develop new ways of telling stories, imagining new approaches to engage us.

The chapter sets out with a reflection on why studying the media as an industry is important for understanding life in media. This is followed by a definition of media industries, including an all-important appreciation that, in the context of the contemporary digital environment, a clear boundary between what a media industry is and what it is not is almost impossible to draw. This opens up tremendous opportunities for new ways to tell stories and experience the world in and through media for media professionals, amateurs, users, and consumers alike, while simultaneously introducing new uncertainties into an already precarious field of business and work. Third, we do a deep dive into the various paradoxes of making media from the perspective of those who work in and for the various media industries—looking at how the industry is organized and managed; how companies, firms, and (groups of) individual media professionals make it work; and finally what working in the various media industries looks and feels like for the people involved. In conclusion, we explore how media industries both structure and respond to our life in media, as new storytelling formats and user experiences are developed and introduced at a sometimes breathtaking pace, affecting the ways in which we can study and understand what making media means.

A crucial consideration of this focus on professionals in the media is the fact that so many who study the media consider pursuing a career in the media. For sure, media firms seem like a wonderful place to work: collaborating with like-minded people on a film or game, chasing down leads and interviewing people to uncover a news story, coming up with a slogan or motto that defines a product that will be recognized by people all over the world—what is not to love? In the decades I have researched media professionals all over the world, the sentiment most often expressed in countless conversations was "I cannot believe I am getting paid to do this!" When we see interviews with media workers or watch "making of" featurettes, we see lots of happy people. When stars receive their industry awards—whether a Shorty (for real-time short-form content on social media), Hugo (for best science fiction magazine), the Game Developers Choice award, a Media Excellence Award (for international broadcasting), a Lion (at the Cannes Festival of Creativity), the Oscar (for the American and international film industry), or any of the other countless national and international prizes for professional media work—recipients inevitably exclaim how wonderful the work is, how lovely their peers and colleagues are, how much they have been able to achieve. It certainly looks like the media industry is a fun, exciting, and overall playful place to be. At the same time, the media as a global industry operates pretty much exclusively on the basis of a so-called nobody knows principle: a deep-seated uncertainty because it is extremely difficult to predict the audience response to a product beforehand, and market success or failure is not easily understood afterward. Despite centuries of management and work, of ongoing growth, of collecting endless data on consumer tastes and market development, and of becoming a truly global industry, the media today are a business built on profound risk, uncertainty, and precarity, much of which gets offloaded onto its practitioners, who often are not paid much (or not at all) for the work that they do. The burden of risk, in other words, is shouldered by everyone throughout the media industries—just not equally.

It is fascinating how an industry that provides a product we do not formally need—unlike food, water, medicine, transport, and electricity, for example—has become one of the most influential and powerful in the world, while also remaining one of the most precarious places to work. This is one of the key tensions we explore in this chapter. Exploring these strains and conflicts is a critical exercise in the study of media industries and production (and a vital part of media studies as a discipline), as it makes us aware of profound sources of inequality and exploitation, as much as it highlights different (and other) ways of doing things. Consider for example the industry-wide paradox between sustaining revenue growth and earning little or no profit. The value of the global media market is well over two trillion US dollars—more than double that of the consumer electronics market (which includes smartphones, personal computers, and television sets) and coming close to the market size of the worldwide car and automobile manufacturing industry. However, beyond such seemingly clear-cut statistics lies a fundamental aspect of the media as an industry: even if it does generate revenue, profitability tends to be elusive as the operating costs of media production often far exceed any kind of income. Typical costs of running a media business include

- publicizing and marketing media products and services to get people's attention;

- incessantly investing in new technologies, talent, and production techniques in the hope of securing future markets;

- acquiring, maintaining, and upgrading expensive advanced equipment, studios, technical know-how and expertise, and all other production facilities and resources; and

- paying off debts to third parties, such as investment firms, technology platforms, and corporate owners.

---

*Like most companies and firms in the industry, practitioners take on substantial risk for the reward of making media professionally.*

---

All these risks carry over to the individual media professional, who faces similar pressures to aggressively self-promote (via professional associations, networking, and social media), to constantly focus on reskilling and upskilling to keep pace with changes in industry practices and new technologies, generally having to cover the cost of necessary equipment and training in order to maintain their career in the media. This is especially the case for freelancers, who tend to form the majority of those working in various media professions. Although extremely few professionals strike it rich in the media—and for many such success tends to be short-lived—most media professionals experience a distinct imbalance between the effort they put into the work and the rewards they reap from it (e.g., regarding earning a decent salary, getting an annual pay raise, enjoying a healthy work-life balance, and predictable career progression). Like most companies and firms in the industry, practitioners take on substantial risk for the reward of making media professionally.

An excellent example of the risky paradox between high revenue and low profit is the American streaming service Netflix, starting out in 1997 as a mail-based DVD rental service, piggybacking on the introduction that year of the first DVD players and discs (pioneered in Japan) in the United States. Netflix's subscriber model was unique as users could keep the DVDs for as long as they liked but could only rent a new movie after returning their existing one. As the company's subscriber base kept growing rapidly, it started development of its own hardware—a Netflix box that would enable people to download and play movies directly from the company's servers. After the introduction and enormous popularity of YouTube in 2006, Netflix abandoned the box and switched completely to video on demand, launching its streaming service in 2007 (expanding to Latin America in 2011, across Europe from 2012 onward, entering Australia in 2015, Russia and Africa in 2016; it is not available in China, and suspended its service in Russia to protest the country's invasion of Ukraine in 2022). During peak times, "the service" (as Netflix is called in the film industry) consumes more online bandwidth than YouTube and Amazon combined. As of 2013, it started coproducing its own original content, next to acquiring licenses to stream existing films, shows, and series. Over the years, the company signed more and more deals with creators around the world to develop original content, subsequently launching its own production hubs in the United States, England, Spain, Sweden, and Canada. With each step, Netflix sought to control costs and solidify its subscriber base—hoping to bring more predictability to the company—while at the same spending extraordinary amounts of money to develop and secure new products, services, and facilities, as well as ramping up massive marketing budgets as it faces increased competition. Competition mainly comes from such streaming services as Amazon Prime, Disney+, and HBO Max, as well as many smaller or regional local video on-demand businesses, including iQiyi in China, Vix for the Latin American market, Britbox in the United Kingdom, Showmax in South Africa, and Takflix in Ukraine. Companies like Netflix and Amazon Prime significantly invest in local content, signing deals and acquiring talent throughout the Middle East, Europe, Latin America, and elsewhere. The bottom line: while Netflix may generate huge revenue from its hundreds of millions of subscribers, it also has to burn more cash on maintaining, marketing, and expanding its products and services. Furthermore, a sudden dip in subscriber numbers can make the service lose significant share value, indicating the fundamental uncertainty governing Netflix's business model.

All this suggests a profound paradox, as the global media industries clearly bring in enormous revenue, producing publicly visible engaging and enjoyable news and entertainment experiences, while at the same time it is an industry riddled with risk, uncertainty, and (corresponding) tension. Understanding the media as an industry requires an appreciation of this quandary. In this chapter, various media industries—journalism, advertising (including marketing and public relations), film and television, digital games, music and recording, and social media entertainment—are considered in terms of a range of these structural dilemmas, contradictions, and conflicts that together color and shape the way professionals and firms operate and "make it work" on a day-to-day basis— whether it is a corporation like Netflix extending its operations all over the world, a local advertising agency working for a client just down the road, or a freelance reporter doing the kind of investigative journalism that she really cares about.

**On Defining Media as Industries**

What makes media an industry is complex and is always the result of some kind of artificial boundary drawing. Examples thereof are generally quantified—such as distribution, profits, number of productions, personnel employed, and so on. The problem with such numbers is that they also serve political goals, intended to make the industry seem productive, successful, innovative, or otherwise constantly busy with exciting new things. As media industries for their financial survival tend to rely on shareholders, investors, commercial owners, partners, or—especially in the case of public service media—government officials, much of the way we see them as industries has a certain image-boosting quality. The numbers produced by the various industries are certainly impressive, documenting global reach, mass market penetration, widespread consumer uptake, and fabulous sales figures. At the same time, we should keep in mind that these statistics are performative, portraying the industry as serious: if companies are able to measure themselves and produce numbers consistently, they can be taken to be well organized and worthy of trust and investment. Generally, such investments do not take note of the oftentimes problematic story those numbers tell—for example, regarding the discrepancies between revenue and profit and about the precarious nature of what it is like to work in the media.

Beyond the performative and somewhat mystifying nature of media industry statistics, a third problem with defining media as an industry in terms of the way it reports on itself is the illusion of coherent and singular organization of activities such numbers suggest, as if all the different aspects of making media—from ideation to acquisition and preproduction, from financing to production, via packaging, design, and promotion to marketing and distribution, and ultimately to distribution and consumption—can be neatly captured in a single narrative or number. For example, to talk about "the film industry" runs the risk of overgeneralization as there are so many different ways of making and distributing film around the world, and local film industries—Hollywood in the United States, Bollywood in India, and Nollywood in Nigeria to name a few high-profile sectors—are both internationally networked (through financing, coproduction, and distribution arrangements) and locally specific (in terms of production cultures and talent pools). Given the overall trend of industry convergence—where various sectors (such as media, telecommunications, and technology companies), channels (such as television, print, and online), and companies collaborate or even become part of the same corporation or network—defining the industry also essentializes it, reducing it to a particular feature without recognizing the many linkages, crossovers, and interdependencies between different media professionals, departments, and firms.

A fourth issue with traditional definitions of the media is the relatively recent entry of social media entertainment as a distinct sector of productivity in the industry. When YouTube cofounder Jawed Karim uploaded an eighteen-second video titled *Me at the Zoo* to his newly formed website (on April 24, 2005) this kickstarted a movement, culminating in a truly global industry of vloggers, influencers, social media entertainers and entrepreneurs, and microcelebrities. Karim's site and upload fit into a much longer history of people sharing stories and scenes from their everyday lives online in more or less spontaneous fashion. In the late 1990s, weblogs emerged out of people's online diaries, which in turn were inspired by newsgroups in the 1980s. An early example of serially sharing slices of life publicly can be found in the late nineteenth-century investigative journalism of the American writer and inventor Nellie Bly. She documented her trip around the world in eighty days between November 1889 and January 1890 for the *New York World* newspaper. During her journey, she sent short progress reports via electric telegraph, filing longer stories by regular mail service. In doing so, she pioneered a genre (and a way of being professional about it) that, more than a century later, has come to define a new type of global industry online.

What sets social media creators apart from traditional notions of media producers and industries is their lack of ownership and control of both the channels they use (such as LiveJournal, YouTube, Instagram, or TikTok) and the intellectual property of their work. Formats, genres, and formulas that determine much of professional media production tend to be largely absent in social media entertainment as well, as almost everything these creators (or *wanghong* as they are known in China) make is determined by the extent to which it seems authentic—not to be confused with real. Although in the context of *mass self-communication* most of us can be considered to be creators of content and experiences online, only very few people manage to turn this into a sustainable business. Successful creators manage to aggregate and engage online communities, which they convert into varying forms of value—for example, through sponsored content and other kinds of commercial partnerships, viewer donations, and receiving revenue from the advertisements displayed around (or within) their published content.

Although genre conventions certainly exist on various online social networks, the rules of production change faster for any routines or formats to sink in. New technological developments, a constant influx of new voices and characters from around the world, and changes in platform policies contribute to this dynamic—which is both exciting and exhausting. Creators have to maintain a permanent stream of updated material, engage in self-branding, and do the relational work of engaging with followers and fans, as well as incessantly finding and securing sources of revenue—all in the context of the algorithmic volatility of media platforms over which they have no control whatsoever.

Some truly fascinating media forms emerge from this creator culture, especially where instances from the mundane everyday interact and mix with stories and genres that have widespread (and possibly commercial) appeal. This is, for example, quite pertinent in the fast-growing popularity of such music genres as *tecnobrega* (coming out of north Brazil since the 2000s) and *amapiano* (emerging out of South African townships around Johannesburg and Pretoria from the 2010s), both in local clubs and around the continent. In both instances, traditional rhythms and melodies are combined with basic beats powered by computers—often acquired through the recycling of electronic waste—by musicians, DJs, and fans alike, gaining massive popularity through social media without the support of the traditional music industry.

Another genre particular to technology-driven, self-sharing, and highly commodified creator culture is the *mukbang*, or eating broadcast, becoming massively popular in the 2010s in South Korea, later on spreading across Asia and North America via YouTube and Twitch. In these shows—at times watched live by millions—someone eats copious amounts of food, earning money through ads on their channels. The success of mukbang tends to be explained by social factors, such as loneliness and people's desire to feel connected, and technological developments, especially the rise of social media.

A third creator trend on social media I like to highlight is the emergence (and popularity) of virtual influencers. In 2018, *Time* magazine named Miquela Sousa one of the twenty-five most influential people on the internet, despite the fact she is not a person at all. Sousa, or Lil Miquela, as she is known on Instagram (where she has millions of followers, known as "Miquelites"), is a computer-generated person, representing a fictional Brazilian American teenager, created in 2016 by the Los Angeles–based content production company Brud. She works as a model and spokesperson for brands like Prada and Samsung and releases her own music videos. Her story, as a young woman establishing herself online with a huge following all the while not being entirely real—which is something none of her fans seems too bothered by—reminds me of the early YouTube phenomenon lonelygirl15. Between 2006 and 2008, a young woman called Bree Avery posted regular slice-of-life videos to the platform, created an account on MySpace as well as a personal website, and rapidly built up a massive fan base. Some of her videos included product placement, and she starred in an antipoverty campaign for the United Nations. At some point it was revealed that she was a fictional character, played by the New Zealand actress Jessica Rose and created by American filmmakers. Although her posts stayed popular, the actress chose not to renew her contract for a new season—something a virtual star would of course never do. The examples of lonelygirl15 and Lil Miquela additionally suggest that many people—followers and fans—do not seem to be particularly bothered by the fact that these characters are not real (see chapter 4 on real and virtual life).

All things considered, it is important to keep an open mind when thinking about the media, as any definition of media as an industry is ultimately a theoretical construct, both useful and limited. Media represent meeting points for multiple industries, which today most notably involve three distinct areas of economic activity:

1. The businesses of news, film, television, games, music, advertising, and social media

2. Telecommunications, including the development and provision of telecom equipment, services, and wireless communication

3. The technology industry, specifically the manufacturing of consumer electronics, and any information technology-related products and services (including software development, web design, and platforms)

The contemporary media industry can be said to compose a part of a larger telecommunications-media-technology (TMT) sector, while the TMT sector itself has become a global media industry. Firms within each of these areas are currently in the business of one or more of the key things that a media company tends to do:

• financing content, generally through private investors and investment firms, grants and subsidies (including tax incentives), sponsorship (including advertising, branding, and product placement), and presales;

- producing (original) content, which can involve employing professionals or inviting anyone to contribute (generally for free);

- distributing content so that it may appear on your screen, speakers, or doorstep in one way or another; or

- exhibiting content directly to people by offering a unique gateway (e.g., a cinema or a dedicated application) to access it.

It is fascinating in this context that companies in the information technology domain tend to insist they are not in the media business. As Mark Zuckerberg of Meta (including Facebook, WhatsApp, and Instagram) has repeatedly stated: "We are a tech company, not a media company. When you think about a media company, you know, people are producing content, people are editing content, and that's not us. We're a technology company. We build tools."[1]

Zuckerberg's counterpart at Alphabet (the parent company of Google, Fitbit, Waymo, and other subsidiaries) Sundar Pichai has made similar statements, despite the fact that both companies do not just curate, distribute, and exhibit content but also invest directly into content production. Platforms shift toward media content acquisition and production to have exclusive material to offer users. Traditional or legacy media firms—such as news publishers, movie studios, and digital game developers—also are likely to strike deals and work together with technology and telecommunications companies to produce, promote, and disseminate their work. The tech sector's reluctance to embrace an identity as media industries stems from its fear of being held legally responsible for the content uploaded and shared on their platforms (as media companies are). Another consideration is their tax rate, which tends to be much lower than that of media companies.

All this collusion and convergence in the TMT sector is in part driven by technological developments. Our lives in media also have something to do with it as we become less loyal to particular channels, media, and franchises, instead each of us assembling our own personalized information environment. The challenge for media makers is not just to come up with compelling stories or to get us to pay attention to their work but to meet us where we are in our digital environment, surrounded by a plethora of options. The industry tends to respond to this challenge by collaborating across the various devices, channels, and platforms that make up the whole of our media experience. On the other hand, mergers and acquisitions have always been the bread and butter of the global media economy. One of the few ways in which media corporations make a profit is by consolidating and bundling their offerings across a variety of media channels (a process called *horizontal media integration*) or by controlling all the elements of the production value chain (known as *vertical media integration*). While various kinds of media integration

---

1. Susmita Baral, "Is Facebook Is A Tech Or Media Company? CEO Mark Zuckerberg Weighs In," *International Business Times*, August 30, 2016, https://www.ibtimes.com/facebook-tech-or-media-company-ceo-mark-zuckerberg-weighs-2409428.

have been a historical phenomenon in the industry, the recent decades have seen an acceleration of this trend, coupled with an equally fast-paced process of media deconcentration as companies seek to divest themselves of unprofitable (or otherwise unwanted) ventures. This feverish dynamic contributes to an overall sense of restlessness as well as excitement, as the media industry constantly changes and the management of such integrated firms is anything but straightforward or uneventful.

Given these developments, it becomes quite difficult to pin media down. For sure, we can point to big media corporations, such as WarnerMedia, NewsCorp, and the Walt Disney Company, that own and partner with thousands of smaller companies and firms, but most of their operations do not involve the making of media anymore. These globally networked companies are copyrights industries, investing mainly in content libraries to be marketed and licensed around the world. Their own productions are, relatively speaking, a small part of their business portfolio. This pattern emerges time and again when we look at each main media industry separately. In advertising, a handful of big holding companies—Publicis Groupe, InterPublic, Omnicom, Vivendi's Havas, Dentsu Aegis Network, and WPP—own or control most of the global market. On the other hand, countless small advertising, marketing, and public relations agencies around the world work mainly as independent contractors to create campaigns. As in all other media industries, these holding firms are gradually losing market share to technology companies, and brands are increasingly producing their own advertising and marketing. The global game industry exhibits a similar pattern, where the largest publishers—Electronic Arts, Ubisoft, Activision Blizzard, Bandai Namco, Capcom, Tencent—tend to license, franchise, and market games produced by thousands of small studios all over the world. Some corporations—Nintendo, Microsoft, Sony, Apple, and Google—are also in the hardware and software business, offering unique opportunities to develop controlling strategies across games, consoles, and operating systems. As in advertising, film, and television, these corporations gain market share primarily by acquiring content-producing companies more so than developing and producing their own unique intellectual property. The music and recording industry tends to be equally dominated by less than a handful of megacorporations—Sony Music, Warner Music Group, and the Universal Music Group—versus countless independent (or indie) record labels. Each of these corporations owns or controls many sublabels that are nominally independent, most importantly sharing accounting and pricing systems.

In every media industry, professionals are most likely to work for a company, agency, or studio with only a handful of colleagues rather than at one of the global behemoths. It gets even more complicated when we consider how all these corporations, publishers, agencies, firms, and studios are connected beyond relatively straightforward structures, such as financing and ownership. It is quite common for films and television series to be distributed in different parts of the world by a competitor of the company that owns the copyrights. In the production phase, firms and professionals working for a variety of larger corporations or holding enterprises can form temporary alliances or *project ecologies* to work on a specific product or service (such as a campaign, film, or news investigation). In journalism, for example, it is slowly becoming commonplace for independent reporters, legacy news organizations, and professional associations to collaborate on covering major international news stories. Among the more high-profile cross-border collaborations are the Panama Papers (2016), Paradise Papers (2017) and Pandora Papers (2021) investigations, all based on leaks of millions of files from the databases of law firms and other companies involved in setting up tax havens for the rich and famous. Journalists—both independents and those working for legacy news organizations worldwide—worked together through the International Consortium of Investigative Journalists (ICIJ) to scrutinize the documents, turning them into an ongoing stream of stories about tax evasion and corruption on a global scale. The ICIJ is a nonprofit network of investigative reporters (from more than one hundred countries and territories) and partner media organizations from around the world, facilitating these kinds of complex collaborations. The implications of these networked investigations reverberate around the world, as details continue to emerge about athletes, celebrities, politicians, and billionaires setting up companies in low-rate (or no-rate) tax places like the British Virgin Islands, where their businesses and identities are hidden from the public.

Temporary alliances in media production are a standard way in which the work of making media gets done, across all industries—adding complexity to any kind of consistent definition of media as an industry. A big brand advertising campaign usually involves many professionals and firms with different contractual obligations, forming teams at various instances of the production process, in many circumstances never meeting in person yet still involved with making such a campaign work. A *tentpole* or *blockbuster* movie—an industry term for a production with a higher-than-average development and marketing budget (in the games industry, labeled as a *triple-A* game)—does not just employ hundreds of professionals; its entire value chain also includes a large number of partner firms. For instance, the 2021 James Bond film *No Time to Die* directly or indirectly involved 107 different production companies, from tiny firms such as the two-person animatics shop MonkeyShine in the United Kingdom and the two-person catering company Eirik & Pedro in Norway to global distributors like Universal Pictures and Metro-Goldwyn-Mayer (MGM).

The impermanent nature of all these collaborations and projects within and across the media, technology, and telecommunications sector raises the question of how we can best capture the media as an industry from the perspective of the people working in it: Is it a *network* of a diverse range of institutions and organizations? An *ecosystem* of interdependent entities? A *field* of people, practices, and values? Or perhaps it is best understood as a *world*, as the production of a media product or service is never exclusively the act of an individual practitioner but rather involves a collective endeavor of various people, organizations, and institutions (temporarily) working together based on mutually shared and understood ideas and understandings that facilitate their cooperative activity.

The media industry, when seen as a world, involves a rather heterogeneous and always changing set of participants, practices, and professional norms and values—all of which are necessary for the production of media content and services. Considering our approach in this book to media as activities, arrangements, and artifacts, the notion of a world of media as industries and work must also include the various technologies, appliances, and material contexts that are part of media production, as these give shape to the work:

- the various *working environments* of media professionals, ranging from newsrooms and other open office-type environments to temporary settings (such as filming locations outside of a studio complex or journalists reporting out in the field), individuals working in shared workspaces (including cafés, hotel lobbies, and libraries), as well as countless people working from the intimate place of their home;

- the full range of *hardware and software* that make specific media production possible, including dedicated computer programs for the organization of television production, content management systems for news publications, and the requisite equipment for designing and developing games, vlogs, and any other digital form; and

- the *material context* of production as determined by the complex contractual, financial, legal, and otherwise economic arrangements under which the work takes place.

All this comes together in a variety of *production cultures*, specific to a place and time, to a particular constellation of people and resources, and to a certain type of project. It is important to include the role of all of us as audiences in this approach as well.

In the context of a life in media, the media as an industry emerges as part of a new economic supersector, where telecommunications, technology companies, and media corporations compete and collaborate to capture the time and attention of media consumers—consumers who are increasingly also in the business of making media. We create media directly by populating and posting on social media, by uploading and disseminating our own content, and by liking, sharing, and forwarding (and thereby promoting) the content made by others. Indirectly, our lives in media produce endless data that fuel the global digital economy. Through our attention to commercial media, we also do the work for media as audiences that are sold to advertisers. Considering media as industries implicates all of us—anyone who uses any kind of media—in an interdependent economic relationship.

Taken together, these considerations for a definition of media as industries suggest an at once global sector dominated by multinational conglomerates and a distinctly local field of professionals, firms, and networks making it work, all of whom invest their time, talent, money, and resources into a risky business where nobody knows whether the project at hand will be successful—let alone commercially viable. Much of the management and organization of media production is therefore based on efforts to reduce (or even control) that risk, which partly contributes to a somewhat performative quality of the various ways the industry presents and talks about itself. The nature of media work is much more messy, temporary, and complex than any definition of the media suggests. Media firms and professionals are quite likely to be working together with people from different fields, even from completely different sectors, including all of us in the process of cocreating content and experiences.

## The Paradoxes of Professionally Producing Culture

At the heart of all this is an industry that produces culture—as the primary value of the products and services of the media have aesthetic, emotional, spiritual, and ethical qualities for us. Although media are inseparable from the way society and everyday life function nowadays, in the strictest sense media are not necessary for our survival as, for example, clean water, sustainable agriculture, and adequate shelter are. The apps, games, films, shows, music, and stories we love so much may not be directly practical, but the value(s) they represent are clearly essential to our sense of identity and belonging. Next to *producing culture*, media industries are a *production culture*—involving a range of practices, routines, norms, and beliefs that are often specific to a generally agreed upon way of doing things of a particular project, team, or organization.

In what follows, we unpack the production cultures of media industries on the basis of a series of paradoxes that codetermine what the experience of making media professionally is like. The choice for such tensions or ambivalences as the starting point for analysis is to pinpoint the sometimes baffling nature of media work and to focus our attention on the continuous struggles that are particular to producing culture and the production cultures of the media. The many ambivalences of media work to some extent feel like constant conflicts and certainly require a lot of shuffling and careful navigation, while also suggesting that there are always other ways of doing things. In other words, media professionals have agency despite (and to some extent, by virtue of) being part of a precarious and sometimes quite exploitative industry, and it is in this resolutely hopeful spirit that the account of what it is like to work in the media is offered.

In total, this chapter covers twelve distinct paradoxes that inform and influence media work, framed along micro-, meso-, and macrolevels of analysis. On the microlevel, we look at the individual media professional at work, where you find tensions that are generally particular to anyone's distinct experience of working in one or more of the main media industries—journalism, advertising, film and television, (digital) games, social media entertainment, and music. This is not to suggest that it doesn't matter who you are or what your background is (including gender, ethnicity, life phase, and ability). Each media professional tends to be confronted by the issues mentioned here, affecting everyone differently and prompting and inspiring many different ways of taking action. On the mesolevel, issues are highlighted that tend to be unique to the ways in which media companies and production projects are organized and managed. Finally, on the macrolevel, we look at the key overall pressures and strains that shape media production, to some extent dictating how the media as an industry is able to make it work both in terms of financial and commercial viability as well as being able to do satisfying and meaningful work.

Of course, the partition in different levels of analysis for understanding media work is a construct, as much as the dual structure of the various contradictions is. Considering the way media work as an industry like this offers a conceptual shortcut, and I hasten to add that these paradoxes can occur on other levels than stated here, mean different things to different professionals working in different industries, and translate into a wide variety of practices. What makes this conceptualization of media work a handy tool is that it allows us to look across the media industry in its entirety, being mindful of the emotional impact and lived experience of working in the media.

Macrolevel: media industries making it work, where the work is

    profitable and loss making
    datafied and creative
    switched on and switched off
    impactful and disconnected

Mesolevel: the organization of work, where the work is

    artisanal and industrial
    informal and militarized
    dynamic and routinized
    collective and fragmented

Microlevel: the individual at work, where the work is

    empowering and exploitative
    liberating and interdependent
    idealistic and homogenous
    pleasurable and precarious

## Profitable and Loss Making

Most media organizations are run as businesses yet tend to include goals beyond turning a profit—such as creating original, beautiful, or otherwise meaningful products and services. Some media are run primarily for idealistic social or cultural purposes, without seeking profit—but they still have to pay the bills. For instance, public service media have generally had a bureaucratic form of organization but with nonprofit social and cultural goals. Most major film studios also operate smaller, "arthouse" studios to experiment with smaller budgets, to earn a reputation and credibility as a nurturer of talent, and to diversify their portfolio of movies—even though these productions tend to be written off (in advance) as loss makers. In this delicate balancing act, most media companies do not turn a profit on the basis of what they make. Despite at times impressive sales figures of digital games, some box office success of films, and huge viewing numbers for shows on

streaming services or for vlogs on video-sharing platforms, all this revenue can quickly disappear when set against the long-term expenses involved in sustaining the business.

This tension between the ideals of creativity and commerce runs through all decision-making in media organizations, inspiring competing logics that govern the production process. Generally speaking, a media company or professional tends to embrace or prefer an *editorial logic*, meaning that they feel creative decisions should be made based on what the media practitioners deem worthy to pursue. However, a definite *market logic* is present, too. In that context, goals are set as determined by such issues as target audience tastes and needs, client demands, and commercial appeal. Success in this context is measured by such metrics as viewing or listening figures, ticket sales, hits and clicks, and time spent. Taken together, these competing goals and logics inspire much of the tension and potential for conflict within media organizations and operations. Other governing logics can be present too, including one that privileges user engagement, interactivity, and cocreation. As many media companies and individual professionals work at the behest of clients, a fourth logic governs decisions on the basis of strict accordance to rules and parameters set elsewhere. In practice, all these logics and goals of media organizations often conflict and at times converge. Different logics also play out on the level of individual practitioners, as many (if not most) media professionals work as freelancers or on temporary contracts, requiring them to constantly balance competing pressures and opportunities. A common formula for navigating all these different pressures and constraints (especially as a free agent) in the media is to strike a balance between *finance*, *fun*, and *fame*:

- Does the project/gig/assignment pay (enough)?
- Does the work inspire/bring joy?
- Would this product or service build or sustain reputation/prestige in the field?

If the answer to any or all of these questions is no, a media professional would do well to seriously consider turning down the project or opportunity on offer.

### Datafied and Creative

Following up on the presence of multiple and at times conflictual guiding principles in the production process, another decision-making logic emerges, given the growing significance of data, metrics, predictive analytics, and key performance indicators (KPIs) in the media industries. Such *data logic* can govern any kind of course of action, from what kind of projects to pursue and determining planning and strategy to playing a role in the creative process itself. At the same time, most organizations tend to privilege intuition and talent as the basis of the media production process, while creativity is prized above all else when articulating the quality of a colleague (or competitor). In a digital environment, more and more aspects of the media production process become datafied, setting the stage for clashes with the discourse and expectations around people's creativity and craftmanship. It is important to note that despite embracing an algorithmic turn, data-driven decision-making at digital-only companies such as Netflix has yet to solve the media industries' nobody-knows dilemma—given the fact that most of its *greenlighting* choices and productions fail to provide significant return on investment.

Data logic across the various media industries tends to play two primary roles: that of demand predictor and content creator. Mining vast databases of people's preferences, subscriptions, clicks, and channel choices is expected to produce accurate predictions on what will be a hit. Increasingly, media workplaces get outfitted with tracking software that shows in real time what the audience is doing and where they are going. The potential for forecasting seems profound. The same goes for content creation through data, which contains the promise of much of the production process to be automated—as already has been applied to copy writing for news organizations or marketing companies and music (for use in digital games and films) created by an artificial intelligence. At the same time, the expectation of creativity and the charisma of certain individuals within the media company or production team often do not take kindly to statistical references, algorithmic influences, and automated solutions to production outcomes. Just as the gut feeling of a creative is never infallible, the results from number crunching are imperfect predictors of success, and the kind of creative work produced by computers tends to lack a certain something—however vague that "something" may be. The tension between these two aspects of all media work does produce interesting clashes and challenges, though.

**Switched On and Switched Off**

Ask around, and anyone will tell you: working on the set of a film, doing *crunch time* (extreme overworking to deliver on deadline) on a digital game, digging deep on an investigative journalism project, recording new music in the studio while trying to stay on time and within budget—it can all be very exciting, meaningful, and intense. Media professionals often describe the work in such instances as going all in, feeling part of a family with their coworkers, purely living on adrenaline (as well as caffeine and alcohol). When it is done, it is not uncommon to celebrate—if anything to let out all the repressed emotions associated with the previously stressful period. On the other hand, the work styles (ways of working and being at work) of media professionals tend to be marked by extended periods of boredom, inaction, and unemployment. Quite a lot of the work in media can be tedious, consisting of sitting at a computer and spending days filling out forms, making phone calls, and performing logistical and other menial tasks, up to and including just getting people coffee or lunch. One professional I once interviewed, a seasoned Hollywood movie assistant director, told me he got his big break in the industry by picking up Steven Spielberg's Christmas presents. In between projects, practitioners can experience tedium as there is not much to do until a new client, project, or account presents itself or is found. For many, if not most, workers this period is unpaid, as crews go on hiatus after a season ends, and many (smaller) companies do not necessarily have new projects lined up—resulting in letting most nonessential staff go.

For most people in the media industry, their time at work is matched with significant periods out of work as nonstandard employment or "atypical" work is the norm in media production—which means that most professionals do not have an open-ended contract with an employer. Roughly speaking there are four types of *atypical media work*:

1. Temporary employment: fixed-term contracts (including project- or task-based contracts), casual work, and daily work

2. Part-time and on-call work: normal working hours (yet fewer than full-time equivalents), marginal part-time employment, on-call work (including zero-hours contracts)

3. Multiparty employment relationships: including temporary agency work, being on call as subcontracted labor

4. Disguised employment / dependent self-employment: work that is not part of a formal employment relationship (including false employment, where a freelancer works for one company exclusively), as well as *speculative work* (meaning work that is unpaid but done in the hopes of securing future gigs, clients, or employment)

As a macrolevel development, atypical work is the dominant mode of employment in the media industries, as it is one of the key ways in which companies manage their financial risk. The tension between being (and feeling) completely switched on and alternately experiencing a switched-off, languishing existence plays out differently yet structurally for many, if not most, media professionals. This is not to say that being switched on is an on the whole enjoyable feeling, as it comes with negative health impacts (such as stress, burnout) and a decrease in the quality of the work and work experience, and the overwork involved is often uncompensated. The same can be said for the uneventfulness of being in between jobs, gigs, or projects, which can be an opportunity to recharge, to retrain, and to reconnect with friends and family.

**Impactful and Disconnected**

A fourth macrolevel friction of media production is being involved with work that intends to reach and affect many, while concurrently working far removed from any meaningful interaction with the public. The products and services of the media industries potentially reach audiences far and wide. Media work at times can make a powerful difference in the lives of people all over the world. Although considerations of the public do not always feature prominently in the ways in which media professionals talk about their work, the media as an industry cannot consider what it does without or beyond the public. In fact, the imagination of an audience—what it needs, what it wants—fuels many aspects of the production process. Some of these reflections are commercial in nature, as decisions can follow a distinct market logic, while the public can also play a role in terms of one's sense of social responsibility and personal motivation to do work that contributes to the common good. However, in the contemporary digital environment, it is safe to say that reaching a mass or otherwise large audience is something few, if any, media productions can ever hope to achieve. Although much of our media use can still be labeled as consuming content, simultaneous multiple media use is now the norm, with people going online with their mobile devices while watching television, and spending time online (specifically with social media) outperforms all other media use. Subsequently, there is a noticeable shift throughout the industry from thinking in terms of distribution and reach toward notions of participation and engagement. It is not just about getting people to notice and pay attention to your product or service—which is difficult enough—but more so to get people to become involved, share their thoughts online, and forward and recommend to each other the new series to watch, the new music to listen to.

There is a specific tension here between getting audiences engaged, while being quite disconnected from society. This disconnect between media producers and consumers is evident in the informal hierarchies of media work, where those that operate furthest removed from interactions with the audience—such as the creatives in advertising, producers and directors in film and television, parliamentary and investigative reporters in journalism, and creative leads in game design—tend to be the professionals with the highest status. Community managers in the games industry, market researchers in advertising and marketing communications, and reader representatives in journalism are lowest on the professional ladder. Furthermore, among peers it is generally frowned upon to use the public as a reference when discussing creativity or quality. Audiences in media industries are imagined (and, to some extent, datafied in surveys, focus groups, and metrics, such as clicks and view counts) rather than experienced. Trying to get access to a building where media work takes place is next to impossible—unless you are there by invitation.

For an industry so dependent on knowing, understanding, and engaging people, its practitioners commonly have little or no actual contact with the public. Industry-audience interactions tend to be staged and managed rather than authentic and organic, focused on extracting data and information rather than soliciting feedback and genuine collaboration. Some professionals and companies choose to act differently, however—investing in cocreation with communities, taking the time to get to know their audience and reconstituting them as club members or constituencies, respecting their engagement beyond the free labor that people do as fans. Examples of this can be found throughout all media industries and in today's digital environment are often the source of inspiration for new companies, such as the American firm Hearken, founded by Jennifer Brandel in 2015 to consult news organizations on ways to better listen to their communities (using a dedicated online platform) and to integrate a diversity of voices into their operations. In 2021, Hearken worked with over one hundred newsrooms in fourteen countries around the world in a direct attempt to bridge the industry's societal disconnect with its desire to do impactful work.

### Artisanal and Industrial

On the mesolevel, or organizational level, of media work, a key strain features in the coexistence of artisanal and small-scale production methods next to producing content on an industrial scale. Much of the work in the media industry depends on craft, talent, and creativity—often using such complex technologies as digital cameras, content management systems, recording equipment, and editing software. Projects tend to be run by teams consisting of people with a variety of skills, backgrounds, and professional roles, requiring careful calibration to make things run smoothly. Much of this fine-tuning is all about building rapport, doing the emotional labor to establish a circle of intimacy and trust. While on a specific assignment or task, the work is meaningful in part because of both its high-pressured intensity and its collaborative spirit. Some of the spectacle of media work is its performative aspect toward colleagues—looking for opportunities to show off your talent and skills, trying to impress coworkers in part to secure future employment and simply to have fun.

These are all the benchmarks of artisan-like processes, relying on skilled workers dedicated to quality (of both the work and the working experience). At the same time, the media industries rely on mass reproducibility of its goods and services, churning out content to accommodate an ever-growing array of channels and platforms. Such an industrial configuration of media work is paramount in generally patterned and routinized production processes and practices, in the development of and strict adherence to genres, formats, and conventions, the rationalization of management (e.g., by outsourcing significant parts of the process to external professionals or firms) and of sticking to well-worn patterns of mass distribution and promotion. This tension makes it possible for media professionals to deny (or remain willfully naive to) the fact that their work is part of a vast factory-like global industry, as much as it enables the media as an industry to privilege intuition, craftsmanship, and artfulness as a primary frame of reference, despite the often highly formulaic and repetitive nature of the industrial production process.

### Informal and Militarized

In a post to his personal weblog (on April 29, 2008), the Chinese-Canadian photographer and digital effects designer Roger Wong describes a typical day at work at the digital effects studio Weta Digital—based in Wellington, New Zealand (founded by Peter Jackson, Richard Taylor, and Jamie Selkirk in 1993 and having won numerous awards for their work on *The Lord of the Rings* and contributing to such major franchises *Avatar*, *The Avengers*, and *X-Men*):

> If you work in our industry then you know what our standard dress code is like, but if you don't work in our industry, the standard dress code . . . is none whatsoever! :) It's pretty much wear what you wanna wear to work. I've seen everything from flip flops and board shorts, to tear away/warm up pants, to jeans and khakis. But the last thing anyone will wear to work is proper business attire . . . no suits, no ties, maybe a dress shirt but it won't be buttoned up or tucked in . . . So yes, here at Weta we have a tradition called "Formal Fridays" . . . An excuse for us to "dress up" . . . Formal Fridays means an excuse to pile into the photo studio to act like school kids at photo time.

This account of a fun ritual where employees regularly "pretend" to be working at an office, complete with official attire and conventions, is rather typical of the generally casual, informal character of the media workplace. Combined with the unusually intense team-based and creative aspects of the work, this explains part of the appeal of a media career. However, this cherished informality also has all kinds of problems. In 2020, numerous reports came out about a toxic working environment at Weta, marked by sexism, bullying, and harassment. The news prompted an independent investigation based on a company-wide survey, interviews with more than 200 current and former crew members, and inviting written statements from everyone involved. The report thereof (published on December 22, 2020) remarked that the company—employing well over 1,500 people—is "energetic, fluid, focused on excellence, but also undisciplined, overstretched, reactive and cliquish." Although it did not find evidence of a "toxic" work environment throughout, the company exhibited the same problems with its culture that can be found in other visual effects companies and in the screen industry more generally. As the author of the report, barrister (and Queen's Counsel) Miriam Dean, concluded: "Weta Digital's crew love their work and the people there, but they do not all love the culture. The company often asks too much of them, it does not consult them, it does not always treat them—or promote them—fairly and it does not always protect them against poor (sometimes even bad) behaviour by others."[2]

Similar stories and reports can be found throughout the media industries. Especially in the digital games industry, numerous high-profile efforts are underway to address issues related to problematic working conditions and office culture. During 2021 and 2022, several lawsuits and investigations unfolded at the American game studio Blizzard Activision—home of such successful titles as *World of Warcraft*, *Guitar Hero*, and *Call of Duty* (with billions of US dollars in annual revenue). Employees complain about different kinds of abuse, such as "cube crawls" in which drunk male employees subjected women to unwanted advances, a lack of women in leadership positions, and unequal pay for women, and accuse the studio of fostering a "frat boy" culture and "rock star" mentality that is a breeding ground for "rampant sexism," including sexual harassment and discrimination against women. In the weeks following the initial lawsuit against the company, thousands of employees organized to stand with their colleagues who experienced mistreatment or harassment of any kind, releasing statements critical of executives, and staging walkouts and work stoppages. Numerous media—including the *New York Times*, *Vice*, IGN (Imagine Games Network), and the *Wall Street Journal*, published their own investigations into the company's lackluster handling of employee complaints. Throughout it all, current and former workers posted their experiences and reflections online combining such references as #MeToo and #timesup with game-industry-specific hashtags #WomenWhoCode and #gamergirls, in conjunction with company-specific tags like #actiblizzwalkout and #ActivisionBlizzard. Over one thousand of their colleagues at the rival French game company Ubisoft pledged support by starting the #ABetterUbisoft campaign, demanding that management do more to end abuse at its own company. Another competitor, Bungie—developer of the hugely popular *Halo* and *Destiny* franchises—announced in September 2021 it would hire a diversity and inclusion director, update

---

2. Miriam Dean, "Independent Review of Workplace Culture for Weta Digital," 1 News, December 23, 2020, https://www.1news.co.nz/2020/12/23/independent-review-of-workplace-culture-for-weta-digital/.

hiring practices, and improve training and tools about social safety, among other efforts to address problems with its working conditions, which include overt sexism, a boys' club culture, and systemic inequalities.

It is not just the games industry that is confronting concerns that partly arise from a working culture where largely informal and intensely collaborative practices clash with an industrialized and high-pressured professional way of doing things. As mentioned in chapter 6, the global #MeToo movement got a major impetus from social media posts of female professionals in the global film, television, and music industry, reporting sexual harassment on the job. These developments coincide with a surge of labor organizing and employee activism within media institutions around the world, including in journalism, advertising, public relations, and television (see chapter 8 for more detail on workers' rights across the media industries).

Online and offline activism (see chapter 6) among media professionals coincide in these highlights of just some of the cases and stories showing how the high-pressure yet informal working environment can be both fun and exciting, as well as harmful and problematic. Part of what explains the particularity of the problems experienced in the media industries is a paradox in its office culture—between lively informality on the one hand and a rather militarized organizational structure on the other. Such structures can be hard to observe in the actual workplace but are quite clear in organizational flow charts of professional roles in media work—consider, for example, the list of credits at the end of a film, the colophon of a print publication, or the ordering of people receiving recognition at industry award ceremonies. It is clear that notwithstanding an easygoing workstyle the media production process is in fact highly structured, where all positions function in distinct hierarchical and dependent relation to each other. This sometimes has severe consequences to those breaching protocol, regardless of how laid-back such rules are formulated (if the rules are made explicit at all). Let's return to the investigation into the problems at Weta in the spring of 2020: "Crew said the organisation was extremely hierarchical; real leadership was scarce; few wanted to rock the boat by speaking up about bad behaviour or heavy workloads because their contracts might not be renewed; and shifting blame to others was common. Favouritism was rife, so promotions were often less about merit than connections; the company reacted rather than acted; strategic planning was lacking, resulting in often chaotic implementation; stress was high because of the heavy workloads and long hours; communication was poor; and technical or artistic excellence . . . trumped all other considerations."[3]

What makes working in the media industries so much fun is partly the reason why the experience can be so frustrating and problematic, especially for people from minority and otherwise marginalized backgrounds. The much-celebrated informality of working arrangements, set against a distinctly hierarchical, interdependent, often male-dominated, and to some extent militarized structure of relations, effectively demands better policies, protocols, and procedures to protect those in vulnerable positions. Thankfully, because of prominent cases and campaigns and an increase of digital as well as on-the-ground organizing, companies and professionals seem to be taking all this increasingly seriously.

**Dynamic and Routinized**

An extension of the pressured and constantly changing working environment is a distinctly dynamic feel to the work. Beyond the periods when workers are on hiatus, between jobs, or without a new client or gig, the work (as the people you are working with) can be genial and quite animated. The context of an always changing digital environment, a near-constant stream of new or upgraded channels and technologies, and all our less-than-loyal uses of media provides additional fuel for the already enterprising nature of media work. In the same breath—and as noted before—much of the production process across the various media industries tends to be tightly prescribed, following well-established conventions, routines, and procedures developed over time to meet the demands of a deadline-driven culture of production. A common phrase in a media company is "that is just the way we do things around here," referring to certain practices to be internalized by every intern, recent hire, or junior employee to fit in. It is the dynamic nature of media

---

3. Dean, "Workplace Culture for Weta."

work that in fact necessitates a somewhat formulaic approach to the production process. Routinization is also a managerial response to an industry marked by uncertainty, as the majority of professionals only get hired on a temporary basis to do the work and often need to fit in on a moments' notice. The increased complexity introduced by our digital environment further pushes the industry into standardized action, especially given the permanent deadline of publishing online.

### Collective and Fragmented

A somewhat counterintuitive tension typical of media work exists between an overall team- and network-based approach to the production process and what is often a fragmented, highly individualized experience of being a media professional. Working in teams suggests collaborating closely with colleagues and peers to get the job done. This collective nature of the work translates to the material context of media production. It is rare to find closed-door workspaces in the buildings of media companies, as open-plan offices tend to be the norm. It is fascinating to see such a relatively uniform design aesthetic dominating media firms around the world, as newsrooms, agency workspaces, or game studios tend to look remarkably alike, preferring shared working environments that privilege coworking over privacy. Yet, given the predominance of atypical working arrangements (as mentioned earlier), the majority of media professionals in fact spend much of their time isolated and alone, working from home or at a coffee shop, only meeting colleagues when they check in with a client or participate on a project for a while. That, in addition to the irregular stop-and-start nature of assignments and careers in the media, suggests a rather disjointed experience.

On top of that, the professional advancement of media workers tends not to follow a neatly linear or steppingstone kind of progression, whereby someone gradually moves up the corporate ladder into more senior positions. Instead, for most practitioners, one's career resembles a spiral staircase structure, where one moves around constantly without necessarily moving forward much nor even finding some kind of job security along the way. Clear lines or protocols for compensation and promotion tend to be notably absent anywhere in the industry. While this may be inspiring as one's work life can be said to be anything but dull, the sometimes-jumbled and unpredictable experience of a patchwork career also causes people (and voices) to leave the profession or to opt out of pursuing work in the media altogether.

### Empowering and Exploitative

When it comes to the microlevel of media work, the individual experience of making media professionally can be deeply meaningful and empowering. This is an industry that encourages and even demands self-expression, prizes autonomy, and organizes the production process in ways that fit those who are comfortable controlling their own time, managing themselves, and handle uncertainty as much as intense pressure well. By contrast, what relies on individual mastery is also an exploitative aspect of the work, as most media professionals are left to fend for themselves in an industry that offers little in the way of open-ended contracts, retirement plans, medical or any other insurances, maternity leave, child-friendly working hours, or any other typical employers' responsibilities. In the media, almost everyone is expected to be entrepreneurial regarding their career, taking on much of the risk associated with the business and becoming a "business of one" in the process.

If you want to be successful as a professional media maker, you may find yourself managing, promoting, and performing yourself as a brand. Media workers, to some extent, have always adopted branding tactics for themselves: building a reputation, carefully managing their emotions in mostly informal working environments, performing this professional identity dutifully to make it work. The prevalence of atypical work and the rise of digital media amplify and accelerate the branding trend, potentially raising the level of self-promotion to extremely stressful levels. Even with a contract in hand, in many media firms, this is now seen as a business-to-business deal in which you, as a business, are contracted to provide temporary solutions to another business's market-specific problems—such as completing a specific project (like wrapping a season of a television show or editing news before it goes online). The digital environment contributes to expectations of permanent availability and has workers stretching their professional identity across a variety of self-branding platforms (including LinkedIn, Instagram, personal homepages, social media contributions, and so on). On the one hand, this can be seen as offering more ways to control one's self-presentation and image, while on the other hand all this puts even more pressure on already struggling professionals.

## Liberating and Interdependent

What binds media practitioners around the world, across professions and disciplines, is a desire for creative autonomy. In many ways, professionals are motivated like fans, in that they primarily want creative autonomy to tell their own stories (or participate in meaningful shared narratives) and would like to be recognized in some way for their work. All this passion can be quite problematic, as it is much harder to develop a substantial critique of the industry you are in if what keeps you at work inside this industry is your own individual passion. You would end up being critical of the very thing that connects most deeply to who you (think you) are. At the same time, the freedom to express yourself, tell your story, and possibly earn a living on the basis of this is undoubtedly a powerful motivator for people to pursue careers in the media industries.

As the digital context offers a growing number of channels, platforms, and ranges of options for storytelling, it should come as no surprise that media work holds much appeal for its opportunities—not just in terms of making money but of having your voice heard and seen. Inversely, the way the media as an industry works is suffused with dependent relations—between different departments and businesses, funders and financiers, employers and clients, professionals from a wide variety of specialized skill sets and disciplines, and a growing range of technologies, channels, and platforms. As mentioned earlier, any work in the world of media is the product of a collective activity that involves numerous people, organizations, and material contexts in interdependent relation. The prized freedom in media work is heavily circumscribed by cooperative and symbiotic associations, where individual voice is both celebrated and erased in the aggregate production process—which can be complementary, collaborative, and harmonious as much as it is marked by codependence, imbalance, conflict, and exploitation.

### Idealistic and Homogenous

On the whole, the workforce of the media industry is overpopulated with relatively young, well-educated individuals who are generally in a socioeconomic position that affords them the option to live in or near thriving urban centers—as this is where most media companies are located—and to spend a significant portion of their career in underpaid or unpaid positions. This situation produces a rather unique twist in the experience of media work, as most of your colleagues and competitors will be similar in outlook on life, often favoring emancipatory ideals and valuing (personal) freedom above all else. The documented informality of production cultures in the media contributes to an overall feeling of being part of a family of like-minded people who all seem to love what they do and who are chiefly motivated to make beautiful and meaningful things that people find useful and enjoy.

You would be forgiven to expect that this lays the foundation for a mostly progressive, diverse, and idealist professional community of peers. However, the majority of studies instead suggest homogeneity rather than diversity among the media rank and file. Pretty much all media industries are still quite male dominated and have little or no representation from the country's minority groups (including people with disabilities). The executive layers of media companies—featuring those with so-called *greenlight power* (referring to the people who have the authority to approve or deny projects to go ahead)—have the least diverse constellation. Among chief editors, executive producers, team leads, and other leadership positions, homophily is still the rule. The life phase of media professionals trends younger, as the precarious nature of media work is less than ideal later in life—when many people, for example, are more likely to be responsible for a family and a mortgage. Combined with the economic cost of making it work in the media—getting degrees at expensive educational institutions, living in costly cities, and taking on internships, apprenticeships, or other forms of underpaid or unpaid labor—this makes for a culturally and socioeconomically largely homogenous workforce. Diversity, while acknowledged as key for a more representative (and economically viable) as well as more creative industry, turns out to be quite an issue throughout the media.

The overall lack of diversity in the media industries is a stark reminder of how intersectionality helps us to understand how exclusion and marginalization works on multiple levels simultaneously. Whether considering the position, role, and experience of (for instance) women in journalism, people of color in advertising, workers in their forties and beyond in film and television, or disabled people in all disciplines, the solution is never as easy as pushing for more diversity hires. The significant problem of the media industries becoming a more inclusive and welcoming workplace runs much deeper than getting more women hired or creating positions specifically for whomever organizations feel is underrepresented on their team. Time and time again, those who are in the minority in the media feel marginalized, sidelined, either not taken seriously, or mainly seen as representing "their" background. This points to an intersectional explanation, where socioeconomic status (or class), gender, ethnicity, age, ability, and even one's level of education, fashion style, and knowledge of popular culture and art all codetermine your "fit" and chance for success at work. Media professionals, generally speaking, perhaps are not necessarily racist, ageist, or misogynist, but the media workforce most definitely has a tendency toward conformist, in-crowd style behavior, preferring shared norms and values over alternate ways of thinking, talking about, and doing things—in part because of the informal nature of the working environment, coupled with the always present pressure to get things done quickly, leaving little room for challenging reflection and exploration. This stands in stark contrast with the idealistic, ambitious, and quite often progressive outlook industry professionals have (or claim to cherish).

Diversity discussions and actions abound throughout the industries on all continents. Consider just a few of the many current initiatives to document and address diversity issues in the various media industries:

- the International Game Developers Association (IGDA) collecting and publishing data on the lack of diversity in the game industry since it started its annual worldwide developer satisfaction surveys in 2014;

- the notable #Oscarssowhite social media and boycott campaign in the US film industry (ignited in 2015 by media strategist April Reign);

- the annual UK screen industries' Project Diamond reports providing long-term monitoring of diversity data (launched in 2016);

- the Walkley Foundation's Media Diversity Australia award (since 2019) honoring journalists who report on diverse people and issues in Australia;

- the World Federation of Advertisers' first-ever global diversity, equity, and inclusion (DEI) census of the marketing and advertising industry launched in 2021; and

- throughout the music industry, labels and professional associations' reports and projects such as the Universal Music Group's Taskforce for Meaningful Change (in 2021), after earlier announcements (in 2020) of the "big three" companies Universal, Warner, and Sony, to set up funds for social and racial justice organizations.

While there clearly seems to be a growing commitment to tackle the lack of diversity in the media across the various industries involved, all these (and many, many more) projects underscore how profoundly problematic this issue is.

### Pleasurable and Precarious

It can be said, not only by practitioners themselves, that working in the media is fun. While there are certainly problematic aspects to it all—as documented in the various tensions discussed here—it can be an absolute thrill to be uncovering injustice on an investigative journalism project, to wrap a new season of an exciting television show, to put the finishing touches on a digital game before it "goes gold" (meaning when it is playable enough to be published), or to see the campaign you have worked so hard on across billboards and screens all over town. The intensity of working toward a deadline, being among colleagues that feel like family, investing your whole self into the work—we should not underestimate how pleasurable and deeply fulfilling all this can be. At the same time, almost every single aspect of media work is suffused with uncertainty—from people's job prospects (or lack thereof) to the success (or failure) of all media products and services, from the decisions made in executive boardrooms to the constantly changing role of technologies in the creative process. Perhaps the relation between pleasure and precarity forms the most fundamental absurdity of professionally making media and explains how, at times, work in the media may feel like an addiction. It provides intermittent happiness and buzz, proves extremely hard to quit, and sometimes becomes harmful to other aspects of people's lives (such as family and friends).

### Media Work for a Life in Media

The gradual transition from either watching television, reading a book (or a magazine or newspaper), or listening to the radio to being concurrently exposed to multiple media in a digital environment—additionally marked by our own productivity as media makers whenever we post, upload, like, forward, or otherwise share something online—proves to be a profound challenge for media industries. As our lives are lived in media, the media as an industry has to come to terms with a profound shift in the storytelling experience, roughly speaking evolving from one-off narratives produced for a particular medium to complex and continuously developing story lines moving across different media. In the process, all the paradoxes and tensions that codetermine media work are jacked up, introducing exciting creative challenges as well as new stressors in an already intense situation.

Envisioning the future of the British Broadcasting Corporation (BBC) in 2004, the company embraced the "Martini media" principle, suggesting that people want to access media anytime, anyplace.[4] The Martini concept is a reference to a series of 1970s European television and radio commercials for Martini, a popular brand of Italian vermouth. The advertisements featured a jingle with the memorable words "capture a moment—that

---

4. Ashley Highfield, "The On-Demand World Is Finally Coming," speech given to FT New Media & Broadcasting Conference 2004, March 2, 2004, https://www.bbc.co.uk/pressoffice/speeches/stories/highfield_ft.shtml.

Martini moment—any time, any place, anywhere—there is a wonderful place you can share—and the right one, the right one—that's Martini" (the original advertising score was composed in 1970 by the British composer Christopher Gunning, later on becoming a hit single in 1977 for the South African singer Danny Williams, adapted and titled "Dancin' Easy"). This Martini media principle, originally referring to content moving freely between different devices and platforms, today extends to the entire media production process as practitioners come to terms with developing story lines that work across a variety of media, while also considering a cocreative role for the audience as people who participate and collaborate in finding, producing, sharing, curating, and remixing content.

From a creative point of view, the evolution toward a Martini media universe is quite exhilarating. A *pervasive* and *ubiquitous* digital media environment inspires new forms of storytelling, combining stories and experiences within and across media in a recognition of people's migratory behavior as we combine and use multiple media devices, channels, and platforms for our daily dose of news, information, and entertainment. Instead of making media for a particular medium, makers now can consider narratives that span the entire digital environment. Beyond the technological challenge this provides, it is important to note that much of the creativity of media making is ensconced in well-established genre conventions, formats, and formulas that are particular to different types of media. For example, a newspaper story customarily follows an inverted pyramid format—whereby the most important information comes first in a story as copy editors need to be able to cut off the last paragraphs of a story (without having to read it) in case there is not enough room on the printed page. In many commercial motion pictures, the main characters and story line need to be established within the first twenty minutes of a film, because market research suggests that audiences otherwise become distracted and are less likely to suspend their disbelief. In this way, every industry has its own set of genres, which get established over time:

- by companies and publishers, as a way of organizing and taming the sprawling and unpredictable economy for cultural products into separate, narrowly defined markets;
- by media professionals, to meet the demands of specific technologies, to develop and deploy available skills and competences, and to meet the expectations of clients and employers;
- by audiences, enabling people to identify themselves along specific styles and traditions (e.g., by stating "I am not into sci-fi" or "I'm a metalhead") to differentiate themselves; and
- by the material context of media production, as specific technologies—such as the printing press, a studio complex, the open-plan office, (proprietary) software and interfaces—impose structural limitations (and possibilities) for storytelling practices.

As the context of cultural production changes from media that have specific boundaries—television as distinct from radio, both quite different from print—to a digital environment where video, images, audio, and text seamlessly interact and flow across a variety of devices and uses, notions of genre and format do not necessarily disappear but are much less subject to consensus and control (either by industries, practitioners, or audiences) than they used to be. In other words, nobody really knows what kind of stories work and in what way they can best be told online—adding yet another layer to the nobody-knows dilemma fundamental to the inner workings of media industries.

In terms of developing compelling narratives that somehow flow and work across multiple media, a distinction can be made between *multimedia*, *crossmedia*, and *transmedia* storytelling practices. A multimedia approach takes a single story, breaks it down using different forms of media, and gets published on a single channel. Crossmedia takes the single story and uses different media and multiple channels for publication. Transmedia storytelling involves the creation of an entire storyworld containing many stories, in a variety of forms, published across different channels. Each of these approaches can be coupled with participatory modes of storytelling, where media users contribute to one or more phases of the production process. This could be already in the financing stage through a crowdfunding campaign, at the preproduction and information-gathering phase by crowdsourcing data and intelligence, or as part of the creative process by offering feedback, up to and including after formal publication by inviting comments and enticing the audience to like, share, and respond using social media.

Although the various modes of storytelling using multiple media are not necessarily new, with the digitization of both work and life comes all kinds of interesting opportunities to connect people, content, and services in ways that were not readily available earlier on. Some examples that have become benchmark cases in different media industries can be highlighted here, each setting the stage for subsequent developments that are very much ongoing and subject to much intervention by companies and practitioners around the world.

In multimedia storytelling, the production of a story can take on many forms—spoken and written word, music, photography and video, animations, illustrations, and infographics—yet generally is produced within one specific channel (e.g., a website, or within a mobile application). In journalism, this is sometimes referred to as "snowfalling" the news—a respectful reference to an influential multimedia reportage titled "Snow Fall" by the *New York Times* published December 13, 2012 (winning the Pulitzer Prize in Feature Writing, as well as a prestigious Peabody Award). The multichapter series of reports on events surrounding a deadly avalanche integrated video, photos, and graphics on a single webpage, with content almost seamlessly flowing across the screen. The package became a benchmark for multimedia storytelling in journalism, successfully replicated by news organizations around the world. Another option for storytellers combines the potential of different media forms with the opportunity to reach and engage audiences across more than just one channel. In effect, such crossmedia storytelling takes a story and tells it in ways particular to the form best suited for the channel it gets published on. The *Star Wars* franchise can be seen as one of the first attempts to use a crossmedia approach to the overall narrative, as no matter what kind of product you would get—the films, books, comic series, board games, video games, and so on—the basic storyline would stay the same, catered to the particular medium you would use. As the franchise crossed into the twenty-first century, it began to include much more complex narrative structures across all its products and even recognized fan-produced work as contributing to the overall storyline. In its contemporary form, *Star Wars* is—much like other popular science fiction and fantasy franchises such as *The Matrix* and the Marvel Cinematic Universe—an example of *transmedia* storytelling, as the idea of a complete and finished full story is abandoned in favor of a complex array of multiple stories, using different media forms, appearing in as many channels, and requiring an active and creative role from the media user to piece together the elements of the comprehensive storyworld that gets created across a variety of media. Beyond big budget franchises, countless examples exist of relatively small scale transmedia, including the *Cheese in the Trap* online comics, website, television series, and film in South Korea (also popular in China), the Argentinian *Aliados* series of television episodes, webisodes and website, mobile applications, a print magazine, live musical, social media accounts, and more, followed by fans throughout Latin America, as well as numerous collaborative transmedia projects by African, American, and European media makers combining radio, animation, and documentary film to explore stories particular to local folklore, cultural histories, and major events in countries such as Rwanda, Nigeria, and South Africa.

Emerging multimedia, crossmedia, and transmedia modes of storytelling inspire much more than the production of a one-off, finished story. Instead, media professionals face the option of designing content and experiences across multiple platforms, integrating user-generated content and other interactive features, and expanding the reach of the story line beyond its initial publication. Ideally this is a chance to tell better stories, using multiple media in complex and engaging ways and integrating the creativity of consumers. From a more critical perspective, combining and integrating multiple media gives media corporations even more reason to merge or acquire different companies to seek more control (and more ways to extract value) of our media experience. An additional complicating factor for corporations and individual media makers alike is the fact that there seems to be no professional or economic monopoly on the potential to reach a large audience by way of the internet, although the search and recommendation algorithms of platform companies act as new gatekeepers up to a point. Furthermore, social media act both as sources, producers, distributors, and promoters of media content, partly because of deliberate strategies by media organizations but generally beyond the control of professional media makers.

*A life in media is, in many ways, a life of mixed realities—albeit one perhaps less virtualized and commodified than the one envisioned by the technology industry.*

Although technological developments can most certainly be described in terms of the rise of new digital intermediaries, platformization, datafication, and algorithms, we should not ignore the tremendous affordances new media also offer to makers. The "underdetermined" nature of digital media enables practitioners to experiment with telling stories in all kinds of ways. Opportunities arise in the field of extended reality (XR), where multimedia, crossmedia, and transmedia storytelling experiences may come together with immersive technological applications that provide so-called mixed reality experiences, where physical and digital objects coexist and interact, such as voice recognition, motion sensing, augmented reality (on mobile devices), 360-degree video, and virtual reality. Claims about an effortless integration of all of these forms of mediated realities in the near future—for example, through a metaverse as heralded by companies like Epic Games (creators of *Fortnite*) and Meta (including Facebook, WhatsApp, Instagram, and Oculus)—tend to be grounded in a fallacy of "perfect" machines (as explored in chapter 4). On the other hand, one could argue that such a metaverse already exists, as our current digital environment composes versions of all these applications that interact through the communication and exchange of people's personal data. Furthermore, life during the global coronavirus crisis shows how video chatting and using virtual hangouts for just about any experience (including weddings and funerals) can quickly become commonplace. Additionally, extending the technology-driven notion of a metaverse to what some call a "pluriverse"—meaning a storyworld where multiple worlds coexist, each offering unique characters, values, and ideas (often informed and inspired by local, indigenous, and marginalized voices)—can be a wonderfully exciting prospect for inclusive storytelling. A life in media is, in many ways, a life of mixed realities—albeit one perhaps less virtualized and commodified than the one envisioned by the technology industry.

Through new (and often relatively cheap and easy-to-use) technologies, more people can participate in making media than ever before, stimulating both the emergence of a global market for media makers and local playgrounds for the inclusion of many different voices and communities. The possibilities to interact with audiences have increased significantly, paving the way for various forms of cocreation. Although traditional (advertising, subscription and sales-based) business models of media organizations are under pressure, the opportunities to monetize media content and products expand, just as the platforms on which content can be shared multiply, potentially providing media makers with more autonomy and creative freedom.

## Make Life

Any description of how the process by which media products come to us gets coordinated and managed has to allow for the ambivalent and sometimes counterintuitive nature of how media production is organized. Who has the power to make decisions and sway things in one way or another and who can become powerful under certain circumstances are key questions that do not always have the same or easy answers. Similarly, the at times intimidating concentration of ownership and size of multinational media corporations belies their vulnerable position in the field (especially vis-à-vis the technology sector), their dependency on talent, and the extent of economic failure throughout the industry. Finally, the nature of work across the media industries is at once fun and inspiring, as much as it can be deeply frustrating and disempowering. One's degree of freedom to move and navigate within these forces is to a large extent determined by a range of forces, varying from the economic, political, material and cultural context of the work; the norms and values of the media organization for (or with) which a media professional works; the particular conventions, rituals, and routines of the field; and the personal characteristics (including level of education, training background, history and experience, socialization, and attitude) of the individual practitioner.

In all the considerations about what it is like to make media professionally, much depends on the question of power. While different affordances, opportunities, and scenarios are always possible in the dynamic, fast-paced context of the media industries, whether people are able to take advantage of such possibilities revolves around having at least some power to act. Power in this context refers to skills, access, position, and motivation, which collectively enable or constrain one's agency in any given situation:

- *skills*—possessing the expertise and ability necessary to make the best use of people, material conditions, and expectations in a given context;

- *access*—having the capability, reputation, and clout to make something happen, including having access to particular people and resources;

- *position*—being in a position where you can make a difference in your activities and circumstances or in those of others as related to one's role in the working environment (e.g., based on workplace seniority and level of experience);

- *motivation*—attitudes and reasons for (not) using the opportunities available to you (as sometimes you can push for some kind of change but are not enthusiastic about doing it).

These four individual-level dimensions of power directly relate to securing work in the media, finding your own creative voice in the media, and getting to do the kind of meaningful, inspiring, and enjoyable work that inspired you to pursue a career in the media in the first place. This is not to say that simply by acquiring such skills every individual media professional gets to do what they want to do. The global media industries are not really known for genuine care and respect for their talent—as exemplified by the countless instances of discrimination and exploitation, as documented in this chapter and throughout the field of media production studies. One way of dealing with this is to develop tactics to creatively navigate the complex and paradoxical nature of working in the media, align with peers you trust, and set clear standards for yourself. A second way of looking at this would be to strategically engage the industry to fight for reform, inspire collective organization, and negotiate for change (we return to this aspect of workers' rights in chapter 8). The industry itself needs to take responsibility too, which, for example, means

- reducing the effort-reward imbalance experienced by media workers, and commit to fair treatment and just pay for everyone involved in the creative process;

- being much more transparent and explicit about hiring, firing, and promotion procedures;

- establishing clear protocols that prevent and combat sexism and any other kind of workplace discrimination; and

- controlling the culture of overwork throughout.

The rationale for wanting to work in the media tends to be universal across all disciplines and professions, as most practitioners—including but not limited to reporters and editors; film or television producers, actors, and crews; buyers, planners, and creatives in advertising and marketing communications; game developers, artists, and programmers; musicians, recording engineers, artist and repertoire managers, vloggers, social media creators, and so on—simply want a chance to tell stories. That story can be about a homeless person you want to help out, about the corruption of a local politician, or about a superhero protecting the planet from an evil superpower, or it can be coming up with a slogan like "Just Do It" (coined by the American advertising executive Dan Wieden in 1988) that everyone around the world remembers. The combination of this drive to create powerful, meaningful, or just successful stories and making it work (which can refer to anything from earning a decent living, aspiring to run a profitable business, achieving stardom, or simply having fun) propels the media as an industry—as a world—into constant action. This anything-but-predictable dynamic should inspire doubt and concern about any kind of blanket statement on how power works in the media.

For sure, multinational corporations strive to control most aspects of the media production value chain, and many media professionals work in precarious and at times exploitative circumstances. At the same time, uncertainty and volatility are rampant throughout the industry, extending precarity from the individual experience of working in the media to the overall way in which the media as a global industry is organized and managed. A continuous stream of technological developments and different kinds of innovation (including product, process, and genre innovations) supercharge this state of flux. Media professionals are thrown into all this, fueled by their passion for storytelling and eager to make a difference. And thrown they are, as work in the media is easy to find while jobs are hard to get. The entire media industry runs on an undercurrent of intrinsically motivated talent, often willing to work for next to nothing as interns, doing speculative or other kinds of unpaid work, or accepting underpayment for their time and services because of the promise of better days ahead. While this suggests a profound power imbalance, it is fascinating to see how the various paradoxes as outlined in this chapter keep opening up the industry to intervention, transformation, and change.

Above all, making media professionally is a form of affective labor: work that elicits an affective investment from its practitioners exceeding conscious deliberation and that is intended to evoke a similarly emotional response among audiences. In an industry that does not produce nor sell goods that we need for our survival, it seems senseless that it relies so much on deep and prolonged engagement among its producers as well as consumers. Why should we care about anything that is going in and with our media? Perhaps it is exactly our lives as lived in media that explain the industry-wide shift from audience orientation to engagement, from mass markets to individual-level targeting, from medium-specific and neatly packaged products to complex, immersive storyworlds that stretch across multiple media, as media quite clearly have become so much more to us as the nonutilitarian goods and services they were seen as providing in the past.

# 8

## Life in Media

Using, making, and studying media have a "dark" side, as much as there is a light side to media. Articulating a good life in media therefore includes an exploration of what makes media problematic, as well as how it brings so much joy. The question is how we can study, understand, and critique media when we are living in media. The answer can be found both in a famous advertising campaign and in the philosophy that inspired a motion picture franchise: think differently.

In August 2021, the American science fiction action-comedy film *Free Guy* premiered, after being in different stages of production for five years. The movie stars the Canadian actor Ryan Reynolds as Guy, a bank teller who gradually finds out he is not human but a nonplayer character (NPC) in a massively multiplayer online game. Embedded in his programming is a piece of code helping Guy to gain consciousness—allowing him to adapt to changing circumstances with character development based on machine learning, making decisions that are not intended by the game (nor the corporation that owns it). Without necessarily wanting to become a hero, Guy ends up rallying his in-game friends to resist attempts by the studio that runs the game to shut their world down to delete unruly NPCs. In many ways, *Free Guy* is a twenty-first-century version of the 1998 film *The Truman Show*, where the Canadian actor Jim Carrey plays Truman Burbank, who lives in a lovely small town called Seahaven, unaware of the fact that he is the star of his very own reality television show. The television show is Truman's life, from birth. His family, neighbors, colleagues, and friends are all actors. Slowly but surely, Truman realizes that he is imprisoned and must flee. Beyond the difference of setting in a TV show or a game world, the films primarily diverge on their solution for the life in media their main characters are experiencing: either opting to escape or fighting to stay.

In a not-so-subtle intertextual reference regarding their names, the characters of Truman and Free Guy rely on their individual ability, as the only "real" people in their respective worlds, to figure out whether all the other people are authentic (or to what extent they act or are programmed) and how far their lives can stretch within the studio or game world. In both cases, their confrontation with what amounts to the limit of their existence

is symbolized by a beach and whatever lies beyond the horizon of this body of water. Love plays a powerful role in both films too. For Truman, his love for an American girl called Sylvia (played by the British actress Natascha McElhone) ignites an interest in his surroundings and what may be possible. Sylvia is an actress in Truman's reality show who plays Lauren Garland, a girl from his high school. At one point in the film, Sylvia tries to tell Truman the truth about his reality (quoting parts from the movie transcript) as they make a run for it to the beach surrounding Truman's hometown:

LAUREN: Listen to me. Everybody knows about it—everyone knows everything you do. They're pretending, Truman. Do you, do you understand? Everybody's pretending.

TRUMAN: Lauren, I don't know what . . .

LAUREN: No, no, no, my name's not Lauren. It's Sylvia. My name's Sylvia.

LAUREN: This . . . it . . . it's fake. It's all for you.

TRUMAN: I don't understand.

LAUREN: And . . . And the sky and the sea, everything. It's a set. It's a show.

TRUMAN: What's goin' on? I really would like to know what's going on!

LAUREN: Get out of here. Come and find me.

The forbidden love—according to the show's creators, Truman is supposed to fall in love with another actress, who he subsequently marries—prompts a lifelong search for Lauren and what happened to her. This is one of many ways in which the film introduces an element of uncanniness in Truman's life, an eeriness that he simply cannot shake off. In *Free Guy*, Ryan Reynolds's character begins to diverge from his rigid NPC script when he falls in love with Molotovgirl—the American avatar of the game developer Millie, played by the British actress Jodie Comer. For the remainder of the film, she acknowledges the thoroughly uncanny feeling that she is falling for Guy, despite the fact that he is an NPC, at one point sighing: "The only non-toxic guy I meet is a robot."

Both *Free Guy* and *The Truman Show* somewhat self-consciously appeal to our experience of living in a comprehensively mediated environment. Our world is similarly characterized by pervasive and ubiquitous media that we are constantly and concurrently deeply immersed in, that we seem to be the stars of, and that influence and structure all aspects of our everyday life. The twin powers of love and the uncanny that provide so much of the fuel that propels the narratives of these reality-within-a-reality movies are also what gives our lives in media energy and purpose. We clearly love (our) media, and our subsequent immersion in media generates continuous experiences of uncanniness—corresponding with a sense of unreality—about the nature of our existence. Importantly, in this world it is also up to each of us to navigate the largely unwritten rules and hidden passages of an ocean of media on our own—a personal charge that seems to be unattainable by most, if not all, of us. Our desire to project a somewhat authentic, whole self gets propelled as much as frustrated by our mediated environment. Media extend and amplify our originality as well as make the project of being a true person forever unrealizable—always chasing the shareable moment in any experience, trying to delete (or retrieve) some particular personal data, and constantly updating and upgrading our hardware, our software, and ourselves (sometimes referred to as wetware).

---

*We clearly love (our) media, and our subsequent immersion in media generates continuous experiences of uncanniness—corresponding with a sense of unreality—about the nature of our existence.*

---

When seen as such, these completely mediatized worlds are rather bleak, conformist, and devoid of life in all its messy, unpredictable, and contradictory forms. When asked how the show can be so successful in convincing Truman that his world is real even though it so clearly features a fake reality filled with fake people, the director of Truman's reality show (named Christof in the film, a not-so-subtle reference to the divine authority of a godlike director) answers: "We accept the reality of the world with which we are presented." It is important to note the implication of this narrative, as it does not seem to be premised on a notion that Truman's world is unreal—it just means that the protagonist of such a world is unaware of being on camera at all times. The televised *Truman Show* and the game environment of *Free Guy* are just other versions of the real, carefully staged, concisely scripted and programmed, and completely mediated. Christof's explanation is taken from Plato's famous essay "Allegory of the Cave" (written as the introduction of the Greek philosophers' seminal work *The Republic*, published around 380 BC). In this story, Plato suggests that all of us mistake appearances for reality—that we live our lives like prisoners locked in a cave, unable to see the true nature of things. Eventually, one prisoner could manage to break free and see the light outside—the kind of person able to take on the role of an enlightened philosopher. Plato's conceit makes a prominent appearance in many other stories (including E. M. Forster's short story "The Machine Stops" from 1909 as discussed in chapter 4), and quite explicitly so in the *Matrix* transmedia franchise, as the human resistance to the world of machines congregates in a massive cave, only to be led out by Keanu Reeves's character Neo (a divinely inspired anagram for "One"). As such, these films seem to promote one of two options for our life in media: either to become "blue shirt guy" or any other stereotypical office or factory worker, barely distinguishable from a robot (or NPC), just going along while pushing the right buttons and following fixed programming and routines; or to join the resistance, become some kind of freedom fighter (or terrorist, depending on your point of view), and either attack, hack, or escape the system.

As discussed earlier (in chapter 4), there are many more strategies and tactics we have at our disposal when considering the long chronicle of media and society entanglement. As a reminder, such approaches can be grouped along three dimensions:

1. We can *fight* the media—either through disconnection and digital "detoxing" activities and destroying the machines around us or by "arming" ourselves with the insights and instrumental skills of media literacy.

2. We can *surrender* to a mediated existence—either by becoming players with information, utilizing the various affordances of our digital environment to get the most out of the media experience, or by truly embracing the uncanny notion that who we are is and always has been interconnected with our technologies and media.

3. We can *become* media—either by adapting carefully (if not critically) to what media want from us or by learning how to reprogram and hack the system, making it a participatory rather than imposed reality.

For Truman and Guy, these options would translate to

• figuring out how the media in their environment work and trying to escape from or take to bits the gaze of trackers, cameras, and programmers, or exploiting all the technologies at their disposal to consciously change the narrative;

• opting to live the (screenwritten, preprogrammed) mediated life to the fullest or to engage other people in the mediated environment (e.g., characters, actors, avatars) in continuous self-reflective debate about the various roles to play in this environment;

- gaining and raising awareness about all the subtle (and not so subtle) ways in which technologies in general and media in particular nudge and direct us, or find ways to completely rewrite the script or code to create a different (yet still mediated) world.

The deceptively straightforward strategy to go to war with the machines is in effect the only way out suggested by the media industry. In the practice of everyday life, all these tactics and strategies overlap, cross over, and influence each other, however deliberately or accidentally so.

It is fascinating to see, after well over a century of mass media, how our public debates as well as depictions in popular culture still do not seem to have come much further than considering media and life as an "either/or" rather than a "not only, but also" relation. In this final chapter, we cut across the historical (as much as contemporary) dichotomy between a completely mediated or nonmediated existence. Instead, I offer a perspective that synthesizes the apparent contradictions between the problematic, or "dark side," of life in media and its lighter side. We need to move to an understanding of media studies as a disposition: a powerful way of observing, analyzing, and understanding media in terms of *possibilities* rather than *properties*. This, for example, prompts us to consider media less in terms of effects, things, and what happens, instead focusing on process, practices, and what can be done. Importantly, this means that neither media nor life are given. Both are always in a process of becoming, influencing, and reinforcing each other and, through their entanglement over time, produce an endless variety of experiences, feelings, and ways of being in the world. Some of these states can be problematic, many are exciting or simply enjoyable, and almost all are quite mundane elements of everyday life and of the way society functions.

## The Dark Side (and the Light)

Sometimes it seems as if the early twentieth-first century offers a unique, singular perspective on the media, as it is rather difficult to ignore the countless speeches, blogs and vlogs, reports, news stories, moral panics, and public outcries on or about (the) media. Media are quite clearly crucial in the eyes of many, which to some extent explains why people project so much of their anxieties and fears—as well as hopeful expectations—onto them. This does not mean media are innocent or inculpable! In fact, quite the opposite should be argued: media play a pivotal part in society and everyday life, and this role is far from flawless—the impact of media is messy and inconsistent, changes often over time, and works differently for different people in different contexts. However, such a nuanced perspective on the general role of media as it affects most of us runs the risk of glossing over some distinct dark sides of a life lived in media that often only affect some of us. In what follows, I highlight three distinct areas of profound concern that should be part of any critical perspective in (media studies for) a life in media. Concurrent to this expose is an exploration of what is or can be done about these various problems with media. It is important to emphasize that this juxtaposition is not intended as offering a false balance (or "both-sidesism" as it is sometimes called in journalism), somehow suggesting that these are both equally powerful developments. In fact, both the dark and light sides of the life in media equation need work (from people and physical infrastructures alike) to become apparent. What highlighting the tension between problems and their potential solutions offers us is a perspective on agency, on the hopeful possibility of doing something about it.

The three key problematic aspects of life in media, and their prospective solutions—structured from the moment media are born to when they die—are

1. The *environmental impact* of (mining for and discarding of) media and efforts to green the media
2. The *workers' rights* aspect of precarious conditions in electronics factories, hardware and software development and engineering, and media production facilities versus corporate social responsibility, fan and shareholder activism, and increased collective organizing (including unionization)
3. The *representational quality* of media (e.g., regarding issues of intersectionality in film, games, and the news and considering the dominance of platforms, surveillance, and algorithmic systems in our digital environment) versus consumer activism as well as a growing media literacy movement around the world.

Not part of this list is the global concern about the impact of disinformation campaigns, propaganda, and fake news. I have to admit that, from a media studies perspective, we have to ask serious questions about the presumed consequences of disinformation on the way people live their lives or on how society functions. First, the research consistently suggests that most people read, hear, or see relatively little disinformation in the media they use. Second, people use multiple media concurrently, and it is doubtful that all the media in their information diet transmit the exact same messages (that in turn are all understood and acted upon in the same way). Third, the fact that so many people espouse suspicion, criticism, or even fully fledged conspiratorial thinking online on just about anything may be less of a consequence of media life and more a window onto a world full of gossip, hearsay, and partial truths that is typical of discussions in the neighborhood café and sports club, only now amplified online for the world to see. The expectation of humanity to come to consensus about reality and truth is problematic to begin with, and concerns about disinformation tend to gloss over this. Of course, this does not mean there is nothing to worry about, and clearly there are circumstances where a dispute over truth as accelerated and amplified in media has dire consequences for the people involved—the genocide of the Rohingya people in Myanmar in 2017 and the suffering of Ukrainians during the Russian invasion of 2022 come to mind, as well as mounting evidence that the spread of misinformation about COVID-19 and its vaccines caused many preventable deaths. What makes disinformation problematic in media studies is not so much the fact that some people sometimes fall for conspiracy theories or half-baked truths but rather the ways in which certain people, narratives, and ideas find their way into media and in turn how we can digest and understand these stories. Chapter 4 speaks extensively to the issue of media and the unreal, whereas the discussion on the representational quality of media below addresses the ways in which we can appreciate and give meaning to the media in our lives. Interestingly, the impact of disinformation could be far greater on those professionally responsible for providing people with the truth—journalists—than the average media user. Journalists tend to fall victim to propagandists who skillfully exploit their professional news values such as reporting on both sides of the story and staying neutral in stories involving conflict. In doing so, reporters and editors of news organizations can unintentionally become pawns in the game of influencing and manipulating public opinion over time. Yet again, we have to stay mindful about overemphasizing unidirectional impact or effects news coverage may have on people, always returning to the most fundamental research question of what someone is actually doing with (their) media.

### Environmental Impact

A significant proportion of the world's metals—such as tin, cobalt, palladium, copper, silver, and gold—are the basic elements that make up our smartphones, computers, cameras, and televisions. Much of these metals is mined in developing nations. Cobalt provides a case in point. It is an essential mineral used for batteries in, for example, laptop computers, tablets, and smartphones. Most of the world's cobalt is produced in the Democratic Republic of the Congo (DRC) in Africa, often by small-scale mining operations that lack the kind of (international) oversight and control large conglomerates to some extent adhere to. Copper, another key element for consumer electronics, is primarily mined in Chile in Latin America. Because of its significance to the economy, the copper industry has been left largely deregulated, resulting in little or no oversight regarding its environmental impact. The life cycle of mining for minerals such as cobalt and copper is damaging its surroundings, as it involves setting off explosions to break and split off rock blocks (a process called blasting or crushing), releasing all kinds of polluting elements into the environment, and using massive amounts of water, electricity, and oil in the mining process.

Once media devices find their way into our hands and homes, their carbon footprint primarily consists of electricity use and how anything with plugs, cords, and electronic components get disposed of and turned into electronic waste (or e-waste). Although the power consumption of a single household may not amount to much, media industries are historically known as heavy users and polluters—consider, for example, poisonous solvents, inks, fumes, dust, and wastewater as the byproducts of printing books, magazines, and newspapers and the manufacturing of film stock, as well as the massive destruction of trees and usage of substantial amounts of land and water resources to print, produce, and distribute media. In today's digital environment, we additionally have to consider the

environmental costs involved with powering and cooling the vast data centers that connect us with the digital products and streaming services offered by the media industries, especially regarding the top three cloud computing companies in the world: Amazon Web Services, Google Cloud Platform, and Microsoft Azure.

To offer an example of how sensitive ethical and environmental issues are to the companies directly involved, we can consider the fate of Timnit Gebru, co-leader of a group at Google that studies the social and ethical ramifications of artificial intelligence (AI). She was let go by the company in December 2020 after coauthoring a paper (at the request of Google) about the downsides of relying on large language models, which are AI trained by using gigantic amounts of text data. Gebru is an Ethiopian American computer scientist much celebrated for her critical work in facial recognition—showing that it is less accurate at identifying women and people of color, which means its use ends up discriminating against them. She also cofounded (with fellow Ethiopian computer scientist Rediet Abebe) the Black in AI network in 2017 to champion diversity in the tech industry. Gebru outlined several key problems for the kind of AI that companies like Google use, focusing on, among other issues, the extraordinary carbon footprint of huge data centers to perform the kind of calculations necessary to run such programs. Managers at Google clearly felt this paper did not focus enough on its efforts to become more energy efficient and demanded Gebru and her colleagues change their research accordingly, which she refused. Thousands of Google employees ("Googlers") and outside experts signed a public letter in support of the paper and Tinmit Gebru's work, to no avail. Gebru went on to found and direct the Distributed Artificial Intelligence Research Institute (DAIR) to continue her work. Concern about the water and electricity use of data centers also plays an important role in the field of *blockchain, cryptocurrencies,* and AI more generally, as all these applications need a lot of computing power. As a side note, blockchain technology also gets touted (by organizations such as the World Economic Forum) as a tool that could help in tackling climate change, for example, by enabling people and companies to respond almost in real time to changes in data patterns about weather conditions.

Electronic waste is one of the fastest growing waste categories in the world. The amount of e-waste doubled in the first decade of the twenty-first century and shows no sign of slowing down. This development gets spurred on by the ever-increasing popularity of media, rising income and literacy levels around the world, increased bandwidth and connectivity, and technological changes. We adopt new media as fast as we discard them, in part because our television sets, personal computers, stereo systems, and smartphones tend to be designed for a limited life span—in industrial terms called the "planned obsolescence" of consumer commodities. Other factors that play into the short life cycle of media are product failures, changing customer needs, and technological transitions (such as from VHS to DVD and Blu-Ray, from one generation gaming console to the next, or from 2G to 3G and onward in mobile connectivity). Our media are also quite complicated to maintain, fix, or repair, as the different parts are often located within glued-shut plastic enclosures, warranties are voided if we attempt to break open our devices, and support for outdated media quickly gets discontinued. What is furthermore striking about our media refuse is that most of these devices are still working (or have many working parts) when we throw them away.

What makes e-waste unique is that it is a hazardous waste that also has significant economic value through the recycling or recovery of valuable metals (e.g., copper, gold, silver, and palladium). As a result, it is traded between the developed and developing world—mostly without any kind of scrutiny or oversight, as the vast majority of e-waste is collected and dumped without any formal documentation. Our unwanted devices generally end up this way in e-waste graveyards across Ghana, India, and China. Agbogbloshie, outside of Ghana's capital city Accra, is the world's largest e-dump, where people—including many children—make a living by taking apart and burning down electronics. The Chinese city of Guiyu and Moradabad (just east of New Delhi) in India have similarly become the main sites of a lucrative media scavenging industry, operating largely beyond the purview of regulators and without much in the way of safeguards for people working there. All these irregular recycling activities are not keeping pace with the global growth of e-waste, in part because of the dangerous nature of the recycling process.

In recent years, efforts have been made by international nongovernmental organizations, governments, and corporations to improve working conditions in mines, manufacturing plants, and recycling industries around the world. A significant effort is underway to make the water and electricity use of the industries involved more sustainable, for example, by switching to renewable energy sources. A wide variety of stakeholders invest in formalizing and standardizing the mining of metals and minerals as much as the recycling of e-waste to both protect workers and reduce harm to the environment. For example, in the Democratic Republic of Congo, the government strives to become the only legal buyer from miners in the informal sector—both to capitalize on the international demand and to end unsafe working practices. International firms active in the region—such as the Swiss-based Trafigura and Chinese cobalt-processing firm Huayou—are furthermore involved in formalizing mining operations to protect workers from accidents and prevent child labor. In Chile, mining corporations such as Anglo American have started using green-hydrogen generators, reusing water and solar energy in an attempt to become carbon neutral. In the production of devices, companies like Apple, Amazon, and Samsung conduct annual audits of the factories they contract to assemble their iPhones, Alexas, and Galaxies, publicly demanding action on any irregularities found. One of the largest such manufacturers is Foxconn (otherwise known as Hon Hai Precision Industry), headquartered in Taiwan and operating electronics manufacturing factories mainly in China, Brazil, Mexico, India, and Malaysia—but also in Hungary, Slovakia, the Czech Republic, Turkey, South Korea, Japan, and the United States. Especially since a series of prominent news reports and documentaries showcasing the poor working conditions at its factories in China in 2006 (and continuing in the years after), the company has been under increased scrutiny by both its corporate clients and government oversight committees. This in part accelerated the company's plans to automate the work in its factories.

It is important to appreciate the significant ways in which local (informal) economies, organizations, and individual people appropriate, make inventive and creative use of, and repurpose materials in the process of mining, developing, and assembling our media. The people who engage in recycling and reusing electronics often do so to pursue a viable and valued trade in affordable secondhand computers, empowering many local businesses as well as artists. An example thereof is the Agbogbloshie Makerspace Platform (AMP), a youth-driven project to promote maker ecosystems in Africa, starting in Ghana—receiving international recognition for its work to help coordinate and organize the many local recycling, making, sharing, and trading initiatives. AMP was launched in 2012 by the Ghanaian materials scientist Kwadwo Osseo-Asare with the French architect and designer Yasmine Abbas, aiming to create and support an "open architecture for crafting space" that includes modular, prefabricated kiosks, kits, and a mobile application for makers. Among its many inspired projects: building a prototype spacecraft.

Within the corporate players in this worldwide system, there is widespread collective organizing and worker resistance—including at the various Foxconn factories around the world. Quite often mining operations sustain the livelihoods of entire communities through several generations, where people are critically involved in managing the process. The people in Ghana, the DRC, Chile, and elsewhere cannot and should not be reduced to passive victims of exploitation. Of course, state repression, corporate control, and long histories of operating in the shadows complicate the agency of the people involved, but it is crucial that we ask the same questions of the mining, assembly, and design of media that we ask of media production and consumption more generally: What is exactly happening, what are people precisely doing, and how does all this make a difference?

A second key insight from an environmental exploration of life in media for the student, scholar, and ardent user of media points to our shared responsibility to take apart and look inside the black boxes of the technologies and electronics that we all love so much. Exploring the building blocks and materials that make up our media does not just help to demystify their inner workings—tracing the various histories of these components helps us understand and appreciate how all new media contain versions of older media (see also the definition of media as outlined in chapter 2). Such "unboxing" and historicizing of media is at the heart of *media archaeology*, which seeks to unpack the often erratic and idiosyncratic technological pasts of specific media. Tracing the genealogy of media—in terms of their materials and components, the production cycle (see chapter 7), and their genres, formulas, and conventions, as well as the ways in which people use and give meaning to them—is one of the foremost ways in which we can take responsibility for the media we use and love so much.

### Workers' Rights

Working in the media at all stages of the life span and product cycle—from mining to manufacturing, from hardware design to software engineering, from ideation to production, from promotion to distribution, up to and including community management and audience engagement—can be stressful, precarious, and even hazardous. To some, this is one of very few ways to make a living. Others consider working in the media as a dream job and would make media even if they weren't paid to do so. In some ways, the miners digging up cobalt and copper, the software engineers coding and building websites and apps, the countless professionals creating media content and experiences, and social media creators and entertainers are in the same league of generally precarious working circumstances with little or no regard for their rights as workers. Of course, we cannot compare the situation of underage minors in the DRC with that of someone working for an advertising agency in the center of a major European city. What is possible, however, is to appreciate the significance of workers' rights at every step of the media lifecycle. At all ends of the broad spectrum of work related to media, the job can be dangerous to your physical and mental health, for a variety of reasons:

- people generally work in largely informal circumstances, generally unregulated and without clear policies and standards (both regarding bodily and social safety);
- the transforming technological context of the work expects workers to constantly learn and adapt to new requirements, skills, and procedures;
- the work tends to be (physically, cognitively, emotionally) involving and demanding, for example, leading to long working hours and having to be "always on" to keep going and make it work;
- jobs are few and far between, often without formal benefits (such as sick pay, medical or legal protections, and scheduled time off); and
- the labor market and culture at work can be quite competitive, high-strung, and full of conflict, characterized by looming deadlines, intense schedules, and pressured productivity.

As mentioned before, workers in the cobalt and copper mines in Congo and Chile (and elsewhere where gold, silver, and other metals and minerals necessary for our media devices are extracted from the earth) are vulnerable to the hazardous and claustrophobic conditions of mining operations and suffer from lung disease, heart failure, and cancer from exposure to the toxic materials used. Similarly, the informal economy that governs much of the recycling of e-waste means that people inhale toxic fumes (from burning down computers and other devices to retrieve precious materials), causing chest pains and persistent headaches, and have little or no access to proper medical treatment. Workers on e-waste sites can experience decreased lung function, skin disorders, and gastric diseases that cause cramps and liver damage.

Reports on the working conditions at the factories of Foxconn and other manufacturers are similarly distressing, documenting excessive overtime work, unhealthy and unsafe working conditions, use of student labor, a militarized disciplinary regime, and workers' suicides. When people get a job at one of Foxconn's factories, the preface of their copy of the Employee Handbook states: "Hurry toward your finest dreams, pursue a magnificent life. At Foxconn, you can expand your knowledge and accumulate experience. Your dreams extend from here until tomorrow." Behind this morale-boosting language stands

Life in Media

the managerial philosophy of Foxconn founder and CEO Terry Gou as expressed in a collection of quotations his managers and workers are required to learn:

Successful people find a way, unsuccessful people find excuses. Growth thy name is suffering.

A harsh environment is a good thing.

Obey, obey, and absolutely obey!

Execution is the integration of speed, accuracy and precision.

As Foxconn (and other companies like it) expand their operations around the world, similar labor conditions and managerial practices emerge in and around the factories it builds or takes over. The role of unions in all this is limited, as unionization is generally weak and the collective organization that does exist often operates at the service of the companies involved—given workers' dependence on these kinds of menial jobs to survive and provide for their families.

Once (digital) media devices leave the factories, they have to be programmed. The operating systems, games, and applications we all use daily are developed and programmed by software engineers. Since the 1980s, reports on the long-term effects of working conditions in software development consistently show that the same conditions that make the work pleasurable—working in teams of like-minded individuals, being creative and building something from scratch in a dynamic environment, meaningfully contributing to a product that people use around the world—also give rise to the more frustrating aspects of such work: not getting along with colleagues (and poorly managed group work), constantly having to learn new technologies (such as specific programming languages and accommodating new hardware standards), experiencing intense pressure to meet high demand. Remote working, already well established in this industry, has become increasingly common throughout the software industry (in part because of the coronavirus crisis), introducing additional stress and burnout factors, such as uncertainties about expectations in the absence of coworkers (to ask questions of or to get direction from) and a blurring of boundaries between work and home, coupled with an inability to switch off.

Although the role of unions is traditionally weak in the new media industries, coders, programmers, and developers do form informal networks, within countries and internationally, as the tech workforce is spread worldwide. An example would be a form of collective organization among employees of a single company, such as Google (as a subsidiary of the Alphabet conglomerate). Tensions within the company between employees and management have grown over such issues as the company's decision to censor its search engine for the Chinese market, its collaboration with US military and law enforcement on surveillance projects, and more generally the way the company struggles to provide a safe and supportive working environment for women and people of color (the case of Timnit Gebru comes to mind). This culminated in a walkout of more than twenty thousand employees in November 2018. Subsequently, workers at Google started to formally

organize, forming the Alphabet Workers Union in 2021—invoking the concluding statement in Google's internal Code of Conduct: "And remember . . . don't be evil, and if you see something that you think isn't right—speak up!"[1] This was followed by a global union alliance called Alpha Global, composed of thirteen different unions representing tech workers in such countries as the United States, United Kingdom, Ireland, Sweden, and Switzerland. In the United States, the national union of the Communications Workers of America launched a Campaign to Organize Digital Employees (CODE) in 2020 targeting tech and video game workers.

One of the key contributing factors in the hurdles to organizing in the information and communication technology sector is an aspect of labor relations that strikes at the heart of all work in the media industries: the fact that many, if not most, professionals are contractors, not employees. This means that the majority works on a temporary basis, moving from project to project within and across multiple clients and companies, therefore often lacking any real power on the job. Furthermore, as the production pipelines of both the tech and media industries stretch across the globe involving countless smaller businesses,

---

1. Google, "Code of Conduct," last updated January 24, 2022, https://abc.xyz/investor/other/google-code-of-conduct/.

offices, and studios, the workforce is fragmented. Historically, management of both multimedia conglomerates and tech companies alike resisted efforts to organize—resistance that in part explains the industry's reliance on outsourcing labor, temporary contracts, and atypical working arrangements. It is fascinating to see how there seems to be real momentum behind efforts to collectively organize—not just in the tech industry but also in the media more generally.

In the advertising industry, a parody video produced by the Canadian ad agency Union Creative after it was shortlisted for an agency of the year award (in 2014) went viral, making headlines all over the world. The video poked fun at the work-life balance of its employees, suggesting that the company's owners celebrated the nomination by letting their employees see their families—if only for a few minutes. As they mockingly conclude the video: "At Union, family and friends come first. After clients and revenue. And awards." Interestingly, much of the coverage this video got in the trade magazines and social media applauded the way Union managed to successfully self-promote its commitment to the work. As one industry observer wrote at the time: "I would venture to guess that every single employee at Union would agree that they don't see their families as much as they'd like."[2] The workaholic mindset parodied in the video features prominently throughout the marketing, advertising, and public relations field. In a 2018 industry-wide survey, two-thirds of professionals in British advertising and marketing said they considered leaving the industry at some point due to work negatively impacting their well-being. A series of industry surveys between 2018 and 2020 conducted by Mentally Healthy—an initiative by people in the Australian creative, media, and marketing industries—documented that about half of the participating practitioners struggle with depression. Aspects of the work that these media professionals felt positively affected their mental health and well-being yet were reportedly lacking in their careers included the following:

- a variety of tasks,
- learning new things,
- not doing the same things over and over, and
- having decision authority on when and how to work.

Beyond the intensity of the workload, the global advertising and marketing field as organized in the World Federation of Advertisers (WFA) is acutely aware of the need for its industry to tackle enduring problems regarding diversity, equity, and inclusion. In 2021 the WFA took its first global census (among professionals across twenty-seven countries) on these matters by asking about people's perceptions of fairness and sense of belonging. A third of respondents reported feeling stressed and anxious at work. The most reported

---

2. Will Burns, "Ad Agency, Union, Proves Its Usefulness in Self-promo Video," *Forbes*, November 7, 2014, https://www.forbes.com/sites/willburns/2014/11/07/ad-agency-union-proves-its-usefulness-in-self-promo-video/?sh=18db2b6d1b98.

forms of discrimination were on the basis of age, followed by family status and then gender. As the WFA indicates, the intersections between these different facets of who people are provide the most damning results in their census, as the professionals who report the lowest sense of belonging in the industry tend to be people, particularly women, with disabilities and an ethnic minority background.

The good news from these and other reports out of the industry is that there is much more understanding and appreciation of stressful working conditions and mental health problems than in the past. Some agencies and networks implement such measures as a dedicated employee hotline to report issues, rolling out in-house mental health programs and employing counselors for one-on-one consultations. Especially during the pandemic, agencies around the world have accelerated these and other initiatives, although there is also some backlash to such offerings as mindfulness and yoga sessions as not addressing the underlying structural problems of excessive overwork, lack of autonomy, and overall precarious working arrangements.

Similar stories can be told about practitioners in film and television, as well as among those in music and recording and digital games. The Film and TV Charity—a British nonprofit fund—started a 24/7 support phone line in 2018, followed up by an industry-wide survey on mental health and well-being. The findings amount to a mental health crisis in the industry, according to the organization: out of well over nine thousand participating professionals, 87 percent experienced a mental health problem, and more than half at some point considered taking their own life related to stress at work. These numbers are even higher for professionals working freelance—who make up the majority in the media industry—and those who identify as BAME (Black, Asian, and minority ethnic; a controversial UK designation for nonwhite minority communities), LGBTQ+, or disabled. This crisis is caused by what Alex Pumfrey, CEO of the Film and TV Charity (and former digital strategist and strategy director at several British media companies), considers as "three Cs" of the media industries: *conditions* of work, the industry's *culture*, and its *capability* to provide support for those who need it. Specifically, the worst aspects of working conditions include unusually high work intensity, little or no work-life balance, and ever-tightening budgets. The industry's culture can all too often be characterized by workplace bullying and sexual harassment (experienced by the majority of women in the industry), with practitioners feeling disposable and expendable on the job, and an expectation to always be tough and "suck it up" to succeed. Thirdly, the industry is seen as lacking the capability to recognize when and how its people are in distress and offering little in the way of opportunities to discuss (let alone meaningfully address) mental health and well-being at work. These findings are not particular to the situation in the UK—earlier well-being studies among film and television professionals in other parts of the world show similar findings, which, for example, prompted the Media, Entertainment and Arts Alliance (Australia's largest union for creative professionals) in 2016 to start its Equity Wellness initiative, including mental health first-aid training. Several companies

Life in Media

in the field offer counseling sessions, giving employees free access to digital meditation and health-related smartphone apps, such as Headspace and Maven, and schedule events around mental health and wellness.

The International Game Developers Association (IGDA), a nonprofit organization of professionals involved in digital game development (founded in 1994), has had its fingers on the pulse of game workers for a long time. A pivotal moment for the industry was the so-called EA_Spouse controversy in November 2004. The American game developer and writer Erin Hoffman posted an anonymous blog to LiveJournal, sharply criticizing the labor practices of Electronic Arts (EA, one of the largest game publishers in the world), as experienced by her fiancé, EA employee Leander Hasty. Her post opens: "EA's bright and shiny new corporate trademark is 'Challenge Everything.' Where this applies is not exactly clear. Churning out one licensed football game after another doesn't sound like challenging much of anything to me; it sounds like a money farm. To any EA executive that happens to read this, I have a good challenge for you: how about safe and sane labor practices for the people on whose backs you walk for your millions?"[3]

Hoffman's post went viral, leading to three class-action lawsuits against EA and some changes throughout the industry at large, especially raising awareness about the treatment of entry-level workers and the expectation of extreme overwork (known as crunch, as discussed in chapter 7). Around the same time, the IGDA launched its first worldwide Quality of Life survey to gain more understanding of the issues that affect working life as a game developer, later on followed by a diversity survey. In subsequent years, these studies were combined and expanded into a regular Developer Satisfaction Survey among its members around the world. The various reports suggest that game developers experience a high degree of employment volatility regardless of whether they are contracted, freelance, or self-employed. Long working hours are part of the job for most, with over one-third feeling the pressure to crunch. Overtime remains poorly compensated, as only very few get paid overtime. Interestingly, between 2019 and 2021 union membership doubled, with the majority saying they would vote in favor of a formal union at either their company or for the sector at large, indicating a growing desire to organize. Overall, the IGDA reports suggest a fast-growing awareness throughout the industry of the pressing need for more equity, diversity, and inclusion: "Many respondents (74%) felt that there is not equal treatment and opportunity for all in the industry. In addition, 56% of respondents perceived inequity towards themselves and 71% perceived inequity towards others based on gender, age, ethnicity, ability, or sexual orientation."[4]

---

3. Ea_Spouse, "EA: The Human Story," LiveJournal, November 10, 2004, https://ea-spouse.livejournal.com/274.html?page=59.

4. Johanna Weststar, Shruti Kuman, Trevor Coppins, Eva Kwan, and Ezgi Inceefe, *Developer Satisfaction Survey, 2021: Summary Report* (Toronto, Canada: International Game Developers Association, 2021), https://igda-website.s3.us-east-2.amazonaws.com/wp-content/uploads/2021/10/18113901/IGDA-DSS-2021_Summary Report_2021.pdf.

Consider this headline from a news story in November 2020 from the Society of Editors (representing about four hundred editors and directors from a variety of news media in the United Kingdom): "77% of journalists suffer from work-related lockdown stress, mental health survey finds."[5] Reports on significant challenges to the mental health of journalists have been paramount—not just during but also before the global coronavirus crisis. We have to consider the consequences of decades of waves of lay-offs, ongoing newsroom restructuring, and the rapid rise of atypical working arrangements in journalism. The risk of being a reporter in part stems from often covering emotionally laden events, conflict, and trauma—factors external yet intrinsic to the work of journalists. The hazardous nature of journalism also comes from factors internal to the profession, given the extraordinary amount of emotional labor journalists have to do on a day-to-day basis, to manage, manipulate, and anticipate the emotions of their colleagues, their sources, an (imagined) audience, and their own feelings. As in other media industries, mental health problems do not just follow from the intense feelings and pressures that practitioners experience at work. Stress and ill effects often result from having to suppress these feelings or from a lack of mental health literacy—as in not understanding when you are (or a colleague is) in distress and not knowing how to offer or find help.

There is some movement on the front of mental health awareness and literacy in journalism—as in the other media industries discussed here. In several countries, former-journalists-turned-therapists offer workshops and counseling to their colleagues, with international professional networks emerging to support working journalists. Examples include The Self-Investigation, a foundation providing online courses, a newsletter and coaching aimed at improving media professionals' well-being, and the Headlines Network, offering podcasts, training (supported by the Google News Initiative), and an online mental health toolkit. Journalism educators have formed the Journalism Education and Trauma Research Group (JETREG), with regional research hubs across all continents, in their own words "to respond to the body of research which shows that practicing journalists are at risk for physical, emotional, moral and psychological injury due to exposure to traumatic events in the course of their career."[6] Several international organizations exist to support journalists who cover (and experience) conflict, crime, and violence. Publishers and broadcasters sometimes offer training for their employees—especially female reporters—on how to handle online harassment. In the United States, numerous newsrooms have unionized in recent years. Freelance journalists in several countries organize themselves—with many national journalism unions now counting independently working

---

5. "77% of Journalists Suffer from Work-Related Lockdown Stress, Mental Health Survey Finds," Society of Editors, November 11, 2020, https://www.societyofeditors.org/soe_news/77-of-journalists-suffer-from-work-related-lockdown-stress-mental-health-survey-finds/.

6. "About JETREG," Journalism Education and Trauma Research Group, https://jetreg.blogs.lincoln.ac.uk/about-jetreg/.

reporters as their largest member segment. Other forms of collective organization include reporters and editors joining forces in various cross-border coalitions, such as the International Consortium of Investigative Journalists (based in the United States), European Investigative Collaborations (based in Germany), the Centro Latinoamericano de Investigación Periodística (CLIP, based in Costa Rica), the Environmental Reporting Collective (with board members in Malaysia, Japan, China, Hong Kong, Taiwan, and Indonesia), and the Global Investigative Journalism Network (featuring 211 member groups in eighty-two countries).

With all this attention for the mental health of people making media, I am not discrediting nor diminishing concerns about the health consequences of problematic media use (as documented in chapters 4 and 5). A key difference is that harmful effects of time spent with media affect some people, in some ways, some of the time—whereas the various reports on the mental health and well-being of media professionals clearly suggest that work-related stress and burnout are features of the work and do not just represent an incidental bug in the system.

What is crucial about all these issues that arise from working in the media is that the professionals in these fields (are expected to) bring their "whole selves" to (the) work. This means that all aspects of who they are—their socioeconomic background, gender, ethnicity and ability, personality and character, tastes and preferences, skills, feelings, and emotions—are considered assets of their professional identity. In media work, just like all other creative professions, this results in people going all in when it comes to their projects, assignments, and jobs. As discussed in chapter 7, such intense engagement amounts to a profound paradox related to workers' rights: what makes the work most alluring—its intensity, the empowering experience of being self-expressive and creative, of building meaningful relationships with colleagues, peers, fans, and audiences—also sets up many of its workers to fail. The permanent pressure to make it work (both commercially, financially, and in terms of cultivating a reputation and professional identity), to meet constant deadlines, to manage one's emotions and those of all the other people involved in the production cycle, and to always perform can be distressing liabilities to one's mental health and well-being.

The work clearly matters to the professionals involved—and that passionate engagement in turn makes everything personal. That means that a compliment from a colleague goes a long way, just as much as any rejection (of an idea, a gig, or project) can make one feel hurt and vulnerable. This is not to say that working in the media is inevitably awful, nor is this long list of potential problems with workers' rights in media production intended to discourage anyone from wanting to work there. Quite the opposite—media work is (or can be) genuinely fun and fulfilling and at its best contributes to society and the greater good in all kinds of meaningful ways. It is perhaps not a stretch to say—with specific reference to those in creative roles - that the pain of working in the media in part makes the work pleasurable. I would argue that students and scholars in the field of media studies have a responsibility to highlight and tackle some of the more problematic issues

about working in the media—not in the least because the majority of students taking courses in media and communication are primarily interested in pursuing a career in the media. This is how the subdiscipline of production studies emerged at the start of the twenty-first century, highlighting a growing interest and awareness among media scholars about the conditions, norms, and values that structure how media get made. In line with how the media as an industry is transforming, audience studies and production studies in recent years of scholarship tend to converge in recognition of a growing interdependency between those who make media and those who consume it—such as between professionals working on a TV series and fans of the show, between developers working on new editions of a game and gamers online, between those working on a film set and the communities where they are filming. As using media to some extent also involves producing media, it seems productive to collapse the categories of production and audience in our studies of media, particularly when we include the role of platforms, online social networks, and other interactive media in our analyses.

**Representational quality**   One of the most powerful contributions that media studies make to public awareness and debate relate to the representational quality of the media—in the various ways in which our media reflect and present the world to us. Traditionally, media representation has been studied as a process, whereby those who make media professionally imbue their work with all kinds of values, ideas, and conventions, and the people who read, listen, and view media interpret and give meaning to all of this. Although makers generally try to do their best to get their imagined audience to "get it" in the way they intended, this is rarely the case—as each of us understands media and information in our own idiosyncratic ways. Media are powerful because they present the world in a specific way; makers use all kinds of skills and techniques to produce a particular version of the world, while audiences actively engage with media to fit or question their own way of looking at the world. This circular, dynamic, and generally quite unpredictable process of representation opens media to a variety of critical questions about the way in which different peoples and places are featured (or marginalized) in the media, how people act on things they learn from the media, and how all this contributes to ongoing debates about the degree of "derealization" felt and experienced as everything is mediated (see chapter 5 for an extended discussion on the *mediation* and *mediatization* of society).

While all this seems straightforward, in our digital environment the representational quality of media is subject to intense discussion. We live in a context where everyone produces media—either professionally, by people uploading and sharing content just for fun, or by everyone when going online (as this produces a digital shadow consisting of all the data you leave behind when visiting sites and using apps). To some extent, this means that we are all implicated in representing the world, in effect continuously shaping and cocreating a world in media to such an extent that we do not live *with* media but *in* media—as is the foundational premise of this book. Just as there is a struggle over authenticity in media life (see chapter 4), there can be said to be a pervasive crisis of representation in the media, a condition amplified by a surge of possible realities produced by AI software and applications (including ChatGPT, Midjourney, DALL-E 2, Synthesia, and many more). It is, for example, fascinating to observe that many, if not most, people, groups, and entire communities feel they (or their issues, hopes, and fears) are not heard or seen in today's media—even though almost everyone operates one or more profiles on social media. Many turn away from the news because it does not represent their lives, although journalists spend more time than ever before doing street interviews, quoting the tweets and other online contributions from regular people in their stories, and featuring grassroots perspectives in the news. Much of contemporary media activism (as reported in chapter 6) focuses on representation as a part of demanding real change in people's lived experience. Especially when it comes to the so-called *cage* of our identity (short for "class, age, gender, ethnicity"; sometimes used as *cages* to include "sexuality" or *caged* to add "disability"), claims about media representation are paramount yet possibly also more complex than they were in the past as we are all involved, in one way or another, in the direct mediated communication and construction of each other's identities.

What all this relates to is a fundamental shift in the development of—and thinking about—media. For much of the twentieth century, media have been mostly thought about in terms of their seminal role in the process of mass communication, referring to messages transmitted to a large audience via one or more media. Media were seen as the (technological and formally organized) means of transmission of such messages. The terminology of *mass* media and communication (originally coined in the 1920s) matters, as this shaped expectations about the power, influence, and effects of media on people. The mass audience was considered large, heterogeneous, and widely dispersed, and its members did not and could not know each other. The context for developments in mass media and communication since the start of the twentieth century has been one of rapid and constant change. It has been a time of growth and concentration of population in large cities, of the mechanization and bureaucratization of all aspects of life, and imperialist expansion (as well as disintegration) by the great powers of the time. It was also a period of profound political change, of large social movements, unrest within nations, and catastrophic warfare between states. Populations were mobilized toward national achievement or survival, and the new mass media played their part in these events as well as providing the masses with the means of relaxation and entertainment. Against this background, the concepts of mass media and mass communication were forged and rose to a dominant status as objects of public concern, which in turn inspired the first studies of the media—for example, about the purported effects of listening to the radio, going to the cinema, or the dangers of people falling for the manipulations of populist politicians and wartime propaganda. Although we now know that this notion of a mass audience never correlated with the reality of mass media—as audiences are industry constructs more so than concrete phenomena in the real world—preconceived ideas about how mass media work and affect people continue to this day. Consider, for example, the global fears about the consequences of disinformation, especially in the context of the coronavirus disease. The early days of fears about all-powerful media certainly seem to have returned, roughly a century later, in the context of our digital environment.

The history of public debate as well as academic scholarship on mass media and communication did not just produce specific ways of thinking about the impact and effects of anything mediated but also fashioned stable concepts and structures through which scholars claimed to understand how media production, content and representation works. Examples of such seemingly stable media and mass communication structures informing much research and theorizing in the field are

- media production taking place in newsrooms, within the film and television studio system, in creative agencies and game studios, organized through publishers and broadcasters, large holding firms, and multinational corporations;
- media content that is based on routinized, scripted, and formulaic industry formats and genre conventions;
- media audiences that are massively aggregated and programmed around release schedules and predictable media events (such as the Olympics, the Eurovision song festival, and a press conference by a country's political leader).

The problem or challenge in a life in media is that these three constituent structures of the mass media and communication process are increasingly unstable and become fluid, in that their constituent elements seem to change faster than it takes new structures to sediment. First, in media production, the dominant trend around the world today (as outlined in chapter 7) is the emergence of multiplatform and multichannel industry structures and value chains (within and across the media, telecommunications, and technology sectors), with production increasingly organized through atypical working arrangements as professionals combine different media, serving different clients and employers and coming up with new storytelling experiences. Furthermore, the "people formerly known as the audience" contribute in all kinds of ways to the media—either through their data or by actively cocreating content—which amplifies the confusion between the respective roles that various people and media play in the production process.

Secondly and correspondingly, content increasingly flows across different media—in part because we copy and paste media across all our accounts and profiles and in part because it deliberately gets designed to do so. This signals the rapid development of a wide variety of multimedia, crossmedia, and transmedia storytelling forms throughout contemporary media productions (see chapter 7). Online, content tends to be a quite unstable category, as stories and campaigns are often produced in multiple versions (to see which one generates the most engagement online) or are produced until further notice as more information, characters, items, and even entire story lines get added down the line. As media users, we collect, curate, remix, share, and forward content all the time, taking information and entertainment out of context and in the process collapsing and converging the various contexts within which media were made. A news story can become the basis for ridicule by turning it into a meme, an advertising campaign is the source of satire through parody, a famous movie scene turns into an inspirational sequence for some kind of activism, and a song becomes the soundtrack of a revolution.

Thirdly, audiencing today—being part of the audience for a particular media product or experience—most of the time involves *concurrent media exposure* (see chapters 1 and 2) as we use media as an ensemble (and thereby experience the digital as an environment) rather than dedicating our time exclusively or deliberately to either a television set, a smartphone, or a print magazine. On top of that, when doing so, we are not just actively interpreting and giving our own meaning to what we consume; in a digital context, we are dedicating a significant amount of effort and emotion creating our own media—in effect turning being part of a media audience into an individual act of media production. This process of mass self-communication clearly unsettles much of our understanding of media users as audiences. Seen as such, one could argue that in our current digital environment mass media and communication converge with interpersonal media and communication—although not all the time and not for everyone equally.

Beyond the ways in which industries and people disrupt age-old notions about the way to study and understand media, the contemporary context demands that we take the role of machines, computing, and data seriously as producing its own unique and distinct *mediatic* process. How we imagine, share, create, transmit, promote, and respond to media transforms under the influence of big data (notably in combination with so-called thick data; see chapter 3), machine learning, and artificial intelligence. Given that data collection (through all our mouse clicks, swipes, and keystrokes), data archiving (in countless databases), and data repurposing (through the use of statistics and algorithms) together form the governing principle of our digital environment, it is safe to say that its role in creating media, distributing content, and turning people into audiences needs to be considered carefully and critically. As we have seen in chapter 7, algorithmically driven decision-making and content-creating tools play a formidable role in the contemporary process of producing media content, albeit not breaking through and dominating to the extent that both their fans and detractors make them out to be.

The most likely explanation for the continued popularity of computer software, big data, statistical analyses, and automation in (the management of) creative work is that it touches upon a deep anxiety about audiences, the profoundly risky nature of the business, and the rapid rise of the technology sector as a source of deep influence in the media industries. All these worries get projected onto seemingly perfect machines of automation, datafication, and artificial intelligence, which in turn come to play a profound role in the process of media making.

Both individual media professionals and large corporations invest in creating one-on-one relations with audience members, using the exchange of detailed information this provides to create unique content and experiences. Journalists turn to subscription-based newsletters and social media channels as a source of income (i.e., crowdfunding), as a platform to tell stories, and as a way to crowdsource newsgathering. Many performers and creators turn to dedicated online platforms for the chance to share content directly with paying subscribers rather than offering their work through established publishers. Such more or less new business models upend the traditional taxonomy of producer, content, and consumption, as every step in the process gets governed by the exchange of complex data (in the form of money, shares and likes, comments, tips and stories, feedback and discussion, and so on), which in turn creates opportunities for new ways to tell, sell, acquire, and participate in stories. On the corporate level, such approaches get explored as well and are taken to new transmedia heights. Consider, for example, the November 2020 launch of South Korean four-member girl group Aespa by SM Entertainment. The "Ae" in their name stands for "avatar experience," as each of the band members—Winter, Karina, Ningning, and Giselle—also has a fully-fledged avatar that fans can download onto their smartphones to interact with in real time. Ultimately, the goal is to have these avatars act autonomously—based on artificial intelligence, programmed with the musician's personality—and develop one-on-one relationships with fans. The group quickly signed on as advertisement models for various industries around the world, including gaming, banking, beauty, apparel, and telecommunications, both appearing in person and giving virtual concerts and online showcases as their avatars. Every single and corresponding video released by the group furthermore continues an overall story line and virtual universe, within which fans can interact and experience the lives of these pop stars, also adding their own fan content through SM Entertainment's dedicated TikTok channel, PinkBlood.

The Aespa example suggests fascinating if not entirely unproblematic futures for media—consider, for example, the authenticity puzzle involved with relating to AI-driven stars, the role and rights of fans as they participate in furthering the narrative, the commodification of every single aspect of an entertainers' personality (including all their interactions with audiences), and the discussion about whether avatars and AI, like humanoid robots, have rights. The American game designer Raph Koster—lead designer of the award-winning massively multiplayer online game *Ultima Online* and founder in 2006 of the virtual world software platform Metaplace—published one of the first formal articulations of the rights of avatars in 2000. In this influential declaration, Koster starts from the notion that avatars are equal in rights, as they are manifestations of actual people. He furthermore assumes: "The aim of virtual communities is the common good of its citizenry, from which arise the rights of avatars. Foremost among these rights is the right to be treated as people and not as disembodied, meaningless, soulless puppets. Inherent in this right are therefore the natural and inalienable rights of man. These rights are liberty, property, security, and resistance to oppression."[7]

While Koster clearly sees avatars as extensions of humans in the context of a virtual world and therefore equal in rights, the overall and ongoing debate about granting rights to avatars, AI, and robots tends to be less about whether these technologies and machines are the same as humans, instead focusing on their continuing evolution, increasing complexity, and growing significance in the way society and everyday life function. Consider, for example, developments in increasingly autonomous drone warfare, in algorithmic operations that become opaque to their programmers and managers (such as in the PayPal example described in chapter 3), and the accelerating climate cost of the tech industry as the key sector that designs and programs automation and AI applications. We can and should expect and stimulate more critical and ethical discussion about rights and responsibilities to evolve particular to emerging hybrid and automated forms of life in media.

While media studies, as a field, contributes in significant ways to discussions about representation in today's complex media landscape, it sometimes seems as if the traditional ways in which we would classify, organize, and interrogate this fast-moving digital, datafied, and virtualizing world may not be all that useful or applicable in the context of a life in media. This real or perceived knowledge gap opens up exciting opportunities for new sense-making approaches, for instance, in the development of new forms of digital, media, and information literacies. Especially since the 2010s, media and information literacy moved to the forefront of debates in political and societal circles about how to

---

7. Raph Koster, "Declaring the Rights of Players," Raph Koster's Website, August 27, 2000, https://www.raphkoster.com/games/essays/declaring-the-rights-of-players/.

handle—and what to do about—our all-encompassing digital environment. Media (or, more specifically, digital) literacy is considered by many stakeholders (including schools, political groups, nongovernmental organizations, foundations, and academics) as a panacea—a general solution for a variety of troubling issues associated with our lives in media. In recent years, significant resources have been pushed toward a variety of media and information literacy (MIL) initiatives around the world. A 2016 mapping report of MIL practices and actions across the European Union by the European Audiovisual Observatory, for example, identified close to a thousand funded organizations, groups, and programs spread across civil society, public authorities, and academia. Similar developments can be observed in other parts of the world, leading to a bewildering proliferation of activities.

It is important to note that media literacy, both as a research discipline and a field of practice, has been around for quite some time. Several governments have consistently supported education about their national press and cinema from the 1960s onward—including countries such as Russia, France, the United Kingdom, Canada, and the United States. The United Nations Educational, Scientific and Cultural Organization (UNESCO) has promoted the development of media education since the mid-1970s, urging member states to promote a critical understanding of media among their citizens. Such programs generally morphed into more comprehensive approaches to media education in the 1980s

and 1990s. Gradually, the focus shifted from learning about the rich history and traditions of national media cultures to more instrumental approaches to media, including learning skills and competences associated with handling technological equipment (such as cameras and computers) and the ability to distinguish between different types of information (such as between news and advertising). Today, the number of programs, initiatives, and centers in some form of digital/media/information literacy, training, and education is truly astounding, prompting initiatives to come up with a global governance structure around media and information literacies, generally focusing on primary and secondary education—yet also mindful of the fact that media literacy involves a lifelong learning process, as media continually transform and evolve.

In the context of a complex and converging digital environment, some inspiring MIL interventions regarding our life in media have been made—for example, by schools in Australia, Colombia, France, Finland, Italy, Portugal, Spain, the United Kingdom, and Uruguay—animated by two related concepts: *transliteracy* and *transmedia literacy*. Transliteracy is an approach to media literacy that takes as its cue a convergence of media literacy, information literacy, and computer literacy, focusing on:

- reading, writing, and interacting with and across a range of platforms, tools, and media and

- navigating through multiple domains (including the ability to search, to evaluate, to test, to validate, and to modify information).

Transliteracy, in other words, is less interested in developing particular literacies about various media—newspapers, radio, television, film, social media, and so on. Instead, this approach engages our concurrent media exposure head on by helping people to figure out what they are doing and how to make sense of this moving between and across different media. Although each medium can be said to have its own literacy—applying a distinction between text, visual, and digital literacy, for example—a transliterate perspective involves the interaction and cross-pollination between these literacies.

Transmedia literacy adds a crucial dimension to transliteracy: informal learning and participatory practices. The central assumption here is that most of us, when we use media, do not just consume news, information, and entertainment—we in fact actively participate in the creation, curation, and circulation of it. Rather than following specific protocols on what people should know about different media, transmedia literacy includes the informal, everyday ways of using and making media as sources of learning and insight. Of particular significance is a notion of collaboration and participation, coupled with an emphasis on play, informal learning, and problem solving. Much of what people learn about media does not happen neatly organized and structured in a classroom or seminar but at home, among friends, and while simply living our lives. This intimate and social aspect of media use has to be integrated with how we study and understand media if we are to make much headway in appreciating what life in media truly means. Media, in other words, are not just technologies, uses, and representations—media are also feelings, emotions, and experiences. As argued quite forcefully in chapter 5, we love media.

In conclusion, some of the most pressing challenges for media studies as outlined here—the environmental impact of media, the way the media as an industry works for the professionals involved, and the complexity of media representation in a digital environment—can be met by a variety of inspiring interventions from within and beyond the field:

- *media archaeology*, opening the black box of media devices to explore their material as well as cultural history and genealogy;

- *production studies*, casting a critical eye on the way media get made, both by professionals as well as all of us as audiences, participants, and cocreators of media content and experiences;

- *transliteracy and transmedia literacy*, developing a way to learn and play with(in) a complex and converging digital environment.

### On the Possibilities of Media (Studies)

What I hope to have shown in this book is that media studies as a field offers tremendous possibilities to help us understand and appreciate the potential challenges, opportunities, and problems of a life in media. The strength of media studies and communication research lies quite possibly exactly therein that these are not disciplines in a traditional sense but rather act as intellectual trading zones where philosophy, literature studies, biology, neuroscience, psychology, sociology, anthropology, political science, and other fields meet in their common quest to study and understand the role of media in society and everyday life. The field of media and communication teaching and research, whether departing from the humanities, the social sciences, or (as is increasingly the case) from an interdisciplinary point of view, has a postdisciplinary character, as it is a permanently impermanent field of study, loosely built on the foundations of many other disciplines while never really coalescing around a more or less consensual paradigm, set of theories, or research methods. Instead of a weakness, this is a strength, making the field flexible, adaptive, and well positioned to take on new challenges such as the ones outlined here. As just about every academic discipline these days makes claims about media, I would argue that a key difference is that while these fields—including, but not limited to psychology, anthropology, political science, and sociology—see media as something that happens to us, while in media studies we tend to consider how people make worlds happen in and through media.

Throughout its history, the scholarly study of (mass) media and communication went through a handful of phases, always inspired and informed by the concomitant rise of new technologies, growing public concerns about the media, and breakthroughs in academic research. Before there were departments of media studies and communication research, scientific exploration of the media focused more or less exclusively on their perceived power in shaping and manipulating public opinion. From the early twentieth century onward, numerous studies were conducted about the potentially damaging influence of radio (in the 1930s), television (in the 1960s), the internet (the late 1990s), and social media and smartphones (the 2010s), and today it is the infodemic and the way people get bamboozled by misinformation online. In each instance, media tend to be considered as all-powerful, or at least as having an impact well beyond what can be considered good for us. A second phase of work in this field coincided with the emergence of dedicated research teams and schools for media and communication and unsurprisingly started from much more nuanced perspectives. Instead of highly influential media messages, attention shifted to the exact nature of mediated messages (using increasingly sophisticated techniques for content and textual and visual analysis) and the various ways in which audiences used, interpreted, talked about, and acted on media. While much work in Europe and North America focused on mass media, research traditions in Africa and Latin America tended to be more oriented toward grassroots and community media, including a specific interest in how people managed to resist the meanings that corporate or state-owned media had with their messages, whether this was advertising, political propaganda, or just the news.

The third phase of work in the field, developing alongside the rapid worldwide growth in mobile communication and the internet from the late 1990s onward, included production much more specifically, especially how media users could also be considered to be productive when they go online. The role of technology became more prominent in media and communication research, to some extent leading to a resurgence of the "powerful media" thesis that gave the field its original impetus. As households, organizations, and institutions become suffused with media, surely this must have some profound consequences. Given the fact that the measurable impact of all these pervasive and ubiquitous media remains elusive, the most recent phase in the development of media studies and communication research (as a postdiscipline) can be characterized as a search for individual and collective agency in the context of a comprehensive, all-encompassing digital environment.

While there is much to be said about the rather hasty and generalized nature of this categorization of the field, it is fair to assume that media studies for a life in media is predominantly conceived from the perspective of our power and agency (or lack thereof). Does a life in media inspire a new culture of autonomy and communication power now that we all participate in the process of (mass) communication, mediation, and mediatization? Or are we just pawns in the age-old chess game where large companies, multinational corporations, and dominant political forces act in ways that keep their privileged positions intact? Perhaps these questions pose an unnecessary dualism, one that does not lead to helpful answers. Instead, I would like to conclude this section with three concrete options that media studies, as a scholarly endeavor, provides to give our quest for agency some necessary impetus.

---

***Does a life in media inspire a new culture of autonomy and communication power now that we all participate in the process of (mass) communication, mediation, and mediatization?***

---

First, it must be clear from the point of view adopted throughout this book that, for us to say anything meaningful about what the consequences of (digital, social, always online) media are, we need to look at what people—as individuals, in groups, and as part of larger communities—are doing with media. This is such a straightforward notion—yet one that is not always followed in research designs nor readily apparent in theoretical constructs. Concerns about ecological validity (in the social sciences) and lived experience (in the humanities) abound as people are generally only asked about their media use, or their media use is documented in artificial settings (such as a laboratory), or the only way in which media is made sense of is how people talk about and interpret media. What is key here is to gather up-close and personal data on what people do with digital media, for example, based on the affordances of the digital (through eye tracking, data scraping, collecting log files, downloading personal account information, and so on), as well as through observation and self-observation (such as through diary keeping, reflective journaling, and a range of creative methodologies). This would add crucial nuance to claims about media influence and effects, showing how people playfully and affectively assemble and mix their media diet—enabling us to assess its nutritional value beyond panicky claims about screen time for children or an infodemic for adults. This idiographic, embodied, and experiential perspective runs throughout this book, framing the various arguments and insights about what media do to us and what we do to media.

A second consideration must be that media use and reflections on what media mean have to be seen in both historical and material context. The same media experience last week (or yesterday) may mean something quite different today, in part influenced by what device, platform, or technology someone is using, when and where they are using it, and in what context all of this takes place. As detailed at various instances in this book, history is important when making any kind of claims about our media today, especially if we take to heart the material, phenomenological and affective dimensions of media—how we experience media and what media feel like. Appreciating the materiality of media—their existence as minerals, metals, plastics, wires, cables, discs, chips, screens, paper, pixels, interfaces, bits, and bytes and so on—is equally significant, not in the least because this material context grounds our life in media in particular places, at specific moments in time, using certain devices, and reminds us of their environmental impact. A historical

and material consideration helps us to avoid the narrow viewpoint of nowism as the habitual pitfall when studying new media. It also allows us to make the role that specific media play explicit—for instance, by looking at the subtleties of swiping or seeing the world through a camera viewfinder—as it helps us to decenter the media, instead focusing more carefully on all the elements of life that play out around, next to, and beyond the devices and technologies we care about so much.

For the third possibility that media studies offers in understanding our lives in media, I would like to acknowledge the inherently (and at times radically) hopeful nature of the field. The scholarship of media studies and communication research is infused with a sense of hope for a change for the better in the present and a prospective of chances given to the future. While hope inspires the actions of most people, it has a unique character in our research and teaching, in that all our arguments about what media are, how media get entangled with (the inner workings of) society and everyday life, and why we should care about this process at all can be seen as deliberate strategies to articulate how things can or should be better—even when faced with daunting evidence suggesting some very real problems and consequences of media. I have strived to make that hopeful perspective explicit in the book by documenting what is and has been going on, while always also outlining what is or can be done about it. This is not some kind of naive celebration of balance nor relativism or optimism but comes from genuine intent to show that there always are different ways of thinking about and doing things—ways for us to commit theoretical violence in a comprehensively mediated context where things sometimes seem to be completely prescribed and predetermined by impenetrable corporations, algorithms, and institutions. There can be a radical element to this hope, as many in their research and teaching of media push for an end to surveillance capitalism, for greening the media, and for a truly diverse, equitable and inclusive media workforce and representation in content—ideals that require a fundamental reorientation of how the technology, media and telecommunications sector has traditionally operated all over the world.

Admittedly, I personally find joy in a digital culture rife with parodies, remixes, memes, and oppositional readings of commercial and political messages, take pleasure in the way people idiosyncratically put together their own unique media life, and revel in the fact that both technologies and human beings are much messier and more unpredictable than our models and theories generally make them out to be. Hope inspires looking for many possible solutions and the development of a moral consciousness that provides ample ammunition to fight whatever evil we encounter. It also propels us into action—which is why hope and love are such crucial elements to study, teach, and theorize the role of media in people's lives.

Looking at what people are doing and feeling, considering all this in historical and material context, and allowing inspiration to come from a perspective of social, even radical hope—these are the possibilities that media studies and communication research most definitely bring to a life in media. This is a point of view that goes beyond offline romanticism or seeing empowerment outside media. At the same time, it is not an ignorant embrace of all things media! It is a grounded, embodied, real-world approach to the way we are in the world, suffused by media—but neither determined by media nor completely in charge of their programming. This leaves us with a final question: How can we live a good life in media?

**The Truman Show Delusion**

In conclusion, I would like to return to *The Truman Show* and *Free Guy* cases mentioned at the start of this chapter. These are not just interesting examples of popular culture providing the playground for intellectual fantasies about life in media—both films can also be seen as offering a perspective that could be tremendously helpful for our way forward. Specifically, the films could be seen as suggesting a way to be authentic and "true" in media that goes beyond either surrendering to omnipresent media or desperately trying to escape.

During the summer of 2008, the American psychiatrists Joel and Ian Gold made headlines around the world with their diagnosis of a new condition in five of their patients, which they coined the "Truman Show Delusion" (TSD). These patients developed the delusional belief that they were the star of a reality television show secretly broadcasting their daily life. In follow-up publications and in an interview I did with Ian Gold (in 2011), the brothers suggest that the combination of ubiquitous and pervasive media and a digital culture where the boundaries between the physical and virtual world are blurring with classic personality disorders, such as narcissism and paranoia, produces this new type of psychosis. Their aim in this analysis was to show that people's mental health and well-being cannot be seen as separate from cultural and technological contexts. Changes in our environment influence and shape our mental makeup, the Golds argue, and some delusions are particularly sensitive to culture and technology. Soon thereafter their diagnosis got confirmed by colleagues elsewhere in the world, identifying three common symptoms of the Truman Show Delusion:

- people feel that the ordinary is changed or different and that there is particular significance in this (consider the discussion of the uncanny inherent to the entanglement of media and life in chapter 4);

- this is coupled with a search for meaning, trying to find out what is really going on, which in this case results in the "Truman explanation," where everyone seems to be fake (to some extent), and you are the only real person;

- an always present feeling develops of unrest and uncertainty about who you are, what is real or fake, what your role is, and how everything makes sense.

As one of their inspirations, Gold refers to the Austrian psychoanalyst Victor Tausk's 1933 paper "On the Origin of the 'Influencing Machine' in Schizophrenia" (as discussed in chapter 4), arguing that the mediatization of everyday life can clearly be deeply problematic to some. However—and this is what I asked Ian Gold during our conversation—could it be that the TSD, in a mild and nonthreatening form, in fact be helpful to us as we try to navigate and make sense of a life in media? After laughing out loud, the psychiatrist reconsidered and suggested that the TSD indeed is something like optimism: an at times useful bias toward life.

A delusion about being cast adrift in an ocean of media where finding truth, originality, and clear direction is a daunting enterprise seems quite an accurate description of what most of us feel in and about media. In a life in media, our world can certainly seem like a television studio or virtual world (as in *The Truman Show* or *Free Guy*), with the significant difference that there is no exit. There is no door that leads out of the studio, no sea to cross to get out, no final epic boss fight to win our freedom. What the characters of the true man and free guy offer us is a perspective beyond escape or surrender—a way of living in media with full knowledge of how media work, consistently claiming our own narrative and choosing to enjoy life while remaining staunchly reflective and critical about their uncanny ability to mediate, to come between whatever is going on and whomever we think we are.

Media studies for a life in media is, in other words, a way to embrace a Truman Show Delusion necessary to appreciate and effectively critique how media affect and direct us and the world we live in, while avoiding the pitfalls of seeing all-powerful media everywhere or simply advocating for finding our way out. The revolution of a life in media is that it is just another step in our evolution. Like all evolutionary moments, it adds much complexity and confusion and does not necessarily lead to a good life or a better world. For that, we need ourselves, armed with the knowledge and insights of media studies and communication research. There is no outside to media, and it is in and with media that we can make our stand for a better world. Media studies helps us to take responsibility for it.

# Appendix 1: Annotated Sources

Unlike most academic works, this book does not have any references embedded in the text. This has been a deliberate choice to keep the overall story of (media studies for a) life in media flowing. However, all the arguments and insights featured in the book are grounded in scholarly research in the field of media studies and communication research, and like all other researchers, I lean heavily on colleagues all over the world for the arguments and words I use. For every chapter, this appendix offers a rough guide through the academic literature and includes some tips for additional widely available resources where relevant. The focus is on readings that are made open access by publishers or otherwise self-archived by authors (on such services as Academia, ResearchGate, arXiv, and Humanities Commons). Although many of the insights in media studies rest on the work of giants—as in any field of study—there is a preference in the selection of materials for current work. This choice partly depends on the relatively recent life in media perspective chosen throughout the book, while it simultaneously satisfies a desire to be more inclusive and diverse when it comes to the range of authors and perspectives represented. In what follows, I start with some preliminary comments and bibliographical notes about the literature. Second, the main premise of life in media as a grounding perspective for this book is outlined. After that, key references are annotated, organized by chapter. These are the main sources of the insights shared in the text, which hopefully inspire you to read on.

During the writing and editing of the manuscript, several colleagues and scholarly friends from around the world have been so kind as to read, offer comments, and make suggestions on draft chapters, and I am deeply indebted to them: Beatriz Becker (Federal University of Rio de Janeiro), Alexandra Wake (RMIT University Melbourne), Zizi Papacharissi (University of Illinois at Chicago), Cecilie Givskov (Copenhagen), Johana Kotišová (Masaryk University in Brno, Czech Republic), Erwin van 't Hof (University of Amsterdam), Stina Bengtsson (Södertörn University in Stockholm), Svetlana Bodrunova (St. Petersburg State University), Ignacio Bergillos (CESAG in Mallorca) and Martha Evans (University of Cape Town). Overall, the book is a reflection of more than two decades of lecturing courses, workshops, and seminars on media, society, and everyday life at numerous universities, reminding me once more how lucky and privileged I am to work with students around the world.

## Disparity in the Field

There are significant calls throughout the field to de-Westernize, decolonize, and more generally to internationalize media studies and (mass) communication research. This does not just mean that we have to be critical and reflective about any sources, cases, and examples commonly used in what we reference—it also assumes responsibility to actively include (native, Indigenous, universal) voices, experiences, and topics in our own work, as much as discuss and cite research published beyond the provincialism of the Western world. A second consideration is that of tackling the white, male-dominated, and heteronormative nature of scholarship in the study of media and communication throughout the humanities and social sciences. Such issues get raised time and time again in the field, as well as via social media using hashtags like #CommunicationSoWhite (inspired by the #OscarsSoWhite campaign started in 2015; see chapter 7). The composition of editorial boards of scholarly journals and the reviewer panels of academic publishers lack diversity as much as the selection of topics to be researched, the (theoretical and methodological) perspectives deployed, and among the authors cited in the papers published there. There is much work to be done here, for sure—and for established scholars as well as students, this can start by critiquing and diversifying the kind of references traditionally used in handbooks and textbooks.

For a comprehensive intersectional review, Alison Harvey's *Feminist Media Studies* (Polity Press, 2020) offers a magnificent introduction that instantly makes one aware of the profound role our gender, race, class, sexuality, ethnicity, ability, religion, and location in the world play in anything we (can) do with media in terms of access, activities, how representation works, what kind of media are produced and analyzed, and who gets to participate in media practices and research. Another remarkable element of Harvey's book is its truly studentcentric perspective, which also inspires the approach in my work. For more detail on these crucial debates in the postdiscipline of media studies and (mass) communication research, I would suggest starting with the papers listed below, offering richly documented examples of the various intersectional disparities in the field (references are organized alphabetically unless indicated otherwise):

- Chakravartty, Paula, Rachel Kuo, Victoria Grubbs, and Charlton McIlwain. 2018. "#CommunicationSoWhite." *Journal of Communication* 68 (2): 254–266.

- Demeter, Marton. 2018. "The Winner Takes It All: International Inequality in Communication and Media Studies Today." *Journalism and Mass Communication Quarterly* 96 (1): 37–59.

- Ekdale, Brian, Abby Rinaldi, Mir Ashfaquzzaman, Mehrnaz Khanjani, Franklin Matanji, Ryan Stoldt, and Melissa Tully. 2022. "Geographic Disparities in Knowledge

Production: A Big Data Analysis of Peer-Reviewed Communication Publications from 1990 to 2019." *International Journal of Communication* 16:2498–2525.

- Ganter, Sarah Anne, and Félix Ortega. 2019. "The Invisibility of Latin American Scholarship in European Media and Communication Studies: Challenges and Opportunities of De-Westernization and Academic Cosmopolitanism." *International Journal of Communication* 13:68–91.

- Sims, Yelana, and Nina Lorenz. 2021. "Looking Back, Thinking Forward: A Digital Humanities Assessment of Equity, Diversity, and Representation in Twenty Years of Publishing in Cinema Journal and JCMS." *Journal of Cinema and Media Studies* 61 (5): 1–30.

- Yep, Gust A. 2003. "The Violence of Heteronormativity in Communication Studies." *Journal of Homosexuality* 45 (2–4): 11–59.

The overall point of view in this book takes its cue from my earlier monographs *Media Life* (Polity Press, 2012) and *Leven in Media* (in Dutch; Amsterdam University Press, 2017). The new book extends this work, offering a way to look at the discipline of media studies from the perspective of media life. Beyond the books, several essays exploring the basic premise of all this work—about considering our lives as lived *in* rather than *with* media— have been published in various journals and edited volumes before (available open access or self-archived online, organized by publication year), for your information:

- Deuze, Mark. 2010. "Survival of the Mediated." *Cultural Science* 3 (2). https://www.researchgate.net/publication/263565923_Survival_of_the_mediated.

- Deuze, Mark. 2011. "Media Life." *Media Culture and Society* 33 (1): 137–148.

- Deuze, Mark. 2014. "Media Life and the Mediatization of the Lifeworld." In *Mediatized Worlds: Culture and Society in a Media Age*, edited by Andreas Hepp and Friedrich Krotz, 207–220. London: Palgrave Macmillan.

- Deuze, Mark. 2017. "No One Listens when Everyone Speaks: On the Future of Media in Media Life." *Mediapolis* 4:11–37.

- Deuze, Mark. 2021. "On the 'Grand Narrative' of Media and Mass Communication Theory and Research: A Review." *Profesional de la Información* 30(1), e300105.

- Deuze, Mark, Peter Blank, and Laura Speers. 2012. "A Life Lived in Media." *Digital Humanities Quarterly* 6 (1). http://www.digitalhumanities.org/dhq/vol/6/1/000110/000110.html.

The argument presented in this work stands, like all other scholarly work, on the shoulders of giants. It fits within a long history of considerations about the role of media as entangled, suffused, and mixed with society and everyday life to such an extent that a holistic, comprehensive perspective on media is needed to help us understand, study, and capture its influence and impact. The origins of recognizing the primacy of technology in general and media in particular as driving the development of human societies and sense-making practices can be traced to the foundational media and communication theories of the Canadian scholars Harold Innis in the 1950s and Marshall McLuhan in the 1960s. Innis attributed the characteristic features of successive civilizations throughout history to the prevailing and dominant modes of communication, each of which has its own "bias" in terms of societal form. McLuhan extended his work into the twentieth century, focusing specifically on the characteristics of electronic media in suggesting that telecommunications, radio, and television all contribute to feeling part of a global village—a world in which communication is instantaneous and distances disappear by technological evolution. McLuhan was rather skeptical about this development, predicting that such instantaneity would ensure maximal disagreement among people, as everyone's personal issues and causes would come to dominate all mediated interactions. For McLuhan, discontinuity, division, and diversity are the central themes in a world characterized by omnipresent media and real-time, instant communication. What furthermore sets McLuhan apart as an important source of influence and inspiration for my approach to media studies is his commitment to classical literature and the arts on the one hand and to engaging all the senses when making sense of—and writing about—the media on the other. Throughout this book, I have therefore sought to bring in historical examples and cases as illustrations of contemporary debates. McLuhan's sensuous approach to thinking and writing furthermore reminds us to appreciate the role of the body (or "bodymind") when using and making sense of media.

From the 1960s onward, numerous scholarly traditions emerged around the world that consider different media forms and channels of communication as an integrated whole when theorizing their impact on the development of identities, communities, and social life. The Brazilian scholar Luiz Beltrão, for example, developed a theory of "folkcommunication" in the 1960s to account for interpersonal and group forms of cultural expression preceding and existing next to mass and industrialized forms of communication, always influencing each other. The Nigerian scholar Frank Okwu Ugboajah coined a somewhat similar notion of "oramedia" in the African context (in the 1980s), emphasizing the interaction between print and electronic mass media with various forms of Indigenous media, including opera, music, dance, drama, poetry, and folktales. Around the same time, Régis Debray proposed the term "mediology" (*médiologie*) in France in 1979 as an entire field of study focused on the fuzzy zone of interactions between technology and culture. In the United States, the discipline of media ecology emerged in the 1970s and 1980s—spearheaded by Neil Postman—to theorize the media as environments within which our perception, understanding, feelings, and values are shaped. Much influenced by Marshall McLuhan and Harold Innis, American communication theorist James Carey subsequently developed an influential perspective on the intersections of culture, ritual, and belief systems with the whole of media and (mass) communication. His German colleague Hermann Bausinger similarly suggested that different media and the role they play (as content, technologies, and the ways people use them) in social life should always be considered as an ensemble. Much like Beltrão, Ugboajah, and Carey, Bausinger (1984, 351) deliberately drew folk culture into the world of technology and (mass) media, articulating his perspective beautifully: "The surrealism of our media world does not merely consist of the content of the media, but includes all the bewildering interplay of intentional and unintentional acts of deliberate and incidental actions related to the media, to people, to the environment—the whole opaque panoply of the everyday."

The American cultural theorist Lawrence Grossberg (1988, 389) also foresaw an emerging "everyday world of media life" experienced every day in terms of people's orientation to omnipresent media—a perspective outlined well before the always-on, digitally networked media of our time. Meanwhile, in Latin America, the Spanish Colombian researcher Jesús Martín-Barbero produced a rich body of work (from the early 1980s on) around a perspective of mediation, seeing culture as a process constituted out of a complex interplay between the evolution of new communication technologies, the way media industries work, and how people use and appropriate media in their everyday lives. What ties all these perspectives together for the argument outlined in this book is their intentional meshing of technological developments, the rise of (mass) media and communication, transformations in social life, and the lived experience of the everyday by people like you and me—all in appreciation that these trends, circumstances, and feelings cannot be understood separately but should be considered as mutually shaping, influencing, and interdependent. Some key early references and primary texts to explore for the origins of the life in media argument are (listed in order of year of publication) as follows:

- Innis, Harold. 1951. *The Bias of Communication*. Toronto: University of Toronto Press.
- McLuhan, Marshall. 1964. *Understanding Media: The Extensions of Man*. Repr. ed. Boston: MIT Press.
- Beltrão, Luiz. 1971. *Comunicação e Folclore*. São Paulo: Melhoramentos.
- Bausinger, Hermann. 1984. "Media, Technology and Daily Life." Translated by Liliane Jaddou and Jon Williams. *Media, Culture and Society* 6 (4): 343–351.
- Ugboajah, Frank Okwu. 1986. "Communication as Technology in African Rural Development." *Africa Media Review* 1 (1): 1–19.
- Martín-Barbero, Jesús. (1987) 1993. *Communication, Culture and Hegemony: From the Media to Mediations*. London: Sage. Originally published as *De los Medios a las Mediaciones. Comunicación, cultura y hegemonía* (Mexico: Gustavo Gilli).
- Grossberg, Lawrence. 1988. "Wandering Audiences, Nomadic Critics." *Cultural Studies* 2 (3): 377–391.
- Carey, James. (1989) 2009. *Communication as Culture: Essays on Media and Society*. Rev. ed. Boston: Unwin Hyman.

With the introduction of the World Wide Web, mobile telecommunications, and portable devices, the study of (new) media proliferated in the 1990s and 2000s, leading to a renewed interest in theorizing media as an interconnected whole. In this period, a tradition of media philosophy emerged in the German-speaking world to account for the ways in which global media and technologies of (mass) communication penetrated everyday life, changing how we think. In the Anglo-American world, interest grew in a phenomenological perspective on media, focusing on the embodied and sensual experience of media beyond a more traditional psychoanalytic perspective. Laura Glitsos's work on how the embodied listening experience to music transforms through interaction with digital media is a key reference here. This redirected attention from the realm of consciousness and reflection alone to include embodied, aesthetic, and ethical senses emerging "in the flesh" as we use and give meaning to various media. Although these schools of thought come at media from wildly different viewpoints—engaging in dialectic between a materialist and technical versus an experientialist and affective perspective—all seek to tackle the challenge of studying and understanding media as profoundly interwoven with the practice of everyday life, where the boundaries between us and our media are less than clear.

Throughout the 1990s and 2000s, numerous scholars developed a domestication perspective on media, monitoring closely how information and communication technologies get to be integrated into people's everyday life and made to fit daily practices, rituals, and social relations (such as parent-child relationships or office hierarchies) and in the process reconfigure the dynamics and arrangements between people and within particular places. In this domestication process, both the (next generation of) technologies and their users change as they gradually adapt to each other. Domestication theory, as introduced by Roger Silverstone in 1992, made its way into twenty-first-century media studies under the header of "digital domesticity" research as a powerful tool to appreciate the comprehensively mediated nature of modern life. Among the many reasons why Silverstone is such a crucial reference in media studies and for the argument in this book is his insistence for students and scholars (as fabulously expressed in his 1999 book *Why Study the Media?*) to always consider the *erotics* of our life in media. Eros is everywhere in media, Silverstone argued, as all our media and all our experiences with media inevitably contain elements that arouse us, that bring us pleasure, that inspire and seduce. We would do well to make these experiences, feelings, and sensations explicit when studying (the) media, as erotics determine so much about what media mean to us, how media affect us, and how we (can) make sense of our media (chapter 5 is deeply inspired by this point of view). In his final book, *Media and Morality* (2006, 111), Silverstone proposed the concept of an all-encompassing "mediapolis" to capture how our lives in media offer tremendous ethical opportunities to take responsibility for the kind of world we want to live in, emphasizing our agency as participants and citizens rather than mere passive consumers of media: "The mediapolis . . . signals the presence in everyday life, both empirically and potentially, of that mediated space within which as participants we confront the world, and where, as citizens, we might confront each other."

Given the ongoing digitalization, convergence, and multiplication of media in our lives, I have found much inspiration in readings that cut up and remix insights from science and technology studies, human-computer interaction, communication research, and cultural and media studies with medium theory, which seeks to understand how elements of particular media as artifacts and infrastructures shape and influence people's experience beyond the role and impact of media as texts. Of seminal importance here is the work by such prominent thinkers as Leah Lievrouw, Lev Manovich, Sherry Turkle, Donna Haraway, John Durham Peters, and Katherine Hayles (all from the United States), Leopoldina Fortunati and Rosi Braidotti (from Italy), and Paddy Scannell, Roger Silverstone, Sonia Livingstone, and Sadie Plant (from the United Kingdom), as all these authors in various ways seek to understand what happens—and what is possible—when we have to consider the agency not just of humans but also of nonhuman actors (such as computers, virtual communities, robots, artificial intelligence, and the cloud) in society and everyday life.

In recent years, elements of all these schools of thought and approaches to research have come together in the global study of media, contributing to a postdisciplinary field that is getting bolder and more confident in how it expresses itself (and articulates its significance) in the world. In their numerous publications together and alone, Nick Couldry and Andreas Hepp have done much to integrate media studies as a theory and research practice with insights from a wide variety of fields, patiently building a comprehensive framework for understanding the mediated construction of reality. Jonathan Gray and Laurie Ouellette (2017) show in their *Keywords for Media Studies* book and website how beneficial an approach to media studies beyond specific media can be, by focusing on a careful selection of key concepts and tracing, defining, and problematizing the terminology of the field (and, in passing, introducing the reader to a wonderful and diverse range of authors). In the work of these and many other colleagues around the world, some inspiring concepts emerge for contemporary media studies that, taken together, benchmark the overall approach outlined in this book:

- *mediatization*, a concept with a long history yet as a core concept recoined co-opted in the 2010s to recognize that media do not just influence society and daily life but as industries and technologies also create routines across society's institutions (such as politics, education, and the family) on their own;

- *polymedia*, introduced by Mirca Madianou (with her coauthor Daniel Miller) in 2012 to shift analyses from a focus on qualities of each particular medium as a discrete technology to a holistic understanding of new media as an entire environment of affordances;

- *transmedia work*, coined by Swedish colleagues Karin Fast and André Jansson in 2019 as an appropriate term for our management, use, and meaning making of the various interconnected media in our environment (a similar approach, focused specifically on how people give meaning to media as they move across, mix, and combine a variety of media, is coined as *intermedial studies* by Jørgen Bruhn and Beate Schirrmacher); and

- *the digital environment*, conceptualized by Pablo Boczkowski and Eugenia Mitchelstein (in 2021) to consider media as enveloping and shaping virtually all major facets of everyday life, experienced by people as an encompassing system of interconnected technical and social possibilities.

To conclude this brief discussion of intellectuals and ideas that inform and inspire the overall premise of this book, some crucial references to consider for further reading to see where the field is going are (chronologically ordered) listed below. A special note has to be made for the four edited volumes Zizi Papacharissi published with Routledge in 2018, collecting work from scholars all over the world to work through the profound changes in the way people make sense of themselves and the world in the context of our digitally as well as socially networked environment. All these books are titled *A Networked Self* (inspired by her earlier book with that title, published in 2011), respectively exploring the themes of love, artificial intelligence, platforms, and (life and) death. In all this work, she continually reminds us to keep an open mind to the always-changing, undeniably creative, and invariably inconsistent ways in which society evolves in the context of a life in media.

- Silverstone, Roger. 2006. *Media and Morality: On the Rise of the Mediapolis*. Cambridge: Polity.
- Hjarvard, Stig. 2008. "The Mediatization of Society: A Theory of the Media as Agents of Social and Cultural Change." *Nordicom Review* 29 (2): 105–134.
- Manovich, Lev. 2009. "The Practice of Everyday (Media) Life." *Critical Inquiry* 35:319–331.
- Turkle, Sherry. 2011. *Alone Together: Why We Expect More from Technology and Less from Each Other*. New York: Basic Books.
- Hayles, Katherine. 2012. *How We Think: Digital Media and Contemporary Technogenesis*. Chicago: University of Chicago Press.
- Scolari, Carlos. 2012. "Media Ecology: Exploring the Metaphor to Expand the Theory." *Communication Theory* 22 (2): 204–225.
- Hartley, John. 2012. *Digital Futures for Cultural and Media Studies*. Hoboken: John Wiley & Sons.
- Madianou, Mirca, and Daniel Miller. 2013. "Polymedia: Towards a New Theory of Digital Media in Interpersonal Communication." *International Journal of Cultural Studies* 16 (2): 169–187.
- Hepp, Andreas, Stig Hjarvard, and Knut Lundby. 2015. "Mediatization: Theorizing the Interplay between Media, Culture and Society." *Media, Culture and Society* 37 (2): 314–324.
- Krämer, Sybille. 2016. *Medium, Messenger, Transmission: An Approach to Media Philosophy*. Amsterdam: Amsterdam University Press.

Annotated Sources 261

- Couldry, Nick, and Andreas Hepp. 2016. *The Mediated Construction of Reality.* Cambridge: Polity.

- Durham Peters, John. 2016. *The Marvelous Clouds.* Chicago: University of Chicago Press.

- Markham, Tim, and Scott Rodgers, eds. 2017. *Conditions of Mediation: Phenomenological Perspectives on Media.* New York: Peter Lang.

- Gray, Jonathan, and Laurie Ouellette, eds. 2017. *Keywords for Media Studies.* New York University Press.

- Fast, Karin, and André Jansson. 2019. *Transmedia Work: Privilege and Precariousness in Digital Modernity.* London: CRC Press / Routledge.

- Kennedy, Jenny, Michael Arnold, Martin Gibbs, Bjorn Nansen, and Rowan Wilken. 2020. *Digital Domesticity: Media, Materiality, and Home Life.* Oxford: Oxford University Press.

- Hjorth, Larissa, Kana Ohashi, Jolyanna Sinanan, Heather Horst, Sarah Pink, Fumitoshi Kato, and Baohua Zhou. 2020. *Digital Media Practices in Households.* Amsterdam: Amsterdam University Press.

- Boczkowski, Pablo, and Eugenia Mitchelstein. 2021. *The Digital Environment: How We Live, Learn, Work, and Play Now.* Boston: MIT Press.

- Bruhn, Jørgen, and Beate Schirrmacher, eds. 2022 *Intermedial Studies: An Introduction to Meaning Across Media.* London: Routledge.

- Scolari, Carlos. 2022. *On the Evolution of Media: Understanding Media Change.* London: Routledge.

## Chapter 1: Media Life

Much of this chapter takes its cue from the central insights developed in my book *Media Life* (2012), as well as Pablo Boczkowski and Eugenia Mitchelstein's *The Digital Environment* (2021). For more detail, I would like to recommend an interview with the authors I had the pleasure to do on February 23, 2021, for my Deuzevlog—a series of conversations with media scholars and professionals from around the world, published somewhat regularly to YouTube (and available as podcasts via Anchor/Spotify and other platforms; all of these interviews can be found using "Deuzevlog" as the search term). It is always a good idea to watch the two *Life in a Day* crowdsourced documentaries. The films, as well as some additional clips, stories, and updates, are archived on YouTube (youtube.com/user/lifeinaday).

A crucial point in this chapter deals with the necessity to consider media from a historical perspective, recognizing how all our so-called new media are, in so many ways, versions of what came before, with both the technologies and the ways we use them having long histories. A great resource here is Asa Briggs, Peter Burke, and Espen Ytreberg's review *A Social History of the Media: From Gutenberg to Facebook*. Another useful historical and comparative analysis of different media systems around the world (and how these interact with a society's politics and culture) are the various books and reviews by Daniel Hallin and Paolo Mancini. As Hallin and Mancini argue—along the lines of prominent media historians including Bridget Griffen-Foley, Jane Chapman, Lisa Gitelman, Susan Aasman, Gianluigi Negro, Fackson Banda, and others—media studies would do well to adopt and integrate more sociological-historical approaches to scholarship, helping us to understand the evolving relations between social formations, the rise (and fall) of new technologies, genres and ways of using media, and the way different people integrate and give meaning media into daily life.

One key insight such a perspective delivers is that nothing (in or about media) is ever really new and that new media mean different things to different groups of people at different points in time. The American documentary filmmaker Kirby Ferguson produced a four-part video series called *Everything Is a Remix* (which originally ran from 2010 to 2012, remixed and updated in 2105 and 2021) which offers a highly entertaining and well-resourced take on this seeming lack of originality in popular culture. Some good scholarly references to get you started on (social and cultural) media histories:

- Briggs, Asa, Peter Burke, and Espen Ytreberg. 2020. *A Social History of the Media*. 4th ed. Cambridge: Polity.

- Chapman, Jane. 2005. *Comparative Media History, an Introduction: 1789 to the Present*. Cambridge: Polity.

- Gitelman, Lisa. 2006. *Always Already New: Media, History, and the Data of Culture*. Cambridge, MA: MIT Press.

- Hall, Nick, and John Ellis, eds. 2020. *Hands on Media History*. Oxford, UK: Routledge (open access).

- Hallin, Daniel, and Paolo Mancini. 2004. *Comparing Media Systems: Three Models of Media and Politics*. Cambridge: Cambridge University Press.

- Hallin, Daniel, and Paolo Mancini, eds. 2012. *Comparing Media Systems beyond the Western World*. Cambridge: Cambridge University Press.

Next to considering media as a (digital) environment and appreciating their social history, a third overall conceptual insight guiding the opening chapter of the book considers how the emerging convergent, all-encompassing media environment not only mediates everything but also unsettles traditional categories we use when making sense of media—such as the distinction between media producers and consumers. In a life in media, we make media as much as we use media, in the process doing the work for media industries (e.g., of promoting their work by liking, sharing, and forwarding it). At the same time,

Annotated Sources 263

these industries become more dependent on our active engagement with their work, creating a kind of mutual dependency that only reinforces the overall experience of living in media. Such authors as David Gauntlett and Henry Jenkins—among many others, including danah boyd, Mimi Ito, Adrienne Russell, Jean Burgess, Divina Frau-Meigs, Carlos Scolari, Sonia Livingstone, and Zizi Papacharissi—strongly suggest that in making media we can find new forms of (individual and collective) agency, offering hopeful perspectives on social learning with media by looking at what people do with their media and helping people to turn their hopes and dreams via media into reality. At the same time, this turns our everyday media practices into products and data to be commodified for the benefit of global corporations. Some excellent references for this critical debate in media studies follow:

- Bird, S. Elizabeth. 2011. "Are We All Produsers Now?" *Cultural Studies* 25 (4–5): 502–516.
- Gauntlett, David. 2018. *Making Is Connecting: The Social Power of Creativity, from Craft and Knitting to Digital Everything.* 2nd ed. Cambridge: Polity.
- Jenkins, Henry. 2008. *Convergence Culture: Where Old and New Media Collide.* Updated ed. New York: New York University Press.
- Jenkins, Henry, Gabriel Peters-Lazaro, and Sangita Shresthova, eds. 2020. *Popular Culture and the Civic Imagination: Case Studies of Creative Social Change.* New York: New York University Press.
- Livingstone, Sonia. 2009. "On the Mediation of Everything." *Journal of Communication* 59 (1): 1–18.
- Silverstone, Roger. 2002. "Complicity and Collusion in the Mediation of Everyday Life." *New Literary History* 33 (5): 745–64.
- Terranova, Tiziana. 2000. "Free Labor: Producing Culture for the Digital Economy." *Social Text* 18 (2): 33–57.
- Wasko, Janet. 2018. "Studying Political Economies of Communication in the Twenty-First Century." *Javnost / The Public* 25 (1–2): 233–239.

## Chapter 2: Your Life

The purpose of this chapter is to develop a comprehensive definition of media and to subsequently explore what we can say about the status and appearance of media today (and the near future) based on their historical development and impact. The approach chosen here is to consider media as infrastructures, providing much of the scaffolding for life. In doing so, I follow the definition of media offered by Leah Lievrouw and Sonia Livingstone in their *Handbook of New Media* (2006, 2; italics in original), a volume in which they collected groundbreaking scholarly work from a wide variety of scholars to introduce students to an exciting new field of study: "Media are infrastructures with three components: the *artifacts or devices* used to communicate or convey information, the *activities and practices* in which people engage to communicate or share information, and the *social arrangements or organizational forms* that develop around those devices and practices."

Lievrouw and Livingstone's approach offers a middle ground between what Andrea Miconi and Marcello Serra (2019, 3457) in their survey of the field map as a proliferation of both weak and strong definitions of media: "According to the weak conception, media are nothing but neutral instruments in the communication process, seen as mere channels that transport information. Alternatively, the strong conception assumes that they have the power of shaping human reality."

Miconi and Serra suggest that the lack of consensus in media studies about what media is, is a sign of healthy intellectual exploration, noting that the digital transformation of our living environment puts the medium back at the core of scientific debate (as it was during the 1950s, 1960s, and early 1970s in the work of Innis, McLuhan, and Debray). An infrastructural definition of media—whereby we consider both the specific role of technology, how devices get used (and appropriated into everyday life) and what all this means from a cultural perspective—owes much to a specific perspective on technology and society as processual, dynamic, and coconstituent. This point of view was developed mainly by European scholars in the 1980s (under such headings as SCOT [social construction of technology] and ANT [actor-network theory]), including Wiebe Bijker, Trevor Pinch, John Law, Bruno Latour, Madeleine Akrich, and Michel Callon. After the 1980s, these ideas found new fertile ground in the emerging scholarship on ubiquitous computing—a term introduced by Mark Weiser in 1991 to recognize and predict the evolution of technologies as those that gradually disappear, weaving themselves into the fabric of everyday life until they are indistinguishable from it. In their 2011 review of ethnographic approaches to ubiquitous computing, Paul Dourish and Genevieve Bell remind us of the importance of acknowledging the complex and contested realities of how people use and interpret everyday technologies, such as personal computers, smartphones, and any other digital devices. The industry's vision of omnipresent and "calm" technologies is, in fact, far removed from the rather restless and messy day-to-day experience of life in media. Some fundamental texts on defining media and technologies in everyday life to consider:

- Bijker, Wiebe, Thomas Hughes, and Trevor Pinch, eds. (1989) 2012. *The Social Construction of Technological Systems.* Anniv. ed. Cambridge, MA: MIT Press.

- Dourish, Paul, and Genevieve Bell. 2011. *Divining a Digital Future.* Cambridge, MA: MIT Press.

- Law, John. 1992. "Notes on the Theory of the Actor-Network: Ordering, Strategy, and Heterogeneity." *Systems Practice* 5:379–393.

## Annotated Sources

- Lievrouw, Leah, and Sonia Livingstone, eds. 2006. *Handbook of New Media: Social Shaping and Social Consequences*. Fully rev. student ed. London: SAGE Publications.

- Miconi, Andrea, and Marcello Serra. 2019. "On the Concept of Medium: An Empirical Study." *International Journal of Communication* 13:3444–3461.

- Weiser, Mark. 1991. "The Computer for the 21s Century." *Scientific American* 265 (3): 94–104.

In this chapter, I use the tripartite definition of media offered by Lievrouw and Livingstone to explore some historical trends in the role and use of media in people's daily lives, borrowing liberally from Jay Bolter and Richard Grusin's work on remediation, which suggests that all media contain versions of previous media. To account for the ways in which media physically and charismatically restructure our homes and how a household functions over time, I recommend the publications of American media historian Lynn Spigel and Italian feminist scholar Leopoldina Fortunati. In their 2011 textbook for media studies, the Australian scholars Graham Meikle and Sherman Young (2012, 10) intriguingly observe that, "for many people, the media are no longer just what they watch, listen to or read—the media are now what people do." As mentioned earlier, the work on mediatization in recent years is an important reference for thinking deliberately about how deep the rabbit hole goes of media's role and influence in society. At the same time, critics of the theory, such as Cristina Archetti, posit evidence-based arguments to prevent hasty conclusions about any kind of inevitability of media in the process and practice of everyday life, reminding us to stay mindful of our bodies, feelings, and social relations as mediators of any kind of effects by media and technologies (compare the reference to her work with Jesper Strömbäck's argument about the mediatization of politics, for example). The key references used here include the following:

- Archetti, Cristina. 2017. "Image, Self-Presentation and Political Communication in the Age of Interconnection: An Alternative Understanding of the Mediatization of Politics." *Northern Lights* 15:89–109.

- Bolter, Jay, and Richard Grusin. 1996. "Remediation." *Configurations* 4 (3): 311–358.

- Fortunati, Leopoldina. 2001. "The Mobile Phone: An Identity on the Move." *Personal and Ubiquitous Computing* 5:85–98.

- Hepp, Andreas, Stig Hjarvard, and Knut Lundby. 2015. "Mediatization: Theorizing the Interplay between Media, Culture and Society." *Media, Culture and Society* 37 (2): 314–324.

- Meikle, Graham, and Sherman Young. 2011. *Media Convergence: Networked Digital Media in Everyday Life*. London: Palgrave Macmillan.

- Spigel, Lynn. 2001. "Media Homes: Then and Now." *International Journal of Cultural Studies* 4 (4): 385–411.

- Strömbäck, Jesper. 2008. "Four Phases of Mediatization: An Analysis of the Mediatization of Politics." *International Journal of Press/Politics* 13 (4): 228–247.

In the final sections of the chapter, I turn to how media and their uses produce (new) social arrangements and how this contributes to different ways of organizing social life. Of course, the literature on how people use and give meaning to media is incredibly rich and varied, yet I would like to single out a few publications that have specifically addressed issues of media use from an integrated, convergent, and environmental point of view. Thorsten Quandt and Thilo von Pape's notion of the home as a "mediatope" is, for example, a really useful concept to define and appreciate the complex entanglement of media and daily practices as these evolve over time. For more than a year, Quandt and Pape followed a hundred German households (through interviews, observations, and surveys), showing how media move through the household in flocks, how the identities of various devices change over time, and how younger and older media fight for survival as they are domesticated and discarded by various family members. This intimate connection between the lives of people and their media "paints a picture of an evolving, living media world within the domestic environment of the household" (2010, 343). Their work fits in a broader international tradition of carefully considering media use as a reciprocal practice between devices, a variety of contents, and different people in different social and material contexts, a "worlding" process that produces particular kinds of reality and experience. The latest iterations of this kind of research focuses specifically on our "natural" uses of newer media through haptics, motion sensing, and other embodied practices.

While most researchers recognize how people inherently combine and mix multiple media, platforms, and channels, it is important to consider work that isolates the unique features of technologies, especially today's omnipresent locative media (such as mobile devices and wireless internet access) in creating what the Brazilian media scholar Adriana de Souza e Silva calls "hybrid spaces" between where we are and how we are connected in media. Ethnographic studies are essential here, as colleagues such as Laura Glitsos, Sarah Pink, Ilana Gershon, David Nemer, Liesbet de Block, David Buckingham, Payal Arora, and many others shed light on how people from a variety of more or less privileged backgrounds and across locations around the world use and give meaning to the different devices, channels, and platforms they appropriate and domesticate.

Beyond the qualitative nature of such work, more quantitative (survey- and experiment-based) research on media use can be a powerful tool to open up ways of understanding and appreciating how media fit in the various ways in which people organize their lives. Ultimately, the way forward is in mixed methods work—something advocated by many yet practiced by few (often due to budgetary and time constraints). Some fascinating examples of such necessary multimethodological research that deeply informs the narrative in this chapter and book come from Anabel Quan-Haase's SocioDigital Lab at Western University in Ontario, Canada (sociodigital.info; see also our interview on Deuzevlog), and Patti Valkenburg's Project AWeSome at the University of Amsterdam in the

Netherlands (project-awesome.nl). Another significant ongoing project to be mentioned here is Global Kids Online, a collaborative initiative between UNICEF, the London School of Economics and Political Science, and the EU Kids Online network. The project, led by Sonia Livingstone and Jasmina Byrne, uses surveys, focus groups, and interviews with children around the world, aiming to bring evidence to important discussions about children's well-being and rights in the digital age (see globalkidsonline.net, as well as my interview with Sonia Livingstone on Deuzevlog). These kinds of projects make us mindful that people perhaps are not all that different in terms of what they desire from their media—while at the same time acting as a reminder that living in (or coming from) a certain place, being of a certain age, and having a particular kind of background really matters for being able and motivated to use, understand, and enjoy media. Some references to get started in this area:

- Arora, Payal. 2019. *The Next Billion Users*. Boston: Harvard University Press.

- Baym, Nancy. 2015. *Personal Connections in the Digital Age*. 2nd ed. Cambridge: Polity.

- Croci, Valentina. 2008. "Natural Methods of Interaction or Natural Interaction in the Everyday Digital World." *Architectural Design* 78 (1): 120–123.

- de Block, Liesbet, and David Buckingham. 2008. *Global Children, Global Media: Migration, Media and Childhood*. London: Palgrave.

- Gershon, Ilana. 2010. "Breaking Up Is Hard to Do: Media Switching and Media Ideologies." *Journal of Linguistic Anthropology* 20:389–405.

- Glitsos, Laura. 2019. *Somatechnics and Popular Music in Digital Contexts*. New York: Springer.

- Helsper, Ellen. 2021. *The Digital Disconnect: The Social Causes and Consequences of Digital Inequality*. London: Sage.

- Kim, Su Jung. 2016. "A Repertoire Approach to Cross-Platform Media Use Behavior." *New Media and Society* 18 (3): 353–372.

- Pink, Sarah, Martin Berg, Deborah Lupton, and Minna Ruckenstein, eds. 2022. *Everyday Automation: Experiencing and Anticipating Emerging Technologies*. Oxford, UK: Routledge (open access).

- Pink, Sarah, and Heather Leder Mackley. 2013. "Saturated and Situated: Expanding the Meaning of Media in the Routines of Everyday Life." *Media, Culture and Society* 35 (6): 677–691.

- Quan-Haase, Anabel, Hua Wang, Barry Wellman, and Renwen Zhang 2018. "Weaving Family Connections on and Offline: The Turn to Networked Individualism." In *Connecting Families? Information & Communication Technologies in a Life Course Perspective*, edited by Barbara Barbosa Neves and Cláudia Casimiro, 00–00. Bristol, UK: Policy.

- Quandt, Thorsten, and Thilo von Pape. 2010. "Living in the Mediatope." *Information Society* 26 (5): 330–345.

- Souza e Silva, Adriana de. 2006. "From Cyber to Hybrid: Mobile Technologies as Interfaces of Hybrid Spaces." *Space and Culture* 9 (3): 261–278.

## Chapter 3: Public Life

At the heart of this chapter lies a poignant statement published in 2010 by the American media researcher danah boyd: "It's just that, in many situations, there is more to be gained by accepting the public default than by going out of one's way to keep things private. And here's where we see the shift. It used to take effort to be public. Today, it often takes effort to be private."[1]

Living in media means living in a global surveillance society, which is not necessarily a society of discipline or control (as suggested by many theorists) as it is also very much a world in which we opt in to live in public. This attitude brings great profit to merchants of personal information (such as platform and streaming service providers, telecommunications companies, and online social networks) as it serves the purposes of all kinds of institutions, activists, communities and individuals alike. At the same time, all this public engagement online feeds the inner workings of algorithms, machine learning, and artificial intelligence, data-driven systems that deeply structure and give shape to the world we see and experience in media (and therefore also to the world AFK, "away from keyboard"). The ongoing and accelerating datafication of all aspects of life—for children as much as for adults, for refugees just as well as for everyone else, for those in the Global North and the Global South alike—requires an appreciation of life in media as both embodied and datafied, as simultaneously human and nonhuman or something Deborah Lupton fittingly calls a "more-than-human" understanding of the reciprocal and interdependent relations between people, media, and information technologies throughout the life course.

A second key insight from the rich literature on surveillance in media studies is the often ambivalent, inconsistent, and contradictory nature of people's relations with media in the context of monitoring behavior. Research shows how people may feel strongly about privacy and personal freedom, while also embracing and even endorsing all kinds of intrusive forms of surveillance. Tama Leaver, for example, points out that the choice not to closely survey and track your children can be seen as a failure of parenting, which contributes to the normalization of intimate surveillance. Likewise, David Lyon theorizes a culture of surveillance that is not only done to us—it is something we do in everyday life. In this context, Jakob Linaa Jensen coined the concept of the "omnopticon": a situation where everyone monitors (or at least potentially surveys) everyone else.

---

1. danah boyd, "Public by Default, Private when Necessary," *apophenia*, January 25, 2010, https://www.zephoria.org/thoughts/archives/2010/01/25/public_by_defau.html.

**Annotated Sources** 269

See the following for some key sources of insights into this technological bias of surveillance capitalism, how surveillance affects and changes our physical environment, and including nuanced appraisals of our interdependent, imagined, ambivalent, incongruent, and overall quite messy expectations and practices regarding surveillance:

- Duffy, Brooke Erin, and Ngai Keun Chan. 2019. "'You Never Really Know Who's Looking': Imagined Surveillance across Social Media Platforms." *New Media and Society* 21 (1): 119–138.

- Hayles, Katherine. 2009. "Waking Up to the Surveillance Society." *Surveillance and Society* 6 (3): 313–316.

- Jensen, Jakob Linaa. 2007. "The Internet Omnopticon." In *New Publics with/out Democracy*, edited by Henrik Bang and Anders Esmark, 351–380. Copenhagen: Samfundslitteratur Press.

- Kazansky, Becky. 2021. "'It Depends on Your Threat Model': The Anticipatory Dimensions of Resistance to Data-Driven Surveillance." *Big Data and Society* 8 (1). https://doi.org/10.1177/2053951720985557.

- Kostelac, Hille. 2000. "'The Gaze without Eyes': Video-Surveillance and the Changing Nature of Urban Space." *Progress in Human Geography* 24 (2): 243–265.

- Leaver, Tama. 2017. "Intimate Surveillance: Normalizing Parental Monitoring and Mediation of Infants Online." *Social Media + Society* 3 (2). https://journals.sagepub.com/doi/10.1177/2056305117707192.

- Lupton, Deborah. 2020. "Thinking with Care about Personal Data Profiling: A More-than-Human Approach." *International Journal of Communication* 14:3165–3183.

- Lyon, David. 2018. *The Culture of Surveillance: Watching as a Way of Life*. Cambridge: Polity.

- Pink, Sarah, Deborah Lanzeni, and Heather Horst. 2018. "Data Anxieties: Finding Trust in Everyday Digital Mess." *Big Data and Society* 5 (1). https://doi.org:/10.1177/2053951718756685.

- Verbeek, Peter Paul. 2008. "Cyborg Intentionality: Rethinking the Phenomenology of Human–Technology Relations." *Phenomenology and the Cognitive Sciences* 7:387–395.

- Zuboff, Shoshana. 2015. "Big Other: Surveillance Capitalism and the Prospects of an Information Civilization." *Journal of Information Technology* 30:75–89.

Beyond the particulars of surveillance, how the various ways in which different actors—states, security and police forces, corporations, platforms and telecommunications companies, institutions, families, and friends—monitor and survey us, and how so many people embrace sophisticated forms of self-tracking (as I write this, I'm self-consciously aware of the smart watch I am wearing on my wrist), it is crucial to think through the implications of surveillance. The most clear-cut discussion about surveillance engages the eternal dilemma of privacy and freedom versus security and control. Increased monitoring tends to be advocated by those promising better (and more) security, which leads those opposed to surveillance to rally for greater personal autonomy and the freedom to be left alone. On the other hand, one could argue that the condition and experience of being free greatly depends on all kinds of formal structures, regulations, and boundaries set and policed by institutions (such as the state). A similar conundrum occurs with the reference to security as the impetus and legitimation of surveillance, as it implies that there is something to be secured—which generally (albeit implicitly) refers to the liberty and self-determination of people. In other words, the relationship between freedom and security inevitably involves a complex negotiation, and a surveillance society produced by our lives as lived in media has a context of massive mutual monitoring that seems voluntary, even desirable.

The delicate dance between security and liberty is the kindling of an important (and recurring) fiery debate. In media studies, scholars tend to be additionally interested in how people experience and give meaning to surveillance and how this shapes our relations with technology and media and each other and how we perceive our role in society. The dominant impetus for scholarship on surveillance in media studies has come through the work of Michel Foucault. Foucault used Jeremy Bentham's eighteenth-century design of a Panopticon to think about how the modern state (and all its institutions, such as schools, factories, and prisons) is able to function so effectively by getting everyone to participate in it—without overt repression. For Foucault, real power is not something simply done to people but something that is always subject to negotiation and contestation. Power is a living force. To get power to consistently flow in one direction, the French philosopher argued, it needs to function almost automatically—which only can be achieved if everyone is convinced of this power without having the means to verify its existence. In the panoptic process, people internalize surveillance and discipline—like the parents do in Tama Leaver's analysis of intimate surveillance and we all do through our mass self-communication online.

Perhaps this is indeed the crux of the surveillance society as exemplified through our life in media: we all know our digital actions and practices are recorded, stored, and datafied, but we do not (and cannot) know how this process works. For Foucault, this produces the kind of disciplined subject exemplified in the identity of a prisoner. It is somewhat bizarre that the most sophisticated tracking technologies today can be both what some would pay to be free of, such as a prisoner's ankle monitor, while others happily spend significant money to be able to wear it all the time, such as a smart watch.

Another French philosopher, Gilles Deleuze, followed in the footsteps of Foucault, suggesting that the end product of a surveillance society built on digital culture was not discipline as internalized by people when participating in all the public spaces and institutions of society but rather total control, as access to just about anything and any place in society becomes governed and supervised through an increasingly complex system of gates,

**Annotated Sources**                                                              271

checks, and monitoring apparatuses. In Deleuze's analysis, there is quite literally no outside to surveillance anymore, and we are all both the victims and enforcers of a system of control. To Deleuze, this produces a contrasting self, a "dividual"—which inspires me to think of who we are as digital selves, that is, information scattered in bits and pieces of data across a multitude of archives and databases, only to be recomposed by algorithms into statistical aggregates on the basis of which an endless variety of decisions are made, from allowing or denying access, offering a discount, extending a warranty, granting a mortgage, and much, much more.

Soon thereafter, Armand Mattelart extended both Foucault's expectation that surveillance would contribute to the exercise of power through self-discipline and Deleuze's conviction that we all have become part of totalizing control societies. Mattelart traces omnipresent surveillance systems back to, for example, the nineteenth-century tradition of anthropometry: the systematic collection and correlation of measurements of the human body, data at the time that were often used to support theories associating biological race with levels of cultural and intellectual development (as a sidenote: the omnipresent Body Mass Index – a deeply flawed measure of someone's health – derives from this questionable practice). The Belgian philosopher uses this genealogy of surveillance to question its fundamental motivation: suspicion. Taken together, these three analyses inspired much of the subsequent literature and research on surveillance and the role pervasive and ubiquitous media play therein. More recently, Shoshana Zuboff bundled and significantly added to these and other prominent perspectives, suggesting how the historical use of personal information to drive political and commercial developments accelerated with the internet and specifically the rise of platforms (such as Facebook, YouTube, and others) to produce a particularly aggressive and intrusive kind of surveillance capitalism. Whereas previous work predominantly focused on inspection and monitoring as a political force, Zuboff focuses on how the mix of technology firms, personal media, and capitalist logic can overpower the very essence of human freedom, as human behavior gets subtly modified—through algorithmic recommendation systems, targeted advertising, and all kinds of nudges, pokes, and jabs—for commercial ends. The genealogy of these approaches to surveillance is one of ever-increasing scope, depth, and detail. As with the overall observation of a life in media, there does not seem to be an outside to surveillance anymore.

Here are some key references to get started with unpacking the various layers of our contemporary surveillance society—next to checking out the various issues of the dedicated open-access scholarly journal *Surveillance and Society*:[2]

•   Andrejevic, Mark. 2002. "The Work of Being Watched: Interactive Media and the Exploitation of Self-Disclosure." *Critical Studies in Media Communication* 19 (2): 230–248.

---

2. *Surveillance and Society*, https://ojs.library.queensu.ca/index.php/surveillance-and-society/index

- Best, Kirsty. 2010. "Living in the Control Society." *International Journal of Cultural Studies* 13 (1): 5–24.

- Deleuze, Gilles. 1992. "Postscript on the Societies of Control." *October* 59:3–7.

- Elmer, Greg. 2003. "A Diagram of Panoptic Surveillance." *New Media and Society* 5 (2): 231–247.

- Foucault, Michel. (1975) 1995. *Discipline and Punish: The Birth of the Prison*. New York: Vintage Books.

- Lyon, David. 2018. *The Culture of Surveillance*. Cambridge: Polity.

- Mattelart, Armand. (2007) 2010. *The Globalization of Surveillance*. Cambridge: Polity.

- Zuboff, Shoshana. 2019. *The Age of Surveillance Capitalism: The Fight for a Human Future at the New Frontier of Power*. London: Profile Books.

Surveillance inspires a lot of excellent journalism and reporting around the world, and it has engendered its own distinct art form in surveillance art, documented and archived at galleries, educational institutions, and media organizations around the world (including the Tate Modern Gallery in the United Kingdom, the Art and Surveillance project at the University of Calgary in Canada, and ongoing collections at Vice, Artsy, and many other online places around the world). Indeed, such artists as the German filmmaker Michael Klier and the American practitioner Julia Scher produced installations and video projects appropriating CCTV cameras, burglar alarms, electronic security systems, and other surveillance technologies in the 1980s and 1990s, well before our current age of social media, Google Street View, and facial recognition software. As mentioned in the chapter, the Oscar-winning 2014 Edward Snowden documentary *Citizenfour*, directed by Laura Poitras, is worth watching, and there are countless films centered on surveillance. In fact, there is so much attention paid to surveillance in both fiction and nonfiction media that it makes one wonder whether this is exactly the power of surveillance in Foucauldian terms: we are all constantly reminded that we are being watched, without ever seeing (and for most of us, directly experiencing) surveillance.

## Chapter 4: Real Life

A crucial inspiration for thinking through the problematic nature of media in society and everyday life as somehow corrupting the possibility of true, perfect communication comes from the seminal work of the American media historian John Durham Peters (1999, 2), whose 1999 book *Speaking into the Air: A History of the Idea of Communication* posits that the privileging of dialogue as the ideal kind of (interpersonal) communication is flawed, in part because it is based on the promise of "a utopia where nothing is misunderstood, hearts are open, and expression is uninhibited." In his review of Peters' work, the

Annotated Sources 273

British media scholar Paddy Scannell (2004, 95) documents how this argument inspired him to rethink all our assumptions about the role media and communication play, seeing how "the premium we place on sincerity and authenticity . . . is precisely indicative of our continued longing for 'true' communication." This is an impossible goal—for, as Truman Burbank in the 1998 movie *The Truman Show* says, "You never had a camera in my head"—meaning that each of us is contained in their very own "heart-shaped box," unable to peer into anyone else's. From this point, Scannell comes to the realization that the communicative infrastructure of the world is inevitably impersonal, available to anyone—and that this is not a depressing or fatalistic view but rather hopeful. It is "indicative of worldly as of divine love—amor mundi, the world's care-for-itself—and our transcendent human, historical essence" (105). With Durham Peters, Scannell concludes that love is the mediating agency between everything and everyone—an insight that profoundly influences my overall perspective for teaching and research and guides every sentence in this book (while chapter 5 is wholly dedicated to the idea of media love).

From this perspective, it is remarkable to see how our field continues to be preoccupied with solving the "problem" of communication in and through media. The notion that media, quite literally, come between us and (our experience of) the world and thereby inevitably introduce some level of distortion in the possibility of real or true experience (and communication) is not particular to the digital environment. Indeed, concerns about this role of media have been voiced throughout history in both philosophical treatises and popular culture. Especially since the sixteenth and seventeenth centuries, as development in all aspects of society—economic, cultural, social, technological—accelerated due to the mechanical (and later on industrial) revolution, we can find much hand-wringing about the nature of reality and the question of what (and who) is "real," and I would strongly recommend sampling some of these older texts and debates to learn about different ways of making sense of these fundamental dilemmas. An additional advantage of such work is that it tends to be out of copyright and is therefore freely available online. The work of authors like Edgar Allen Poe, E. M. Forster, E. T. A. Hoffmann, Jorge Luis Borges, Robert A. Heinlein, Mary Shelley, Philip K. Dick, various short stories in the collection of ancient Arab folk tales One Thousand and One Nights involving encounters with various robots, and many others—often referenced in this book—are warmly recommended.

A second strand of references for the various arguments and life in media strategies—fight, surrender, and to become media—comes from some of the classics of modern science. It is both important and fun to trace the genealogy of contemporary discussions and arguments about the nature of our reality as it relates to technology, machines, and media—if anything, to appreciate that we have been here before. What makes us human, whether there ever was (or will be) consensus about reality and the truth, how machines are set to replace us (or not)—all these and many other current debates have a long history in the world's literature and arts. Some good references used for the argument in this chapter include

- Gottfried Leibniz's *The Monadology* (written in 1714);

- Charles Darwin's *On the Origin of Species* (published on November 24, 1859);

- the French sociologist, criminologist, and science fiction author Gabriel Tarde's work, linking the monadology, Darwin's theory of evolution, and the technological changes of his time, for example, his *The Laws of Imitation* (1890, published in English translation in 1903) and *Monadologie et sociologie* (1893);

- Victor Tausk's paper "The Origin of the Influencing Machine in Schizophrenia" (first published in 1919, translated from German to English in 1933 by Dorian Feigenbaum and printed in the journal *Psychoanalytic Quarterly* that year); and

- Sigmund Freud's *Civilization and Its Discontents* (first published in German in 1930).

When it comes to concrete discussions on the elements of our mediated reality, three strands of scholarly literature stand out. From such disciplines as media theory and media philosophy, we get crucial arguments about the material dimension of media. Media theory considers how mediated messages mean different things to different people as determined by the different channels used to communicate them, whereas media philosophy tends to advocate an even more fundamental role for media as providing the ontological basis for knowing and understanding reality.

- Baudrillard, Jean. (1981) 1998. *Simulacra and Simulations*. In Mark Poster, ed., *Jean Baudrillard, Selected Writings*, 166–184. Stanford, CA: Stanford University Press.

- Debray, Régis. (1994) 1996. *Media Manifestos: On the Technological Transmission of Cultural Forms*. Translated by Eric Rauth. London: Verso.

- Floridi, Luciano 2009. "Against Digital Ontology." *Synthese* 168:151–178.

- Hansen, Mark. 2006. "Media Theory." *Theory, Culture and Society* 23 (2–3): 297–306.

- Hayles, Katherine. 1999. How We Became Posthuman. Chicago: University of Chicago Press.

- Kittler, Friedrich. 2009. "Towards an Ontology of Media." *Theory, Culture and Society* 26 (2–3): 23–31.

- Vandenberghe, Frédéric. 2007. "Régis Debray and Mediation Studies." *Thesis Eleven* 89:23–42.

A second strand of academic research connects the way media represent the world and the various ways in which people communicate and give meaning to (mass) mediated communication regarding what reality is (or seems to be). Media, considered in terms of what people do, in this process inevitably produce a reality that is not subject to consensus. It is a reality that is always contested, negotiated, and full of conflict and contradiction. As the research shows, all of us tend to have complex, multidimensional ideas about what constitutes realness (in media). Furthermore, when news media and documentaries report on reality, the practitioners involved tend to be committed to being truthful and factual, while their representations end up creating realities of their own. Such an analysis is neither relativistic nor postmodern—as many scholars would point out, it really matters what we, our media, and the way we communicate consider real, true, and authentic. Pursuing these discussions in and through media and mediated communication perhaps brings us closer to reality than simply observing the real—as the late Chilean cyberneticist and biologist Humberto Maturana remarked: "Everything said is said by an observer to another observer that could be him or herself."[3] The world is not simply "out there" for us to either access or ignore with our media. We produce data and information for and about ourselves by making all kinds of choices in media and in so doing cocreate the world we live in. Some helpful entries into the literature on how media as activities challenge notions of truth and reality:

- Aguado, Juan Miguel. 2009. "Self-Observation, Self-Reference and Operational Coupling in Social Systems: Steps towards a Coherent Epistemology of Mass Media." *Empedocles* 1 (1): 59–74.
- Chalmers, David. 2022. *Reality+: Virtual Worlds and the Problems of Philosophy.* New York: W. W. Norton.
- Enli, Gunn. 2015. *Mediated Authenticity: How the Media Constructs Reality.* New York: Peter Lang.
- Funkhouser, Ray, and Eugene Shaw. 1990. "How Synthetic Experience Shapes Social Reality." *Journal of Communication* 40 (2): 75–87.
- Hall, Alice. 2003. "Reading Realism: Audiences' Evaluations of the Reality of Media Texts." *Journal of Communication* 53 (4): 624–641.
- Kepplinger, Hans Mathias, and Johanna Habermeier. 1995. "The Impact of Key Events on the Presentation of Reality." *European Journal of Communication* 10 (3): 371–390.
- Luhmann, Niklas. (1996) 2000. *The Reality of the Mass Media.* Cambridge: Polity.
- Maturana, Humberto R. 1988. "Reality: The search for objectivity or the quest for a compelling argument." *The Irish Journal of Psychology* 9 (1): 25–82.

The third scholarly area of exploring media and reality relations can be classified as considering the social arrangements of media in society and everyday life. Arguably this is the richest and most developed kind of research tradition in media and communication studies, as scholars in the field ultimately aim to say something about how people and

---

3. Humberto R. Maturana, "Reality: The search for objectivity or the quest for a compelling argument." The Irish *Journal of Psychology* 9, no. 1 (1988): 27.

humanity change (or could be changed) in the context (or as a consequence of) media. The references I tend to turn to for inspiration—most of which are documented more extensively in the various editions of the late Denis McQuail's seminal handbook, which I have the privilege to take on as coauthor and editor—question and complicate assumptions about the direction of such changes, the quality of communication, and the universal applicability of theories and methods about the media. Importantly, it is especially here that humanities-based traditions of media studies find common cause with social scientific approaches to communication research, as most scholars in these fields argue that mixed, integrated, triangulated, or otherwise collaborative approaches are necessary to grasp the complexities of life in media. The earlier mentioned work of Patti Valkenburg and her team is a key reference here. The few truly interdisciplinary projects that do exist tend to conclude that the key to understanding the role and impact of media in everyday life lies in idiographic explanations, appreciating individuals in their particular social, material, cultural, spatial, and economic context. Despite the fact that this chapter is somewhat mediacentric in its analysis of our relationship with the real, I remain mindful of Zizi Papacharissi's conclusion throughout her work that while it is crucial to acknowledge how specific technologies and certain media shape us and our worldviews, it is ultimately up to us to connect and create shared narratives.

- Deuze, Mark, and Denis McQuail. 2020. *McQuail's Media and Mass Communication Theory*. 7th edition. London: Sage.

- Fortunati, Leopoldina. 2005. "Is Body-To-Body Communication Still the Prototype?" *Information Society* 21:53–61.

- Hepp, Andreas, Andreas Breiter, and Uwe Hasebrink, eds. 2018. *Communicative Figurations: Transforming Communications in Times of Deep Mediatization*. New York: Springer.

- Papacharissi, Zizi. 2016. "Affective Publics and Structures of Storytelling: Sentiment, Events and Mediality." *Information, Communication and Society* 19 (3): 307–324.

- Peters, John Durham. 1994. "The Gaps of which Communication Is Made." *Critical Studies in Mass Communication* 11 (2): 117–140.

- Risam, Roopika. 2018. *New Digital Worlds: Postcolonial Digital Humanities in Theory, Praxis, and Pedagogy*. Chicago: Northwestern University Press.

- Spence, Patric. 2019. "Searching for Questions, Original Thoughts, or Advancing Theory: Human-Machine Communication." *Computers and Human Behavior* 90:285–287.

- Sundar, Shyam. 2020. "Rise of Machine Agency: A Framework for Studying the Psychology of Human-AI Interaction (HAII)." *Journal of Computer-Mediated Communication* 25 (1): 74–88.

Annotated Sources

- Valkenburg, Patti M. 2022. "Social Media Use and Well-Being: What We Know and What We Need to Know." *Current Opinion in Psychology* 45 (10): 1294.
- Valkenburg, Patti M., Jochen Peter, and Joseph B. Walther. 2017. "Media Effects: Theory and Research." *Annual Review of Psychology* 67:315–338.

When it comes to ever-increasing intimacies between people and media (including all variations of blending of humans, robots, and artificial intelligence), popular culture offers countless inspiring references beyond high-profile Hollywood movies, for example, in the works of such contemporary science fiction authors as Zen Cho, Marge Piercy, Becky Chambers, Robin Wasserman, Octavia Butler, and Nnedi Okorafor. As always, many of the themes involved can be found in earlier works, both in literature, film, games, and elsewhere. Some notable mentions of titles that explicitly explore the bewildering consequences of realities suffused by a variety of media that have served me well in articulating the perspective in this chapter and book include

- the novel *La invención de Morel* by the Argentine writer Adolfo Bioy Casares (published in 1940) about a man losing himself in a parallel virtual world, made into several film versions in Italy and France and serving as the main inspiration for the American television show *Lost*, running from 2004 to 2010;
- *Simulacron-3* by Daniel F. Galouye (published in 1964), about people oblivious to the fact that they are living in a computer-generated city simulation, adapted as a German television film by Rainer Werner Fassbinder in 1974 and as the motion picture *The Thirteenth Floor* by Jozef Rusnak in 1999;
- Philip K. Dick's novel *VALIS* (from 1981), about a vast network of satellites (including the title character, Vast Active Living Intelligence System) projecting images and information directly into people's minds, also serving as a reference for the *Lost* series (in its seventy-sixth episode, characters can be seen reading Dick's book as well as Casares's novel);
- the science fiction novel *Snow Crash* by the American novelist Neal Stephenson (published in 1992), about a world where a computer virus simultaneously affects people's real and virtual bodies (in an online world called the Metaverse, also the name for Facebook's initiative into virtual reality and the inspiration for its name change to Meta); the book is credited for popularizing the term *avatar* and influencing the development and design of Google Earth and such virtual worlds as Second Life, as well as Microsoft's online multiplayer gaming network Xbox Live;
- the comic book series *The Invisibles* by the Scottish artist Grant Morrison (appearing between 1994 and 2000) and the *Matrix* film franchise, originally written and directed by Lana and Lily Wachowski yet claimed by Morrison as stealing the idea

and design from his series about a secret organization battling an alien system of control over humanity;

- the 2020 digital game *Cyberpunk 2077* (part of the *Cyberpunk* game universe introduced by the American designer Mike Pondsmith in 1988), developed by the Polish studio CD Projekt Red; the story world focuses on humanity coming to terms with a technological future of hybrid human-machines, featuring cybernetic prosthetics, organic computer circuitry, and direct human-machine interfaces; the 2020 game has a starring role for the Canadian actor Keanu Reeves, who also plays the protagonist (Neo) in the *Matrix* story world; the cyberpunk genre has its roots in 1960s and 1970s science fiction (notably including the novels of Philip K. Dick), with a distinct Japanese variant emerging in the 1980s—especially in the work of the manga artist Katsuhiro Otomo, in turn inspiring the groundbreaking 1989 film *Tetsuo: The Iron Man* (as mentioned in chapter 2).

**Chapter 5: Love Life**

Love is a central organizing principle and the mediating agency in all arguments throughout this book—not in the least because people clearly love (their) media. It is this kind of love, as well as the role media play in our understanding and enacting of love, which is at the heart of this chapter. To focus specifically on love as a subset of emotions and media is still relatively rare, despite the fact that people clearly love (their) media. A significant reference in this chapter is the work of Roger Silverstone, especially his highlighting of the erotics of media. Catherine Stimpson (2009, 14) reminds us how "the humanities are being snobbish and condescending about love. These attitudes lead to a neglect of teaching and learning *about* love and *with* love in all its varietals, permutations, vicissitudes, necessities, values, and victories."

Following Stimpson, bringing love into media studies is a trifurcated exercise: we need to explore how whatever we study is a lover or capable of loving, what it says about love, and how we can love it. This inspired my reading of media love in terms of a taxonomy of media as practice, mediation, and mediatization.

Emotions enjoy increased attention in the study of media and (mass) communication—in the twenty-first century linked to an "affective turn" in cultural studies more generally and an "emotional turn" specific to media studies. Love, however, is much less explicitly articulated with media studies, even though in recent years there seems to be a surge of scholarship across the social sciences and humanities regarding theories of love. Noteworthy is the emergence of love studies as a coherent field of theory and research with roots in comparative literature, linguistics, and social and feminist theory. As Leopoldina Fortunati (2009, 14) poignantly concludes: "Theories with heart are needed in order to understand properly processes so complex as body-to-body and mediated communication." Some explicit references to get us started on love and media:

- Blum, Virginia L. 2005. "Love Studies: Or, Liberating Love." *American Literary History* 17(2. pages 335–348.

- Deuze, Mark. 2022. "Media Love: On the Media(tiza)tion of Love and Our Love for Media." in *Mediatization of Emotional Life*, edited by Katarzyna Kopecka-Piech and Mateusz Sobiech, 26–40. London: Routledge. (This chapter was a first draft of what is now chapter 5 in this book.)

- Fortunati, Leopoldina. 2009. "Theories without Heart." In *Cross-Modal Analysis*, edited by Anna Esposito and Robert Vích, 5–17. New York: Springer.

- Jónasdóttir, Anna. 2014. "Love Studies: A (Re)New(ed) Field of Knowledge Interests." In *Love—a Question for Feminism in the Twenty-First Century*, edited by Anna Jónasdóttir and Ann Ferguson, 11–30. London: Routledge.

- Malinowska, Ania. 2022. *Love in Contemporary Technoculture*. Cambridge: Cambridge University Press.

- Rusu, Mihai Stelian. 2018. "Theorising Love in Sociological Thought: Classical Contributions to a Sociology of Love." *Journal of Classical Sociology* 18 (1): 3–20.

- Scannell, Paddy. 2004. "Love and Communication: A Review Essay." *Westminster Papers in Communication and Culture* 1 (1): 93–102.

- Silverstone, Roger. 1999. *Why Study the Media?* London: Sage.

- Stimpson, Catherine. 2009. "Loving an Author, Loving a Text: Getting Love back into the Humanities." *Confrontation* 104:13–29.

Interestingly, the current interest in affect, feelings, and emotions is partly due to developments in cybernetics and new insights about human-machine relations (see also chapter 4). This kind of work, following the course of the late twentieth century with the advance of personal computing and the omnipresence of digital devices, suggests that the neat divisions people like to draw between technology and humanity—between media and life—are anything but clear when we, for example, carefully study the way organic and inorganic beings make decisions, relate, and communicate, finding that both living and nonliving systems can have purpose. Another distinct element of the affective turn is its emphasis on (and attention to) the body as an essential part of how people live and act in the world and make sense of it. Such nondualist thinking about mind and body can be traced back to Charles Darwin's work—specifically where he emphasizes the evolutionary link between emotions and instincts. While inclusion of the body is perhaps rather straightforward in some other scientific disciplines, in the history of media studies the body (in all its complicated messiness) has been rather neglected. Not anymore though.

Media and communication scholars today tend to focus on how emotion and affect are produced by media, the way they are communicated through media, and the kinds of emotion people develop when using media. Jens Eder, Julian Hanich, and Jane Stadler

(2019, 91) offer a helpful categorization of the different ways in which media studies engages the emotions today:

- Emotion representation: How are different emotions represented and expressed in media, and through what means?

- Emotion elicitation: Which emotions do media evoke in users, and by what forms and structures?

- Emotion practice: In which practices are emotions integrated, and how are they interwoven with particular media uses and characteristics of specific media?

- Emotion culture: Which socio-cultural causes and effects do media emotions have in certain cultures and epochs? How are they linked to power, ethics and politics, and how do they change over time?

Some typical contemporary examples of media studies where (strong) emotions feature prominently in the research are studies on how young people express and experience sexual intimacy via social media and dating applications, on the affective nature of people's political participation (and polarization) online, and on all the practices related to people's fandom and fannish behavior. In other words, the majority of studies tend to dwell on the issue of what kind of emotions different media provoke rather than looking at (for example) how our emotional traits and states influence and inspire media design, use, and function. Research that connects these emotional dimensions opens up much more fundamental ways to appreciate their significance in everything we do when it comes to (our) media. Inspiring examples are Misha Kavka's consideration of reality television as a "technology of intimacy," danah boyd's work on the complicated reasons why youths love social media, and Perry Parks's proposal for joy as a news value, while Mirca Madianou and Daniel Miller recognize such media as mobile phones and internet cafés as "technologies of love" for transnational families that try to keep in touch as they are scattered across the globe.

Overall, a turn to emotion and affect helps us to build a bridge between what often seem mutually exclusive domains or unnecessary dualities, such as between culture and nature, inside and outside, online and offline, human and nonhuman, the body and the mind, male and female, and so on. It is for this reason that a life in media perspective needs to engage emotions head on, starting with perhaps the most powerful feelings of all: those associated with love. As with all work that deliberately considers emotions, this can best be done with a component approach that includes how people experience media (including bodily sensations, such as sweating, heavy breathing, changes in heart rate, goosebumps, and so on) and how we describe, share, and evaluate such feelings and appreciates the broader cultural and social context within which all these emotions take shape. Love, as a broad concept or category of emotions, in media can be seen both as a universal human experience and as a range of feelings and ways of making sense that are particular to a situation and unique to each individual. Foundational texts to hit the ground running with media love include the following:

- boyd, danah. 2015. *It's Complicated: The Social Lives of Networked Teens.* New Haven, CT: Yale University Press.
- Clough, Patricia Ticineto, ed. 2007. *The Affective Turn: Theorizing the Social.* Durham, NC: Duke University Press.
- Eder, Jens, Julian Hanich, and Jane Stadler. 2019. "Media and Emotion: An Introduction." *NECSUS European Journal of Media Studies* 8 (1): 91–104.
- Gregg, Melissa, and Gregory Seigworth, eds. 2010. *The Affect Theory Reader.* Durham, NC: Duke University Press.
- Kavka, Micha. 2008. *Reality Television, Affect and Intimacy: Reality Matters.* London: Palgrave.
- La Caze, Marguerite, and Henry Martyn Lloyd. 2011. "Philosophy and the 'Affective Turn.'" *Parrhesia* 13:1–13.
- Lünenborg, Margreth, and Tanja Maier. 2018. "The Turn to Affect and Emotion in Media Studies." *Media and Communication* 6 (3): 1–4.
- Madianou, Mirca, and Daniel Miller. 2012. *Migration and New Media: Transnational Families and Polymedia.* London: Routledge.
- Papacharissi, Zizi. *Affective Publics: Sentiment, Technology, and Politics.* Oxford: Oxford University Press.
- Parks, Perry. 2021. "Joy Is a News Value." *Journalism Studies* 22 (6): 820–838.
- Wahl-Jorgensen, Karin. 2019. "Questioning the Ideal of the Public Sphere: The Emotional Turn." *Social Media + Society* 5 (3).

For the analysis specific to this chapter, I turned to work by Alice Mattoni and Emiliano Treré, who used three fundamental concepts of media studies—media practices, mediation, and mediatization—to build a conceptual framework to study social movements and the media. Their inspired work returns in the next chapter on media activism. Understanding media as practice assumes that what we do and how we interact with our environment in turn shapes and changes our experience of social life. As a theoretical approach it is inspired by what is called the "practice turn" in the social sciences and humanities (of the late twentieth century), which is a response to the frustration among researchers that phenomena were either explained by looking at the individual or by pointing at broad structures in society and social systems. In media studies, this, for example, meant scholars traditionally tended to consider the agency and sense making of a specific media user or looked at the media at large for their explanatory frameworks. In recent scholarship, the earlier mentioned affective turn in media studies gets combined with a focus on practice and experience to take seriously all the different ways in which people make sense of themselves and their environment—including the cognitive dimension (of the mind), the sensorial (visual, touch, sound, smells, and so on), the emotional, embodiment (movement, posture, cravings, aches, pains, etc.), and the imagination. Some key readings in (media and) practice theory:

- Ahva, Laura. 2017. "Practice Theory for Journalism Studies." *Journalism Studies* 18 (12): 1523–1541.

- Archetti, Cristina. 2022. "Researching Experience in Journalism: Theory, Method, and Creative Practice." *Journalism Studies.* https://doi.org/10.1080/1461670X.2022.2061576.

- Bräuchler, Birgit, and John Postill, eds. 2010. *Theorising Media and Practice.* New York: Berghahn Books.

- Couldry, Nick. 2004. "Theorising Media as Practice." *Social Semiotics* 14 (2): 115–132.

- Mattoni, Alice, and Emiliano Treré. 2014. "Media Practices, Mediation Processes, and Mediatization in the Study of Social Movements." *Communication Theory* 24:252–271.

- Schatzki, Theodore, Karin Knorr Cetina, and Eike von Savign, eds. 2001. *The Practice Turn in Contemporary Theory.* London: Routledge.

Mediation and mediatization are well-established concepts in media studies, with mediation theory particularly inspired by the work of Roger Silverstone and Jesús Martín-Barbero.

- Martín-Barbero, Jesús. 2006. "A Latin American Perspective on Communication/Cultural Mediation." *Global Media and Communication* 2 (3): 279–297.

- Couldry, Nick. 2008. "Mediatization or Mediation? Alternative Understandings of the Emergent Space of Digital Storytelling." *New Media and Society* 10 (3): 373–391.

- Silverstone, Roger. 2002. "Complicity and Collusion in the Mediation of Everyday Life." *New Literary History* 33 (4): 761–780.

Mediatization theory is a more recent scholarly intervention, spearheaded by work done in the Nordic countries by Stig Hjarvard and Knut Lundby and in Germany by Andreas Hepp, Maren Hartmann, and Tanja Thomas. See also earlier references to work by Jesper Strömbäck and Cristina Archetti mentioned in this appendix. Some argue that mediatization should be seen as one of the metaprocesses in the transformation of societies around the world, on par with globalization, individualization, secularization, and marketization. Others regard it more as a sensitizing concept, keeping us aware of how media not only influence individual and social life but also have become significant actors

Annotated Sources                                                                                    283

in their own right as industries. The media (as part of the cultural industries) today wield formidable economic and political power, beyond their appeal as artifacts, activities, and social arrangements.

- Corner, John. 2018. "'Mediatization': Media Theory's Word of the Decade." *Media Theory* 2 (2): 79–90.

- Hepp, Andreas. 2019. *Deep Mediatization*. Abingdon, UK: Routledge.

- Hjarvard, Stig. 2008. "The Mediatization of Society: A Theory of the Media as Agents of Social and Cultural Change." *Nordicom Review* 29 (2): 105–134.

- Krotz, Friedrich. 2007. "The Meta-Process of 'Mediatization' as a Conceptual Frame." *Global Media and Communication* 3 (3): 256–260.

- Lundby, Knut, ed. 2009. *Mediatization: Concept, Changes, Consequences*. New York: Peter Lang.

- Lunt, Peter, and Sonia Livingstone. 2016. "Is 'Mediatization' the New Paradigm for Our Field?" *Media, Culture and Society* 38 (3): 462–470.

- Schulz, Winfried. 2004. "Reconstructing Mediatization as an Analytic Concept." *European Journal of Communication* 19 (1): 87–102.

Overall, there is no shortage of media about love, and our love for media also knows plenty of art forms worthy of exploration—many of which are referenced in the chapter. It is particularly fascinating to watch older science fiction films (or read such books) in comparison with contemporary work in popular culture in terms of the role technology, media, and machines play in the context (or as the subject) of the narrative.

To conclude, any discussion of media love should include a thoughtful consideration of sex and media. Luckily, there is some phenomenal work in media studies on the topic – including a subdiscipline of porn studies - of which I warmly recommend (as it has informed my work) the following:

- Attwood, Feona. 2006. "Sexed Up: Theorizing the Sexualization of Culture." *Sexualities* 9 (1): 77–94.

- Attwood, Feona, and Clarissa Smith. 2014. "Porn Studies: An Introduction." *Porn Studies* 1 (1–2): 1–6.

- Faustino, Maria João. 2008. "Rebooting an Old Script by New Means: Teledildonics—the Technological Return to the 'Coital Imperative.'" *Sexuality and Culture* 22:243–257.

- Jacobs, Katrien, Marije Janssen, and Matteo Pasquinelli, eds. 2007. *C'Lick me: A Netporn Studies Reader*. Amsterdam: Institute for Network Cultures.

- Tulloch, John, and Belinda Middleweek. 2017. *Real Sex Films: The New Intimacy and Risk in Cinema*. Oxford: Oxford University Press.

## Chapter 6: Change Life

Given the hopeful nature of media scholarship, it should come as no surprise that there is a vibrant field of research available on (the role of) media and social change. For the purposes of this chapter, I divide titles into five distinct categories: historical reviews of media (and digital) activism, international case studies, contemporary theoretical interventions, hybrid warfare, and the link between activism and popular culture. This bibliography starts with works that offer a historical perspective on media, activism, and social movements, with a bias toward titles published in the twenty-first century given the extraordinary role online, social, and networked media play in processes of social transformation and change.

- Carpentier, Nico. 2011. *Media and Participation: A Site of Ideological-Democratic Struggle*. Bristol, UK: Intellect.

- Castells, Manuel. 2012 *Networks of Outrage and Hope: Social Movements in the Internet*. Cambridge: Polity.

- Gerbaudo, Paolo. 2017. "From Cyber-Autonomism to Cyber-Populism: An Ideological History of Digital Activism." *tripleC* 15 (2): 477–489.

- Karatzogianni, Athena. 2015. *Firebrand Waves of Digital Activism 1994–2014: The Rise and Spread of Hacktivism and Cyberconflict*. London: Palgrave.

- Kaun, Anne, and Julie Uldam. 2018. "Digital Activism: After the Hype." *New Media and Society* 20 (6): 2099–2106. This is the introductory overview essay of a journal special issue on digital activism, including articles on cases in China, Russia, and Italy.

- Özkula, SuayMelisa. 2021. "The Problem of History in Digital Activism: Ideological Narratives in Digital Activism Literature." *First Monday* 26 (8). https://doi.org/10.5210/fm.v26i8.10597.

- Mutsvairo, Bruce, and Massimo Ragnedda. 2019. "Does Digital Exclusion Undermine Social Media's Democratizing Capacity?" *New Global Studies* 13 (3): 357–364. The authors offer a cautionary tale of all-too-easy celebrations of the democratizing potential of social media, especially in the context of considering using media for social change.

- Valera-Ordaz, Lidia, and Guillermo López-García. 2019. "Activism, Communication and Social Change in the Digital Age." *Communication and Society* 32 (4): 171–172. This is the introductory overview essay of a journal special issue on digital activism including articles on cases in Spain, Brazil, and Mexico.

A second set of titles broadens the scope of the literature to deliberately include cases and studies from different parts of the world, which is in many ways a typical aspect of scholarly research on media activism.

- Bosch, Tanja. 2017. "Twitter Activism and Youth in South Africa: The Case of #RhodesMustFall." *Information, Communication and Society* 20 (2): 221–232.

- Custódio, Leonardo. 2017. *Favela Media Activism: Counterpublics for Human Rights in Brazil*. Lanham, MD: Lexington Books.

- Khamis, Sahar, and Katherine Vaugn. 2013. "Cyberactivism in the Tunisian and Egyptian Revolutions: Potentials, Limitations, Overlaps and Divergences." *Journal of African Media Studies* 5 (1): 69–86.

- Lee, Francis, and Joseph Chan. 2018. *Media and Protest Logics in the Digital Era: The Umbrella Movement in Hong Kong*. Oxford: Oxford University Press.

- Martens, Cheryl, Cristina Venegas, and Etsa Franklin Salvio Sharupi Tapuy, eds. 2020. *Digital Activism, Community Media, and Sustainable Communication in Latin America*. London: Palgrave. Of interest to note, Etsa Franklin Salvio Sharupi Tapuy is a researcher as well as an Amazonian leader of Quijos and Shuar heritage.

- Podkalicka, Aneta, and Ellie Rennie. 2018. *Using Media for Social Innovation*. Bristol, UK: Intellect. This is an open-access volume featuring numerous activist initiatives, case studies, and action research on the use of (new) media in and by Australian indigenous communities.

- Sreberny, Annabelle. 2015. "Women's Digital Activism in a Changing Middle East." *International Journal of Middle East Studies* 47 (2): 357–361.

- Treré, Emiliano, Sandra Jeppesen, and Alice Mattoni. 2017. "Comparing Digital Protest Media Imaginaries: Anti-Austerity Movements in Spain, Italy and Greece." *tripleC* 15 (2): 404–422.

A third set of titles I highlight because of their profound theoretical engagement with the mutual constitution of media and activist projects, events, and processes of social change. Although much of the scholarly literature warns against technological determinism, and very few (if any) academics would consider media only as an afterthought in their analyses of contemporary protest movements, work that brings both ends of the intellectual spectrum of thinking through technological and social change together (within the specific context of activism) is still relatively rare.

- Cammaerts, Bart. 2021. "The New-New Social Movements: Are Social Media Changing the Ontology of Social Movements?" *Mobilization: An International Quarterly* 26 (3): 343–358. This work suggests that the mix between the opportunities and challenges social media offer to activists amounts to an emerging ontology of social movements, a framework that augments the argument in this chapter.

- Dahlberg-Grundberg, Michael. 2016. "Technology as Movement: On Hybrid Organizational Types and the Mutual Constitution of Movement Identity and Technological Infrastructure in Digital Activism." *Convergence* 22 (5): 524–542.

- George, Jordana, and Dorothy Leidner. 2019. "From Clicktivism to Hacktivism: Understanding Digital Activism." *Information and Organization* 29 (3). https://doi.org/10.1016/j.infoandorg.2019.04.001. This article offers a useful categorization of different kinds of digital activism.

- Hancox, Donna. 2017. "From Subject to Collaborator: Transmedia Storytelling and Social Research." *Convergence* 23 (1): 49–60. This piece combines activism online and offline with storytelling and performativity, aptly including a discussion of KONY2012.

- Jackson, Sarah, Moya Bailey, and Brooke Foucault Welles. 2020 *#HashtagActivism: Networks of Race and Gender Justice.* Boston: MIT Press (open access).

- Stefania, Milan, and Lonneke van der Velden. 2016. "The Alternative Epistemologies of Data Activism." *Digital Culture and Society* 2 (2): 57–74. This article sets a research agenda on data activism focused on our agency vis-à-vis datafication.

The chapter contains an addendum on the role of media in war, admittedly inspired by the fact that during the writing and editing of this book, Russia invaded Ukraine—a war that affected me both professionally and personally. In 2021 the Academic Council of the Faculty of Journalism at Moscow State University (MSU) kindly conveyed the title of "Foreign Honorary Professor of the Faculty" to me. After Russia's illegal invasion in February 2021 and MSU's public support for this illegal war (as expressed in a formal statement as well as in a letter from the Russian Union of Lectors that echoed President Putin's propaganda), it was clear that keeping a formal institutional relation intact was untenable, and I relinquished this title upon due deliberation. What kept me from immediately rescinding my position at the university are the good relations I have with colleagues and students in Moscow.

When it comes to media and war, much scholarship focuses on the role of journalism and the news coverage of war, as well as on war propaganda. In his review of numerous books on war and media, McQuail argues that it is clear that war in the twenty-first century enters a new situation, requiring information dominance and public diplomacy. He concludes that "there is no doubt that the media have become more involved in issues of war and peace, but they are not the main source of the problem and they do show some potential to contribute to solutions."[4] In the context of pervasive and ubiquitous media, research now adds more ethnographic approaches, mindful of the fact that everyone participates in using and making media in times of armed conflict. In doing so, war becomes a shockingly banal lived experience rather than something that happens to people exclusively because of artillery and armies. It is exactly at this everyday level that media come in as a powerful agent affecting the social fabric of conflict.

---

4. Denis McQuail, "On the Mediatization of War," *Gazette* 68, no. 2 (2006): 118.

# Annotated Sources

- Budka, Philipp, and Brigit Bräucher. 2020 *Theorising Media and Conflict*. New York: Berghahn Books.
- Fridman, Ofer. 2017. "Hybrid Warfare or Gibridnaya Voyna?" *RUSI Journal* 162 (1): 42–49. Fridman also published a subsequent monograph on Russian hybrid warfare that informs my discussion of the concept in this chapter.
- McQuail, Denis. 2006. "On the Mediatization of War." *Gazette* 68 (2): 107–118.
- Murray, Williamson, and Peter Mansoos. 2012. *Hybrid Warfare: Fighting Complex Opponents from the Ancient World to the Present*. Cambridge: Cambridge University Press.
- Ogunyemi, Ola, ed. 2017. *Media, Diaspora and Conflict*. New York: Springer.

Finally, I would like to add a fifth set of bibliographic references for this chapter, particularly regarding the relation between popular culture, activism, and participation in media. Of interest is the Civic Imagination Project run out of the University of Southern California's (USC) Annenberg School of Communication and Journalism in Los Angeles (see civicimaginationproject.org). A disclaimer for this source is that I used to work at USC as a Fulbright scholar back in 2003. What makes this project so relevant here is how it provides both an intellectual framework and a tool to work with communities to "identify, map, and analyze stories that have inspired social action." It is one of arguably countless cases from around the world where grassroots issues, media, popular culture, and people jumping into action all come together, in a way operationalizing all the arguments about life in media. If media is coconstitutive of conflict, it can also coproduce a better world.

- Brough, Melissa, and Sangita Shresthova. 2012. "Fandom Meets Activism: Rethinking Civic and Political Participation." *Transformative Works and Cultures* 10:1–27.
- Carpentier, Nico, and Henry Jenkins. 2022. "What Does God Need with a Starship? A Conversation about Politics, Participation, and Social Media." In *The Social Media Debate*, edited by Devan Rosen, 203-21. London: Routledge.
- Ito, Mizuko. 2010. *Hanging Out, Messing Around, and Geeking Out: Kids Living and Learning with New Media*. Boston: The MIT Press.
- Jenkins, Henry, Sangita Shresthova, Liana Gamber-Thompson, Neta Kligler-Vilenchik, and Arely Zimmerman. 2016. *By Any Media Necessary: The New Youth Activism*. New York: NYU Press.
- Jenkins, Henry, Gabriel Peters-Lazaro, and Sangita Shresthova. 2020. *Popular Culture and the Civic Imagination: Case Studies of Creative Social Change*. New York: NYU Press.
- Mina, An Xiao. 2019. *Memes to Movements: How the World's Most Viral Media Is Changing Social Protest and Power*. Boston: Beacon.

288 Appendix 1

- Moreno-Almeida, Cristina. 2021. "Memes as Snapshots of Participation: The Role of Digital Amateur Activists in Authoritarian Regimes." *New Media and Society* 23 (6): 1545–1566.

**Chapter 7: Make Life**

To be honest, this should be the "easiest" chapter for me to provide an annotated bibliography for, yet it is also the hardest. Arguably, most of my academic life I have spent researching how media get made professionally. I was a freelance culture and arts reporter before joining the university and have carried my passions and frustrations about the work with me to my research and teaching agenda ever since. Thanks to the support of colleagues and students and the participation of so many media professionals, I have been privileged to be able to publish several articles and books on media work, appearing alongside the work of academics I admire and respect. That kind of personal connection also makes it difficult to list the works that inform the argument in this chapter and that could set the reader on their way to dig deeper into the phenomena of media work and production studies. Not to argue that my work in this area should be anyone's starting point, but merely to state for the record where I base my claim of expertise for the arguments presented in this chapter, here are some references to the work that I have done in this area:

- Deuze, Mark. 2007 *Media Work*. Cambridge: Polity.
- Deuze, Mark. 2009. "The Media Logic of Media Work." *Journal of Media Sociology* 1 (1–2): 22–40.
- Deuze, Mark, ed. 2011. *Managing Media Work*. London: Sage.
- Deuze, Mark, and Nicky Lewis. 2013. "Professional Identity and Media Work." In *Theorizing Cultural Work: Transforming Labour in the Cultural and Creative Industries*, edited by Mark Banks, Stephanie Taylor, and Rosalind Gill, 161–174. London: Routledge.
- Deuze, Mark, Gemma Newlands, Johana Kotišová, and Erwin van 't Hof. 2020. "Toward a Theory of Atypical Media Work and Social Hope." *Artha Journal of Social Sciences* 19 (3): 1–20. This essay was part of a special issue on work in the creative realm.
- Deuze, Mark, and Mirjam Prenger, eds. 2019. *Making Media: Production, Practices and Professions*. Amsterdam: Amsterdam University Press.
- Elefante, Phoebe, and Mark Deuze. 2012. "Media Work, Career Management, and Professional Identity: Living Labour Precarity." *Northern Lights* 10 (1): 9–24.
- Van 't Hof, Erwin, Deuze, Mark. 2022. "Making Precarity Productive." In *Precarity in Journalism*, edited by Linda Steiner and Kalyani Chadha, 189–202. London: Routledge.

**Annotated Sources**

It is important to note that media production was not really a distinct field of study (within media studies) until well into the twenty-first century. However, there was a vibrant twentieth-century literature on media work before it coalesced into a field. Many excellent works have been published dating as far back as the 1950s on what working in the media is like, for example, in Hollywood, within local newsrooms, and in television, film, and music production cultures around the world. An approach consolidating such efforts; integrating media industries, management, and production; and considering both political economy perspectives and the lived experience of workers across various media professions was lacking until David Hesmondhalgh's seminal book on the cultural industries, the first edition of which was published in 2002. Ever since, a rapidly growing list of works appeared—mostly as books and edited volumes, as a dedicated scholarly journal for the study of media production was not introduced until 2014 with the launch of the *Media Industries Journal*.

The references suggested here are organized along the lines of doing media production studies, general works on media industries and work, some key titles in the area of media management and organization, and concluding with what are some key issues of critical interest for studies on making media: the role of data, algorithms, automation, and platformization on the one hand and the mental health and well-being of media professionals on the other hand. It has to be said that attention for the well-being of workers is a relatively recent phenomenon in media studies—but then again, almost all the topics raised in this chapter and in the bibliography offered here are to some extent.

Studying and understanding how media get made, what the lived experience of media professionals looks like and feels like, and how all this ties into broader explanations of how the media as an industry work requires a great deal of theoretical and methodological legwork. Production scholars tend to combine insights and approaches from a variety of disciplines, using this creatively to overcome unique obstacles when doing fieldwork—for example, a lack of access and facing distrust from practitioners (or their employers, clients, or funders). Additionally, we have to be careful when documenting the work lives of media professionals as this can lead to what Vicki Mayer (2008, 145) describes as "bizarre forms of complicity," with the researcher becoming enmeshed in the very control chain and subsequent uneven power relationships they are critiquing. To get us going, here are some reflections on the intricacies of production studies:

- Arsenault, Amelia, and Alisa Perren, eds. 2016. *Media Industries: Perspectives on an Evolving Field*. Self-published, Amazon CreateSpace.

- Banks, Miranda, Bridget Conor, and Vicki Mayer, eds. 2016. *Production Studies, The Sequel! Cultural Studies of the Global Media Industries*. London: Routledge.

- Caldwell, John, Vicki Mayer, and Miranda Banks, eds. 2009. *Production Studies: Cultural Studies of Media Industries*. London: Routledge.

- Havens, Timothy, Amanda Lotz, and Serra Tinic. 2009. "Critical Media Industry Studies: A Research Approach." *Communication, Culture and Critique* 2 (2): 234–253.

- Hesmondhalgh, David, and Sarah Baker. 2011. *Creative Labour: Media Work in Three Cultural Industries.* London: Routledge.
- Mayer, Vicki. 2008. "Studying Up and F**cking Up: Ethnographic Interviewing in Production Studies." *Cinema Journal* 47 (2): 141–148.
- Paterson, Chris, David Lee, Anamik Saha, and Anna Zoellner, eds. 2015. *Advancing Media Production Research.* London: Palgrave.

Research documenting what working in the media is like proliferates, often organized around specific cases (such as a newsroom, movie set, or advertising agency) and particular countries. Although it is encouraging to see more research being done outside of the United States and United Kingdom, it would be too much to list all such excellent entries in production studies of, for example, Nollywood, Hong Kong's creative industries, or eastern European screen cultures. Instead, my argument in this chapter seeks to address issues and ambivalences that cut across countries and professions, and the following references are tremendously useful in doing so. I would like to highlight the importance of Mark Banks's (2017) work, calling on those of us who study media production to pursue and push for "creative justice," which means that we should respect all the "internal" benefits, capacities, and pleasures media work provides, without discounting the "external" structures and pressures (such as exploitation, alienation, low pay, and stress) that can make media work deeply unfair and unjust; that we should always advance social arrangements that allow for the maximum range of people to enter and participate in the work, in which they will be fairly treated and justly paid and rewarded for their efforts; and that our studies should ideally contribute to reducing the physical and psychological harms and injuries inflicted by media work, making sure that practitioners are treated fairly and justly as dignified and deserving human beings.

- Banks, Mark. 2017. *Creative Justice: Cultural Industries, Work and Inequality.* Lanham, MD: Rowman & Littlefield.
- Christopherson, Susan. 2006. "Behind the Scenes: How Transnational Firms Are Constructing a New International Division of Labor in Media Work." *Geoforum* 37:739–751.
- Curtin, Michael, and Kevin Sanson, eds. 2016. *Precarious Creativity: Global Media, Local Labor.* Oakland: University of California Press.
- Freeman, Matthew. 2016. *Industrial Approaches to Media: A Methodological Gateway to Industry Studies.* London: Palgrave.
- Hesmondhalgh, David. 2010. "Media Industry Studies, Media Production Studies." In *Media and Society*, edited by James Curran, 45–163. London: Bloomsbury.
- Hesmondhalgh, David. 2018. *The Cultural Industries.* 4th ed. London: Sage.

# Annotated Sources

- Hill, Erin. 2016. *Never Done: A History of Women's Work in Media Production.* New Brunswick, NJ: Rutgers University Press.

- Jenkins, Henry. 2006. *Convergence Culture: Where Old and New Media Collide.* New York: NYU Press.

- Laaksonen, Salla-Maaria, and Mikko Villi, eds. 2022. "New Forms of Media Work and Its Organizational and Institutional Conditions." Special issue of *Media and Communication* 10 (1).

- Maxwell, Richard, ed. 2017. *The Routledge Companion to Labor and Media.* London: Routledge.

- Miller, Toby, and Marie-Claire Leger. 2001. "Runaway Production, Runaway Consumption, Runaway Citizenship: The New International Division of Cultural Labor." *Emergences* 11 (1): 89–115.

- Perren, Alisa, and Jennifer Holt, eds. 2009. *Media Industries: History, Method, and Theory.* Oxford: Blackwell.

- Winseck, Dwayne, and Dal Yong Jin, eds. 2011. *The Political Economies of Media: The Transformation of the Global Media Industries.* London: Bloomsbury.

Although media studies tend to move quickly from the macrolevel of media and society relations to the microlevel of individual experience and sense-making practices, the in-between or mesolevel of organization and management is of crucial importance when interpreting media production. Most research in this area consistently finds that organization-level dynamics, such as ownership structure, workplace environment and production culture, and professional socialization, to a significant extent determine what happens and how things work—more so than who the individual workers are or within what kind of overall media and social system they operate.

- Albarran, Alan, Bozena Mierzejewska, and Jaemin Jung, eds. 2018. *Handbook of Media Management and Economics.* 2nd ed. London: Routledge.

- Cottle, Simon. 2003. *Media Organization and Production.* London: Sage.

- Johnson, Derek, Derek Kompare, and Avi Santo, eds. 2014. *Making Media Work: Cultures of Management in the Entertainment Industries.* New York: NYU Press.

- Küng, Lucy. 2017. *Strategic Management in the Media.* 2nd ed. London: Sage.

- Lampel, Joseph, Theresa Lant, and Jamal Shamsie. 2000. "Balancing Act: Learning from Organizing Practices in Cultural Industries." *Organization Science* 11(3): 263–269. This paper inspired my approach to paradoxes (called "polarities" in the article) in how media are managed to create value.

- Lowe, Gregory Ferrell, and Charles Brown, eds. 2016. *Managing Media Firms and Industries.* New York: Springer.

Specific mention must be made of contemporary scholarship that looks specifically at the role of data, algorithms, automation, and platformization in media industries and production, as such "technologization" of the creative process has profound consequences for the work of practitioners across all media professions (including journalism).

- Arsenault, Amelia. 2017. "The Datafication of Media: Big Data and the Media Industries." *International Journal of Media and Cultural Politics* 13 (1–2): 7–24.

- Bilton, Chris. 2017. *The Disappearing Product: Marketing and Markets in the Creative Industries.* Cheltenham, UK: Edward Elgar.

- Napoli, Philip. 2014. "On Automation in Media Industries: Integrating Algorithmic Media Production into Media Industries Scholarship." *Media Industries Journal* 1 (1): 33–38. This article conceptualizes the notion of data as content creator and demand predictor in media production.

- Poell, Thomas, David Nieborg, and Brooke Erin Duffy. 2021. *Platforms and Cultural Production.* Cambridge: Polity.

In conclusion, work on the mental health and (subjective) well-being of media professionals needs to be mentioned, given the extraordinary pressures these workers face on the job. This selection is also inspired by the fact that so many students in media and communication programs worldwide seek careers in the media and need to prepare effectively not just for the skills and competences involved but also for how they can practice self-care and pursue creative justice for themselves and others while doing so. The following works either directly or indirectly address these issues in more detail.

- Baym, Nancy. 2018. *Playing to the Crowd: Musicians, Audiences, and the Intimate Work of Connection.* New York: NYU Press.

- Bulut, Ergin. 2020. *A Precarious Game: The Illusion of Dream Jobs in the Video Game Industry.* Ithaca, NY: Cornell University Press.

- Deuze, Mark, and Tamara Witschge. 2018. "Beyond Journalism: Theorising the Transformation of Journalism." *Journalism* 19 (2): 165–181.

- Ertel, Michael, Eberhard Pech, Peter Ullsperger, Olaf Von Dem Knesebeck, and Johannes Siegrist. 2005. "Adverse Psychosocial Working Conditions and Subjective Health in Freelance Media Workers." *Work and Stress* 19 (3): 293–299.

- Heyman, Lucinda, Rosie Perkins, and Liliana Araújo. 2019. "Examining the Health and Well-Being Experiences of Singers in Popular Music." *Journal of Popular Music Education* 3 (2): 173–201.

- Kotišová, Johana. 2019. "Devastating Dreamjobs: Ambivalence, Emotions, and Creative Labor in a Post-Socialist Audiovisual Industry." *Iluminace* 31 (4): 27–45.

- Liu, Huei-Ling, and Ven-hwei Lo. 2018. "An Integrated Model of Workload, Autonomy, Burnout, Job Satisfaction, and Turnover Intention among Taiwanese Reporters." *Asian Journal of Communication* 28 (2): 153–169.
- Mayer, Vicki. 2011. *Below the Line: Television Producers and Production Studies in the New Economy.* Durham, NC: Duke University Press.
- McRobbie, Angela. 2002. "Clubs to Companies: Notes on the Decline of Political Culture in Speeded Up Creative Worlds." *Cultural Studies* 16 (4): 516–531.
- Monteiro, Susana, Alexandra Marques Pinto, and Magda Sofia Roberto. 2016. "Job Demands, Coping, and Impacts of Occupational Stress among Journalists: A Systematic Review." *European Journal of Work and Organizational Psychology* 25 (5): 751–772.
- Rimscha, Bjørn von. 2015. "The Impact of Working Conditions and Personality Traits on the Job Satisfaction of Media Professionals." *Media Industries Journal* 2 (2). https://doi.org/10.3998/mij.15031809.0002.202.
- Šimunjak, Maja, and Manuel Menke. 2022. "Workplace Well-being and Support Systems in Journalism." *Journalism.* https://doi.org/10.1177/14648849221115205

## Chapter 8: Life in Media

In this concluding chapter, all the different arguments and lines of exploration come together, and the sources used for the various points are detailed in previous annotations. Media studies for a life in media is an approach that seeks to combine the various affective, practice, and material turns in the scholarship on the role of media in society and everyday life and infuse it with a distinct ethical and hopeful perspective. At the end of the book, I therefore want to share some key resources in media studies regarding the environmental impact of media, an appreciation of workers' rights in the various media industries, and how the field responds to rapid changes in our digital environment. The fourth and final category of references for the book and this chapter is the area of media (and information) literacy as a place where the hopeful promise of media studies meets the practice of media education.

When it comes to ethical and sustainability issues, media studies as a field still has some catching up to do. Beyond Richard Maxwell and Toby Miller's seminal book on greening the media, most of the work in this area is published outside of the discipline. In the context of the International Panel on Social Progress (its report launched in 2018), media scholars from around the world discussed the role of media and communications in contributing to social progress, arguing that effective media access is a core component of social progress and signaling various ethical and regulatory roadblocks along the way—important work, although the panel did not discuss the environmental impact of such a life in media. When it comes to teaching (about) media, I would advocate always integrating ecological aware-ness of what happens when media are born, produced, used, and when media die.

- Berkhout, Frans, and Julia Hertin. 2004. "De-materialising and Re-materialising: Digital Technologies and the Environment." *Futures* 36 (8): 903–920. This presents a nuanced analysis that suggests that the diffusion and use of information and communication technologies lead to both positive and negative environmental impacts.

- Couldry, Nick, Clemencia Rodriguez, Göran Bolin, Julie Cohen, Ingrid Volkmer, Gerard Goggin, Marwan Kraidy, Koichi Iwabuchi, Jack Linchuan Qiu, Herman Wasserman, Yuezhi Zhao, Omar Rincón, Claudia Magallanes-Blanco, Pradip Ninan Thomas, Olessia Koltsova, Inaya Rakhmani, and Kwang-Suk Lee. 2018. "Media, Communication and the Struggle for Social Progress." *Global Media and Communication* 14 (2): 173–191.

- Craig, Geoffrey. 2019. *Media, Sustainability and Everyday Life*. London: Palgrave. This book offers a textual analysis of the way a variety of media (social media, television, newspapers, and advertising campaigns) present sustainability primarily through commercial contexts and processes of consumption.

- Lopez, Antonio. 2021. *Ecomedia Literacy: Integrating Ecology into Media Education*. London: Routledge.

- Maxwell, Robert, and Toby Miller. 2012. *Greening the Media*. Oxford: Oxford University Press. This is a key source for many of the arguments in this chapter.

The resources referenced for chapter 7 already cover most of the issues covered in this final chapters' point on the rights of workers in the global media industries. It is interesting to note that media professionals are increasingly likely to self-organize, either formally through unions and other professional associations or informally via online communities using Whatsapp, Telegram, Signal, and Facebook groups. Some early scholarly work in this area:

- Cohen, Nicole, and Greig de Peuter. 2020. *New Media Unions: Organizing Digital Journalists*. Abingdon, UK: Routledge.

- McKercher, Catherine. 2002. *Newsworkers Unite: Labor, Convergence and North American Newspapers*. Lanham, MD: Rowman & Littlefield.

- Rossiter, Ned. 2006. *Organized Networks: Media Theory, Creative Labour, New Institutions*. Rotterdam: Nai Publishers.

On the issue of representation and how media studies can tackle this issue in today's polymedia, intermedial, or transmedia context, I would reference earlier mentioned work by such scholars as Mirca Madianou, Andreas Hepp, Karin Fast, Jørgen Bruhn, Beate Schirrmacher, Patti Valkenburg, and others who have offered different approaches to

# Annotated Sources

theorize and empirically address doing research on people and their media as an ensemble. Over the last few decades, numerous interventions have been made to invigorate, challenge, push, and tickle the field of media studies to change its core premise, to reconceptualize, to rethink, and to engage more deliberately in public debates—since so many of the issues the world is facing (the parallel pandemic and infodemic, hybrid warfare, and even the climate crisis) are mediatized. Some key texts that inspire the argument as outlined in this book are listed below.

- Cardoso, Gustavo. 2008. "From Mass to Networked Communication: Communicational Models and the Informational Society." *International Journal of Communication* 2:587–630.

- Castells, Manuel. 2007. "Power and Counter-Power in the Network Society." *International Journal of Communication* 1:238–266. This paper introduced the concept of "mass self-communication" into the field.

- Chaffee, Steven, and Miriam Metzger. 2001. "The End of Mass Communication?" *Mass Communication and Society* 4 (4): 365–379. This article suggests that mass communication, as a concept, should be replaced with media.

- Hallin, Daniel. 2020. "Comparative Research, System Change, and the Complexity of Media Systems." *International Journal of Communication* 14:5775–5786.

- Fuchs, Christian, and Jack Linchuan Qiu. 2018. "Ferments in the Field: Introductory Reflections on the Past, Present and Future of Communication Studies." *Journal of Communication* 68 (2): 219–232. This is an introduction to a special issue discussing the state and possible future of the field.

- Jensen, Klaus Bruhn. 2018. "The Double Hermeneutics of Communication Research", *Javnost / The Public* 25 (1–2): 177–183. This article reminds us that studying media and communication is a practical discipline: we study something that people do and, in doing so, make ways of using and thinking about media possible.

- Kraidy, Marwan. 2018. "Global Media Studies: A Critical Agenda." *Journal of Communication* 68 (2): 337–346. This article urges us to step out of our comfort zones (of research and academia) to deliberately engage in public debates.

- Krajina, Zlatan, Shaun Moores, and David Morley. 2014. "Non-Media-Centric Media Studies: A Cross-Generational Conversation." *European Journal of Cultural Studies* 17 (6): 682–700. This essay reminds us that studying media requires a genuine interest in social life, including embodied practices that are always meaningful to people in the everyday.

- Matassi, Mora, and Pablo Boczkowski. 2021. "An Agenda for Comparative Social Media Studies: The Value of Understanding Practices from Cross-National, Cross-Media, and Cross-Platform Perspectives." *International Journal of Communication* 15:207–228.

- Mihelj, Sabina, and James Stanyer. 2019. "Theorizing Media, Communication and Social Change: Towards a Processual Approach." *Media, Culture and Society* 41 (4): 482–501. This article draws a distinction between media and communication as an *agent* of social change and as an *environment* for social change, advocating an integrated approach for all research in the field.

- Miller, Toby. 2009. Media Studies 3.0. *Television and New Media*, 10 (1): 5–6.

- Poell, Thomas. 2020. "Three Challenges for Media Studies in the Age of Platforms." *Television and New Media* 21 (6): 650–657.

- Waisbord, Silvio. 2019. *The Communication Manifesto*. Cambridge: Polity. Similar to Marwan Kraidy, Waisbord calls on us to engage in public scholarship.

In the end, it is fascinating to consider the rapidly growing field of media and information literacy, both in terms of research and practical applications. Media scholars get increasingly involved in these debates, and the field develops in interesting ways accordingly—focusing more and more on creative, playful, informal, and imaginative ways in which we can learn to use media rather than simply appreciating media as critical consumers.

- Frau-Meigs, Divina. 2012. "Transliteracy as the New Research Horizon for Media and Information Literacy." *Medijske Studije* 3(6): 14–27.

- Gauntlett, David. 2018. *Making Is Connecting*. 2nd ed. Cambridge: Polity.

- Hartley, John. 2007. "There Are Other Ways of Being in the Truth: The Uses of Multimedia Literacy." *International Journal of Cultural Studies* 10 (1): 135–144. This article signals the importance of read-write literacy beyond read-only literacy.

- Jenkins, Henry. 2009. *Confronting the Challenges of PC: Media Education for the 21st Century*. Chicago: John D. and Catherine T. MacArthur Foundation Reports on Digital Media and Learning.

- Livingstone, Sonia. 2004. "Media Literacy and the Challenge of New Information and Communication Technologies." *Communication Review* 7 (1): 3–14.

- Meyrowitz, Joshua. 1998. "Multiple Media Literacies." *Journal of Communication* 48 (1): 96–108. Meyrowitz makes a distinction between content literacy (media as transmitters of messages), medium literacy (seeing media as environment), and media grammar literacy (considering production variables for each medium).

- Scolari, Carlos, Maria-José Masanet, Mar Guerrero-Pico, M.-J. Masanet, M. Guerrero-Pico, and María-José Establés. 2018. "Transmedia Literacy in the New Media Ecology: Teens' Transmedia Skills and Informal Learning Strategies." *El profesional de la información* 27 (4): 801–812.

# Appendix 2: Key Concepts in Tweets

1.  Active data: data about people's movements and activities
2.  Affective data: data about people's thoughts, ideas, feelings, and emotions
3.  Affordance (vs. functionality): the potential action that is possible by a given object or environment (versus the ability of a particular device or technology)
4.  Algorithm: a series of instructions telling a computer how to transform data into useful knowledge
5.  Ambient intimacy: being able to keep in touch with people with a level of regularity and intimacy enhanced by (social) media
6.  Anti-surveillance: behavior intended to either avoid monitoring altogether or make observation more difficult to achieve
7.  Artificial intelligence: a self-learning group of algorithms
8.  Atypical work: work that people get paid for but without the benefits usually associated with formal employment (e.g., a contract, health care, pension plan)
9.  Augmented reality (AR): a direct or indirect view of a physical environment merged with or enhanced by virtual computer-generated imagery (see also: Mixed reality)
10. Authenticity puzzle/contract: the delicate negotiation between media producers and consumers about what is fake or real
11. Avatar activism: the appropriation of popular culture—such as film franchises, games, rock and pop music—for civic and protest purposes
12. Blockbuster (see also Tentpole, triple-A): industry term for a best-selling media product such as a film or digital game
13. Blockchain: a digital ledger of transactions duplicated and distributed across a network of computers instead of centrally organized by (for example) a bank
14. Charismatic technologies: media that contribute to processes of personal transformation, identity formation, and expression

15. Circuit of culture: concept that highlights interdependent relations between people and media, focusing on production, identity, representations, regulations, and consumption

16. Clicktivism: the use of online media (such as social networks, websites, blogs, and vlogs) to publicize, promote, and support causes for social change

17. Close reading of media: a systematic analysis of all aspects of a specific text, documenting the process in great detail to find out what it is trying to say

18. Command line interface: an interface whereby users interact with a computer by typing in text commands to run programs and manage files

19. Concurrent media exposure: being exposed to multiple media at the same time, whether people are aware of such exposure or not

20. Convergence culture: the parallel integration of multiple media (in corporations and storytelling practices) and the cultures of media production and consumption (by inviting audiences to cocreate stories)

21. Convergence logic: A creative decision-making process (in the media industry) that considers people as cocreators of products and services.

22. Coveillance (see Synoptic surveillance, Inverse surveillance)

23. Cross-media storytelling: to publish or push the same story using multiple forms of media

24. Crowdfunding: raising money to finance projects and businesses from a large number of people via online platforms (such as Indiegogo, Kickstarter, Patreon)

25. Crowdsourcing: obtaining work, information, ideas, or opinions from a large group of people who submit their contribution online (via websites, social media, or dedicated platforms)

26. Crunch time: working extreme overtime to get a media production project finished on deadline (term mainly used in the games industry)

27. Cryptocurrencies: digital currency organized through a blockchain (see Blockchain) and protected by cryptography rather than a bank.

28. Cultural analysis of media: analysis of media texts as a source of meaning for (and about) a particular culture or community.

29. Cyborg: a cybernetic organism, partly human and partly machine

30. Dark participation: various forms of malicious online participation, including offensive speech, hate speech, cyberbullying, producing and distributing fake news and conspiracy theories

31. Data empathy: enriching the statistical analysis of big data with personal stories, backgrounds, and context that give meaning to the material

32. Data glut: an overwhelming and ever-increasing amount of information gathered and stored (by individual people, organizations, or institutions)

33. Data logic: a creative decision-making process (in the media industry) that is primarily oriented toward data, metrics, predictive analytics, and key performance indicators (KPIs)

34. Data mining: finding patterns and relations in large data sets using statistical methods

35. Data portability: controlling personal information based on open software standards allowing people to reuse their data across different applications online

36. Dataveillance: the use of information and communication technologies in the surveillance of people, where data is primarily gathered (and analyzed) by computers

37. Digital commons: a form of communal ownership of data, information, culture, and knowledge (possibly including underlying technological infrastructure and services)

38. Digital culture: an emerging value system and set of practices and expectations as expressed in computer-mediated communication

39. Digital democracy: democracy enhanced by information and communication technologies, enabling people to find information, engage in deliberation, and participate in political decision-making online

40. Digital disconnection/detox: a range of practices related to disconnecting or disengaging from (online) media for different purposes, including protest and activism, self-regulation and health, freedom and sustainability

41. Digital divide: the gap between people with a high degree of access to information and communication technologies and those with limited access or no access at all

42. Digital inequality: differences between people in the material, cultural, and cognitive resources required to use information and communication technology

43. Digital shadow: the information you create about yourself and the information others (including algorithms) create about you online

44. Disinformation: the dissemination of false or misleading information with the deliberate intent to manipulate and deceive

45. Double articulation (of media and society): media influence established processes in society, as well as independently creating routines within and across society's institutions

46. Editorial logic: A creative decision-making process (in the media industry) that is primarily oriented toward (perceived) peers, colleagues, and competitors

47. Egodocuments (or self-life writings): all forms of voluntary and involuntary autobiographical writing, such as memoirs, diaries, letters, travel accounts, blogs, vlogs, and social media posts

48. ELIZA effect: people's tendency to anthropomorphize machines (often specifically referring to an artificial intelligence)

49. Environment / ensemble / manifold / communicative figuration / intermediality (see also Concurrent media exposure): Different concepts to describe the sensation and experience of living with (and using) multiple media somewhat simultaneously

50. Ephebiphobia: an irrational fear of adolescents or teenagers

51. Everydayness: everyday experience is made up out of cycles (day and night, life and death, etc.) and repetitive behaviors, making life both mundane and always changing

52. Frictionless sharing: steps platforms take to reduce friction and get people to spend more time online using their products and services; friction refers to deliberate choice in media use

53. Functionality (vs. affordance): the ability of a particular device or technology (vs. the potential action that is possible by a given object or environment)

54. Gender-bending: acting in away (online or offline) that defies or challenges traditional notions of gender, especially with respect to dress or behavior

55. Graphic user interface: an interface that allows people to interact with electronic devices via visual indicators (such as menus, icons, pointers, and windows)

56. Greenlighting: giving permission for a project to go ahead (generally involving financial expenditure)

57. Hacking: the reconfiguration or reprogramming of a computer system to function in ways not intended by the producer(s)

58. Hacktivism: the act of hacking (see Hacking) for politically or socially motivated purposes

59. Halo effect: the "what is beautiful is good" stereotype

60. Horizontal integration (of media industries): media companies consolidating and bundling their offerings across a variety of media channels

61. Hourglass structure (of media industries): the media consist of a handful of big corporations, few middle-sized companies, and many tiny companies and single contractors

62. Hybrid warfare: when a nation's government and military combine information warfare and cyberwarfare with conventional warfare

63. Infodemic: a rapidly spreading large amount of information about a problem that is typically unreliable or the product of a disinformation campaign

64. Interface (see Command line interface, Graphic user interface, Natural user interface)

65. Intertextual(ity) / intertextual referencing: parts of a media text (i.e., any work, object, or event that communicates meaning) that refer to other texts, for example, when media industries integrate characters and story lines across a variety of media

66. Inverse surveillance (see Synoptic surveillance, Coveillance): a form of surveillance where the many observe and monitor the few

67. Lean-forward media: media that (physically, emotionally, or cognitively) engage the user directly, requiring people to pay close attention

68. Life in media / media life: media disappear, media are what people do, and people love media

69. Market logic: a creative decision-making process (in the media industry) that is primarily oriented toward (intended or imagined) consumers, audiences, and markets

70. Martini media: (making or using) media that are available anytime, anyplace, anyhow

71. Mass self-communication: the circulation and reformatting of any digitally formatted content, generally consisting of egodocuments (see Egodocuments) posted online

72. Material access to media: having access to media as determined by media artifacts (i.e., hardware and software)

73. Materiality: the assumption that the physical properties of an artifact have consequences for how it is or can be used (see also Affordances, Functionality)

74. Media activism (also, digital/cyberactivism): a form of activism that either has the media as the object to be reformed or uses a variety of media to further its goals of social or political transformation

75. Media activities: the activities and practices involved when people use media

76. Media archaeology: a way to think about material media cultures in a historical perspective (e.g., by taking media artifacts apart and tracing the genealogy of their constituent parts)

77. Media arrangements: how people organize and coordinate their lives with and around media

78. Media artifacts: the devices (i.e., hardware and software) people use to live in media

79. Media as practice: studying the open-ended range of practices focused directly or indirectly on media (see also Media activities)

80. Media multitasking: deliberate use of multiple media at the same time

81. Media repertoire: a collection of media sources that people regularly use or a particular way in which people manage and use various media

82. Mediatime: all the time people spend concurrently exposed to media

83. Mediagenic: when something or someone fits the criteria for selection by media

84. Mediation: the circulation and appropriation of information and ideas via media (as institutions)

85. Mediatization: the process by which the media takes a prominent role in society, both as an industry and as the function that media have in the daily life of people and institutions

86. Mediology: replacing ontology as the primary source of how and what we know about the world

87. Meme(s): an idea (digitally represented in a phrase, image, video) that spreads from person to person by replication and adaptation (and is copied or changed along the way)

88. Metaverse: (hard definition) a network of virtual worlds accessible via headsets or (soft definition) the notion of a seamlessly integrated media experience

89. Metavoicing: reacting to other people's online presence and posts (particularly used in the context of activism)

90. Microrebellions: acts of protest and resistance by one person (or only a few individuals), documented and shared on popular social media

91. Motivational access to media: personal reasons for wanting to engage or not engage or participate with the media

92. Natural user interface: an interface that makes you use electronic media using touch, gestures, or voice, in effect naturalizing the interface (and making it invisible)

93. Neo-Luddite (or reform Luddite): someone who opposes (the indiscriminate use of) technology or believes the use of technology has problematic consequences

94. Nowism (see also Presentism): excessive focus on the present or on immediate gratification

95. Omnopticism (see Sousveillance): a situation where everyone monitors (or can monitor) everyone else

96. Open source: an approach to software design where anyone can freely view, edit, modify, and distribute the source code

97. Panopticism (see Vertical surveillance): the systematic ordering and controlling of people by their perception or knowledge of being under constant surveillance

98. Participatory surveillance: the extent to which people willingly submit to having their personal information collected and tracked

99. Passive data: data about people at a particular time and place

100. Personal information economy: an economy where value is primarily extracted by the gathering of (or providing access to) personal data

101. Pervasive media: media are always on, impossible to (completely) turn off

102. Platform cooperatives: a collectively owned and democratically governed digital platform designed to provide a service or sell products

103. Platformization: the penetration of infrastructures, economic processes, and governmental frameworks of digital platforms in different economic sectors and spheres of life and the reorganization of practices and imaginations around platforms

# Key Concepts in Tweets

104. Polysemy: media content always has multiple meanings

105. Precarity: employment that is extremely unpredictable, where jobs tend to be irregular, often temporary, flexible, and casualized (see also Atypical work).

106. Presentism (see also Nowism): the view that only the present exists

107. Reciprocal surveillance (see also Omnoptic surveillance): a type of mutual surveillance, where people are monitoring each other potentially on a mass scale

108. Relational labor: the work involved (for creative professionals, such as performers, artists, journalists, etc.) to build and sustain connections with audiences

109. Remediation: all media are interdependent, in that media are always a remix of older media forms and newer ones

110. Self-surveillance: the attention one pays to one's behavior when facing the actuality or virtuality of immediate or mediated monitoring

111. Shadow profiling: using data collected about nonusers to create profiles of people who have never signed up for online social networks

112. Sharenting: the practice of parents continually publishing and publicizing content about their children online (generally without consent)

113. Shiny (New) Toy Syndrome: always wanting or being oriented toward the latest or newest piece of technology or gadget

114. Singularity: the moment when humans merge completely with intelligent machines

115. Sit-back media: media that require a low level of active (physical, emotional, or cognitive) engagement

116. Skills access to media: media access shaped by the level of knowledge of how to manage and use media

117. Slacktivism (see also Clicktivism): participating in media activism doing little more than messaging, following, or sharing hyperlinks online

118. Social media: websites or devices that offer content and experiences generated by the users of those sites and devices

119. Sousveillance (see Omnoptic surveillance): a form of reciprocal surveillance or coveillance, where everyone is (or can be) monitoring everyone else

120. Speculative work: work that is unpaid but done in the hopes of securing future gigs, clients, or employment

121. Splinternet: a trend whereby the global internet is fragmenting based on such factors as technology, commerce, religion, and divergent national interests

122. Supersaturation (of media): the notion that media completely saturate the average household and in doing so become unremarkable

123. Surveillance capitalism: collecting and mining personal data for profit

124. Surveillance: the systematic monitoring of the many (i.e., people's behavior, actions, and communications) by the few (e.g., companies and government agencies)

125. Synoptic surveillance (see Coveillance, Inverse surveillance): a type of surveillance where the many monitor the few

126. Synthetic media: any kind of media that is produced through automated means

127. Tactical media: "do-it-yourself" media of all kinds used by groups and individuals who feel aggrieved by or excluded from the wider culture

128. Technology bias: technological objects are not just passive or neutral instruments but actively connect us with the environment in which we live and in doing so shape our lived experience in particular ways

129. Technomyopia: people tend to overestimate the short-term impact of technology while simultaneously underestimating the long-term potential

130. Teledildonics: using technology to mimic and extend sexual interaction

131. Teleparenting: a parenting style where parents are in almost constant electronic touch with their children as they grow up

132. Thingness: the actual state of being a (real, material) thing

133. Transmedia storytelling: storytelling across multiple forms of media, with each element making distinctive contributions to people's understanding of the story

134. Transmedia literacy: developing media literacy collaboratively across a variety of media based on people's existing knowledge, skills, and creativity

135. Transliteracy: an approach to media and information literacy that helps people navigate and make use of a variety of media.

136. Truman Show Delusion: feeling that the ordinary has changed, constantly searching for meaning, and experiencing a fluidity of the basic sense of identity.

137. Ubiquitous computing: technology that works in the background, is intuitive and easy to use, is available anywhere, and is interconnected

138. Ubiquitous media: the notion that media are everywhere, impossible to (completely) escape from

139. Usage access to media: media access in terms of the time and types (e.g., uploading, downloading) of media use

140. Vertical media integration: a media company buying up firms providing services in the same line of production (e.g., a film studio buying a movie theater chain)

## Appendix 3: Ten Tips for a Life in Media

In February 2018, at the invitation of a Dutch network organization for media literacy education, I published a list of ten tips for a good life in media. It was translated into English and posted to the online publishing platform Medium. For the purpose of this book, I went back to that list and updated it. It is printed here for your information, in the hope that it helps.

For us, media are like water for fish. In this appendix, I offer ten tips for a good life in media—a happy and healthy media life. Media in general and digital media in particular are inevitable and indispensable. What makes media so special is the opportunity that technology offers us to give shape to the world in which we live. What we do in and with media has consequences for reality and our world offline. That gives us power. The question now is what we will do with this power (as with great power comes great responsibility).

Media are so ubiquitous and pervasive in our lives that we cannot really switch them off or do without anymore. The question is therefore less whether media and information technologies are good or bad for us but rather what it is that we want from our world and how we take responsibility for that desire. We have to use media as a tool with which we can work, instead of as a crutch on which we depend.

Media can contribute to a good life—a life where we have fun, take responsibility, and have a chance to tell the story of who we are and what matters most to us. It also means that the great issues of our time—catastrophic climate change, war and conflict, the refugee crisis, the instability of the global financial market, pandemics, and the increasing scarcity of drinkable water—are, in part, mediatized and therefore rely on media to be addressed effectively.

From this action-oriented and hopeful perspective, I offer ten tips for a life in media. My advice appeals to our shared and individual responsibility regarding living a happy and healthy life. This certainly does not mean that education, government, and business do not have a role to play in creating a productive, responsible, and safe media environment. Critical media literacy and digital life skills should be required in all types of formal education. Governments need to work together to enforce safeguards for our privacy and identity online, and businesses should take appropriate steps toward establishing transparency and accountability mechanisms based on clearly defined public values. Beyond these necessary steps, we all have a daily life to lead, and for this purpose, these ten tips have been conceived.

### 1.  Forgive Yourself

Media are difficult. Just as we mastered the manual and workings of a specific device or platform, an update or alternative comes out with which everything starts again. Media are tempting, and creators and developers are very good at tempting us.

Media are "experience machines" in which we can easily lose ourselves as we connect with friends and loved ones, play cool games, get lost in a YouTube playlist, or watch well-crafted television series. Such behavior generally does not have any greater consequences than that you sleep a little less or occasionally do not pay enough attention to what is going on around you. Sometimes we click on links without thinking, share or like without reading, forward without checking. And for some people, in some particular contexts, all this listening, watching, clicking, and liking makes them really anxious, feeling isolated and alone despite being connected to so many others online.

It is unavoidable to make mistakes with (and in) media. Forgive yourself! If you do worry, try to plan on losing yourself in media (e.g., by binge-watching a series or pulling an all-nighter playing your favorite game) a bit more carefully. Just realize that there is no way of doing it right, so you may as well come to terms with the way you do it.

### 2.  Technorealism

Life in media is not better, nor is it worse, than life without or outside of media. Experiences in media are not fake or unnatural simply because the connection, knowledge, and information in media comes to us through technology. The opposite is also true: something you experience in media is not necessarily more intense or immersive than what you can sense and perceive without using any devices. Techno-utopianism and dystopic viewpoints can be useful to really consider and think through the potential impact of technologies in the long term but do little to help us in navigating our daily lives in media.

A grounded attitude begins with recognizing that everything we know about ourselves, each other, and the world around us is shaped and formed in one way or another by media. Media on the one hand and humanity on the other are inseparably linked by an endless feedback loop. From the earliest cave paintings to the wall in our social media timeline where we scribble notes about ourselves and the people we care about, media are part of who we (think we) are.

Our interdependence with media does not mean that nothing or no one can be trusted anymore. We all sense that no one is entirely honest when they present themselves in media—whether that is the politician vying for our vote on television or a friend posting selfies of a successful life on a social networking platform. However, we must guard against a "suspicion society" where the way we relate to each other is based on mutual mistrust, inspired by the omnoptic surveillance context we live in—a world where everyone is (or can be) monitoring everyone else via a complicated network of mutual, mediated surveillance, mass media phenomena, and 24/7 activities online.

It is better to keep looking at each other deliberately, with an open heart—especially at people with different opinions and people with backgrounds different from yourself. Sure, we must reclaim conversation, both online and offline—based on the fundamental insight that the distinction between both ways of living has vanished (and perhaps never existed).

### 3.  Media Are Magic and Messy

Most of us do not really know how media work. How does the news come about? How is a film made? What is involved in developing a computer game? What about the hardware and software of the internet and online social networks? What is inside your smartphone, where do those electronics come from, and how does this charismatic device we love so much work? Most of us do not know. It's like magic.

Media are magical: we put so much into them—our emotions, expressions, and expectations—and, as if by magic, out comes an awesome array of experiences, content, services, and knowledge and information. At the same time, it is helpful to remind ourselves of the fact that the black box of media never works perfectly: the technology is messy, fickle, charismatic, and unreliable—just like we are.

Smartphones stall, and laptops crash; television signals disappear, and internet connectivity drops for unclear reasons. It is in those moments—when media break down—when we suddenly realize we are indeed living in media. So do assume media have power, and enjoy the wild ride; just never assume computers, media, and all associated technologies are in any way perfect machines. They are not coming for us. People are messy, and technology is too.

### 4.  We Have Been Here Before

New media do not exist. All newer media are versions of older media. To some extent, we all suffer from shiny new toy syndrome: the tendency to fall in love, over and over again, with updates or new releases of cool consumer electronics and software. I mean, all those extra pixels and gigabytes must do something, right?

Yet newer media are not necessarily better nor all that much different than older devices and applications. We are all part of a culture of planned obsolescence: we replace our technologies every couple of years or so, partly because that is how the industry works, partly because what consumers desire changes so much. That means we are not powerless in the context of the upgrade culture that typifies our life in media. It just suggests that oftentimes reusing, recycling, and repurposing existing technologies is the best solution for you and your family—especially because it is in fact quite unlikely that we do something different with all our new stuff.

Media do not program us; nor do media make us do things we would otherwise—as human beings—never do. Media amplify and accelerate what we think about, how we feel, and how we act in life. The conversation that society has with itself in media is like the buzz in a crowded café—only with thousands, millions, or even billions of participants. The themes of that conversation—especially in social media—are as old as humanity itself and can therefore be seen as crucial for us as human beings, as being part of a society: anger, contempt, fear, disgust, happiness, sadness, and surprise. Media coconstitute our humanity.

Technomyopia—overestimating the short-term effects of media, while underestimating their long-term influence—is a predictable and very human response to technology. It is a trap for us to think that a new device, platform, or service will change everything in the short term. On the other hand, in the long term, media play an extremely important role in the way we make sense of the world, how we organize our lives, and how we relate to each other. Some historical awareness is very useful for a good understanding of media. We have been here before.

### 5.  Don't Overdo It

Try things out! Download cool games, click and read everything, let your child play with a tablet, discover new music, just . . . don't overdo it. There is generally no such thing as using "too much" media. Sure, there is such a thing as excessive and problematic media use, but media addiction—regarding social media, video games, or any other medium— tends to be grossly overstated, medicalizing a condition that most of us have at times experienced. Consider, for example, screen time rules for kids: it seems like good parenting, but if it becomes an excuse not to watch, read, converse, or play with your child, then such rules are useless. Generally speaking, parental media monitoring works well—but we must beware of simplistic generalizations when it comes to media: every household is different.

What seems to make sense is to take a little media vacation every now and then. Give yourself a break from the news once in a while. If possible, do a digital detox. On the other hand, the deceptive attraction of this type of advice comes from a position of profound privilege: What if your family is scattered all over the world and media are the only lifeline to your loved ones? What if media are your technologies of love, enabling and sustaining relationships across the world in times of global migration?

As a rule of thumb, mindful and wholehearted media use should go hand in hand. Next to doing everything online, why not also ask someone out without using an online dating service, send an email instead of an app, write a letter instead of email, call instead of writing, or strike up a conversation rather than stare at a screen?

Media only become problematic when they replace and push out other actions and experiences. Media only become powerful when we take them for granted.

### 6.   Realize You Are a Maker

In the past, until the end of the last century, media use for most of us was synonymous with media consumption. That is to say, we listened to, read, and watched the media. Nowadays almost all our media use is productive: we make media when we use media. This is the era of convergence culture, where the production of media takes place alongside the consumption of media. As media industries converge to produce increasingly compelling experiences for consumers, we—as users—come together online to share our own stories and participate in all the wildly varied ways we are informed and entertained. What is amazing is that we do all this often without realizing the communication power we have.

The verbs we use today for our media use are all active, forward leaning, and engaging: we check, surf, click, link, share, like, favorite, forward, upload, download, install, scan, comment, and post. That means on the one hand that we share a lot of personal information in media—which makes quite a few companies rich and contributes to a global surveillance society. On the other hand, we now have much more media power than ever before. It is time to use that power.

The stories we tend to tell in media—our mass self-communication—are generally about ourselves. However, we are more than the bits and bytes that make up our lives: we care about (certain) others, about (some) social and political issues, about the planet. What stories are we sharing about those people and things, the issues that do so much more than just generate data for companies? What world did you make today?

### 7.   Be Mediawise

Being media literate and becoming mediawise are survival-relevant skills in today's information society. After many years of advocacy and lobbying efforts by academics, foundations, industry watchdogs, policy makers, and teachers, media education is slowly but surely beginning to make it into the classroom.

In the United States, media literacy is included in common teaching standards for English language arts. In the Netherlands, the government has made digital literacy a key element in secondary education required by law. Similar developments are underway in Canada, Australia, and England. At the European Union level, media literacy education is heavily promoted to the member states. Latin America, the Middle East, and Africa are the fastest growing regions in the world in terms of internet penetration, and organizations are scrambling to provide access and literacy education in such countries as Rwanda, Nigeria, South Africa, Algeria, Chile, and Venezuela.

Overall, there is a growing awareness about the profound role (digital) media play in the daily life of all (including the aging population). We should prevent this common quest for digital skills and critical literacy to be based on instrumentalist button pushing, fear, and distrust. What we can and should do is enjoy our media life by becoming mediawise.

Check local and international reports and resources, such as those produced and collected by UNESCO, the Center for Media Literacy (United States), MediaSmarts (Canada), Mediawijzer (Netherlands), Ofcom (United Kingdom), ACMA (Australia), the Association Development of Education in Africa (ADEA), the GO Digital ASEAN initiative of the Asia Foundation, and all the organizations, schools, and programs involved in the rich Latin American tradition of *educomunicación*, and many others. Contact community centers and advocacy groups; search for media scholars, trainers, and educators, and ask questions. Ignorance is no excuse.

## 8. True Play

It is true that in a digital environment, much is determined by algorithms: statistical formulas that, based on your personal information and correlated with the data of countless other people, provide you with products and services (or deny access and participation). Algorithms—especially those continually tweaked by social media firms—are generally just as opaque as media are to us in general. This lack of transparency, coupled with the often rather ridiculous rhetoric around artificial intelligence and machine learning, inspires quite a bit of hand-wringing and fear-mongering.

Recommendation systems, automated playlists, and news feeds would inevitably lead us to a life locked inside our own personal telecocoon, blind to others, where all information we get comes to us through our very own filter bubble. This is a questionable argument. Not only are all these assumptions underresearched and overhyped; studies and common sense suggest that people in the real world do not rely on just one source or platform for all their news and information. This does not mean algorithms are inherently benevolent, let alone neutral—it just suggests we can use statistics-driven platforms to get what we want and should consider being more deliberate about it. For example, use sites like Yelp, Booking, and Momondo to orient yourself, but book your reservation with the restaurant, the hotel, and the airline directly if you want your money to go to the people who cook your food, make your bed, and get you to your destination safely.

Take some time now and then to look for other things, ideas, and people than you are used to, not just because that is fascinating and fun but also so that the advertisements and recommendations you see everywhere on the internet clearly do not know what to do with you anymore.

Whatever you do, do not forget to have fun and embrace true play: not everything we do and share in media has to be serious; not all information we seek out is supposed to have educational value; not every comment we make or message we send must be full of well-argued, thoughtful insights. The often meaningless, silly, and otherwise mundane forms of everyday communication we engage in online are in fact what promote social harmony—what make us human.

## 9. Study Media

In recent years, numerous books, documentaries, and reports have appeared around the world warning people about the dangers of (the) media. Despite addressing legitimate concerns, the level of fear and panic engendered by these publications is problematic. In an effort to focus more on the evidence and on (digital) empowerment, media scholars around the world have made efforts to make their research and insights available (often for free) to everyone.

Some excellent examples can be found in the work of my colleagues Patti Valkenburg and Jessica Taylor Piotrowski (at the University of Amsterdam), Payal Arora (at Erasmus University Rotterdam), Tanja Bosch (at the University of Cape Town), Sonia Livingstone (at the London School of Economics), Lynn Schofield Clark (at the University of Colorado), Anabel Quan-Haase (at the University of Western Ontario) and danah boyd (at New York University). These are all experienced authors with an impeccable status in the field, having conducted countless studies, experiments, and interviews with children, teenagers, and young adults around the world about the formative role media play in their daily lives.

Time and time again, the conclusion from these (and other) scholars in the field is that media are complicated, that people have similar reasons and motivations for using media but their ways of doing so are wildly differing, and that all these media most certainly have consequences for our lives, both positive and negative, and although it is next to impossible to generalize conclusions, people (young and old) tend to be both resilient and critical when it comes to media. The kids are all right.

Media studies and communication research have a long, rich tradition of helping us to understand and bring nuance to concerns about the role of media in shaping public opinion and affecting people's lives. Today, many other academic disciplines—from management and economics to philosophy and psychology—talk about media. What media scholars do especially well is offer a critical view of media while still being mindful of the central role media play in people's lives. In media studies, we (should) start our research and theories from a place of curiosity and wonder rather than fear and judgment.

For all of us, the important thing is to keep reading, discussing, and sharing media scholarship that is grounded in everydayness. If you need help finding, reading, and understanding the research, just ask. We are here to help.

### 10. Media Art(s)

Artists are often the best sources for perspectives, possibilities, and opportunities concerning what our life in a media world looks like—or could look like. Artists show us a truth that is difficult to see in daily life, that escapes the confines of a carefully crafted research project, that moves beyond the insights of a deeply philosophical take on media. Likewise, artistic methods make headway in scholarly work as a profound source of insight into the embodied or "somatechnic" experience of a life in media—beyond rationalization and deliberation.

There are numerous amazing festivals around the world where the mix of art, media, technology, and creativity is central to the works on display, and I always find it illuminating and inspiring to explore these kinds of exhibitions and performances. There will be an event, exhibition, performance, festival, display, or installation near you.

# Index

Abbas, Yasmine, 229
#ABetterUbisoft, 204
Access, 217
#Actiblizzwalkout, 204
Active data, 77, 297. *See also* Data
#ActivisionBlizzard, 204
Activism, 153–154, 171–174. *See also* Media
   activism; War and conflict
 analog media in, 165
 co-optation of, 176, 180
 cyber, 161–162
 employee, 167
 environmental, 178
 inclusivity and exclusivity, 168, 170
 as mainstream, 181
 media's role in, 165–166
 online vs offline, 162, 166–167, 205
 platforms for, 149, 159–160
 representation and, 239
 use of term, 157–158
*Activist, The* (TV show), 168
Advertising
 actors, use in, 14
 audiences sold to, 21
 campaigns, 194
 market control in, 193
 Martini media principle, 211–212
 Mass Observation project, 14
 reality-unreality in, 121–122
 work/life balance, 233–234
Aespa (girl group), 242
Affect, 78, 218, 278–279, 281–282, 297
Affordances, 28, 132, 165, 170, 248, 297
Agbogbloshie Makerspace Platform (AMP), 229
Agency
 collective, 247, 263
 consumption and, 258
 love as a mediating, 273, 278
 media and, 1, 60, 196, 263
 in media studies, 281–282
 nonhuman, 259
 power and, 217

Aging, 57
Ahwaa (discussion platform), 172
Airline industry, 66
Algorithms. *See also* Artificial intelligence
 data use by, 41
 disproportionate outcomes of, 56
 humans circumventing, 67
 media production and, 242
 positivity prioritized by, 120
 preferences of, 174
 sameness of, 173
 social media, 22
 surveillance and, 55–56, 78
Alibaba, 66
Alphabet (tech company), 9, 65, 192, 232. *See also*
  Google
Alpha Global, 232. *See also* Work and labor: unions
Amazon, 66, 71, 226
Analog media. *See* Old media
Android, 74. *See also* Google
*Animatrix, The* (anthology film), 108
Anonymous (hacker group), 173
Anthem (health care company), 62
Anthropometry, 271
Apple, 29–30, 35, 65, 71, 96. *See also* iPhone
Appropriation, 175, 180
Apted, Michael, 14
Arab Fund for Arts and Culture, 173
Arab Spring, 153
Archaeology. *See* Media archaeology
Archetti, Cristina, 265, 282
Aristotle (baby monitor), 70–71
Arora, Payal, 311
ARPANET (computer network), 114
Arrangements of media, 27, 44–49, 115–124, 266
Artifacts, 27–36, 112–113, 151. *See also* Devices
Artificial intelligence, 297. *See also* Kurzweil, Ray
 in *2001: A Space Odyssey,* 104
 aging and, 57
 algorithms and, 41
 DALL-E, 121
 ethical aspects, 243

**Artificial intelligence** (cont.)
  facial recognition, 227
  God based on, 111
  social/ethical aspects, 226–227
Artists, 312
Attachment to media, 30–31
Attention, 12
Atypical work, 200–201, 297. *See also* Work and labor
Audiences, 240–241
Augmented reality (AR), 297
Authenticity, 128, 221, 250, 297
Autobiography, 77
*Avatar* (Cameron), 180
Avatars, 166–167, 180, 242–243, 297
Avery, Bree, 191
Azzopardi, Chris, 9

Banking, 76
Banks, Mark, 290
#BantheScan, 59–60
Battell, John, 51
*Battlestar Galactica* series, 110
Baudrillard, Jean, 108–9
Bauman, Zygmunt, 57
Bausinger, Hermann, 256–257
BBC (British Broadcasting Corporation), 211
"A Beautiful Young Nymph Going to Bed" (Swift), 101
Becoming media, 106–24, 112–124
Bell, Genevieve, 264
Beltrão, Luiz, 256
Bentham, Jeremy, 54, 178. *See also* Panoptic surveillance
BeReal (social network), 175
Berkhout, Frans, 294
*Big Brother* (tv show), 73
Big data, 66–67. *See also* Data
Bil'in (Palestine), 179–180
Bioy Casares, Adolfo, 108–9
*Bits and Pieces of Information* (Wachowski and Wachowski), 108
Black box fallacy, 29
Black in AI (network), 227
Black Lives Matter movement, 172
*Blade Runner* (Scott), 104
Blizzard Activision (game studio), 204. *See also* Video games
#BLM (BlackLivesMatter), 152
Blockchain, 227, 297
Bly, Nellie, 189
Boczkowski, Pablo, 260–261
Bodies, 29–30, 35, 40, 62, 166–167
Body Battery (smartwatch), 78
Body Politic (support group), 167
Bollywood, 188
Bolter, Jay, 265

Borges, Jorge Luis, 108–109
Bosch, Tanja, 311
Boyd, danah, 268, 311
Bradshaw, Peter, 8
Brandel, Jennifer, 202
Branding, 122
Brazilian Spring, 153
Breazeal, Cynthia, 71
*Brokeback Mountain* (Lee), 30
BTS (musical group), 14
Bullying, 74–75
Bungie (games developer), 204–205
Burnett, Leo, 122
Bush, George W., 123
Business and industry, 64–68, 141–142, 197–198, 204–205, 207. *See also* Work and labor
Butler, Samuel, 86–87, 96. *See also The Matrix*
Buycott (app), 149
Byrne, Jasmina, 267

*Caché* (Haneke), 95
Cameron, James, 180
Camover, 61
Campaign to Organize Digital Employees (CODE), 232. *See also* Work and labor: unions
Capek, Karel, 89
Capitalism, 55, 160, 167. *See also* Work and labor
Carey, James, 256
Carlsberg, 167–168
Casares, Adolfo Bioy, 277
*Cats* (Hooper), 104
CBS, 5
CCTV (closed-circuit television), 58, 61, 69, 272
CD Projekt Red, 278
Celebrity branding and endorsements, 14–15, 154
*La Cenicienta* (tv show), 73
Censorship, 83, 119, 170, 175, 178, 231
ChatGPT, 239
Charismatic technologies, 37–38, 297
Children, 70–71, 90, 96, 267
CIA, 61
*Circle, The* (Eggers), 75
*Citizenfour* (Poitras), 58, 272
Civic engagement, 171
Civic Imagination Project (University of Southern California), 287
Clarke, Arthur C., 29, 104
Clickbait, 43
Clicktivism, 166, 179, 298, 303
Clinton, Hilary, 72
Clip (camera), 75
Coalizão Negra Por Direitos (Black Coalition for Rights Brazil), 172
Cobalt, 225
Cognitive Assistant that Learns and Organizes (CALO), 114
Colbert, Stephen, 57, 120

# Index

Collaborations, 15
Come to the Street movement, 153
Command line interface (CLI), 34, 298
Communication, 25, 89–94, 133, 162, 256, 301
#CommunicationSoWhite, 254
Community, 82, 123, 133–134
Concurrent media exposure, 11, 24, 39, 47, 114, 136, 224–225, 298
Confidence in media, 19–20
Congo, Democratic Republic of the (DRC), 225, 230
Conspiracy theories, 133
Consumption, 39–40, 106, 123, 130–132, 308–309
Content
  cross-platform, 212, 241
  financing of, 191
  media defined by, 25, 27
  polysemy, 128
  regulation of, 174
Convergence, 47, 188–189, 298
Copyright, 180
Coronavirus crisis
  contact tracing, 105
  depictions of, 9
  digital surveillance and, 20, 64, 70
  disinformation about, 116–117, 224–225
  influencers on, 181
  as infodemic, 116, 124
  life in media and, 3
  long covid, 167
  media love and, 133
  media's danger and, 96
  mediatization of, 116
  work conditions during, 216, 231, 234
Couldry, Nick, 259
CounterSocial (social network), 175
Coveillance, 74
Craig, Geoffrey, 294
Creation myths, 107. *See also* God
Creator culture, 189–190
Crimea, 119
Crossmedia, 213, 298
Crowdfunding campaigns, 166, 214, 242, 298
Crowdsourcing, 18, 298
Crunch time, 200. *See also* Time
Cryptocurrency, 227
Cybernetics, 279
Cyberpunk, 278
Cyberspace Administration of China, 59
Cyborgs, 110, 298

DALL-E, 121. 239
Dark participation, 74, 298
Darwin, Charles, 29, 86–87, 279
Darwin, Erasmus, 87
Data
  access to, 79–80
  activism, 151
  analysis, 61
  archiving, 242
  carbon footprint of, 227
  collection, 242
  content creation and, 200
  as digital air, 65
  everyday life and, 268
  forms of, 66, 77–78, 297, 302
  glut, 60, 298
  in health care, 62
  logic, 199–200, 299
  mining, 299
  ownership of, 63
  portability, 299
  predictive uses of, 67
  repurposing, 242
  state's use of, 60
  surveillance and, 78–80, 299
Dating apps, 5, 134, 136, 143, 280. *See also* Romantic relationships
Dean, Miriam, 204
Death and dying, 9, 73
Debray, Régis, 256
Deepfakes, 121
Deleuze, Gilles, 270–271
Descartes, René, 57
Devices. *See also* Artifacts; Materiality; Subtlemob
  attachment to, 30
  disconnecting from, 10–11, 96, 151, 299
  domestication of, 40
  expression enabled by, 143–144
  international use, 37
  media defined as, 25, 27, 44
  programming, 231
  sales, 36–37
Dialogue, 272–273
Dick, Philip K., 104, 277
Digital
  activism, 162, 173–175
  commons, 88
  as default, 55
  democracy, 299
  detoxing, 96, 151, 173, 222, 299
  economy, 175–176
  health care, 133
  inequality, 299
  insurance, 73
  selfhood, 271
  shadow, 299
Digital commons, 176, 299
Digital inequality, 17
Digital literacy, 17, 20, 106
*Digital Services Act of the European Union,* 97
Digitization, 112, 118
Disconnection, 49, 96, 151, 173, 202, 222, 299
Disney Company, 193

Distributed Artificial Intelligence Research Institute (DAIR), 227
Domestication theory, 258
Dongria Kondh tribe, 180
Dopamine, 83
Dourish, Paul, 264
Drones, 56, 61–62
DynaTAC 8000X (Motorola), 33

Eck, Otto van, 77
Economics and economies
  content creation, 191
  cryptocurrency, 227
  digital, 175–176
  e-waste, 227–228
  media industries, 186, 195
  surveillance and, 81
Eder, Jens, 279–280
Editorial logic, 9, 198
Education, 68–69, 97–98, 106, 158
Edufactory, 69
Eggers, Dave, 75
Eirik & Pedro, 194
Election campaigns, 72, 122–123
Electricity, 225–226
Electronic Arts (EA), 235. *See also* Video games
Electronic waste, 227–228, 230
#EleNão, 152
*ELIZA effect,* 103, 114, 300
Embodiment, 29–30, 131
Emotions
  categorizing, 279–280
  expression of, 145
  responses to media, 113, 125–126, 131, 143
  scholarship on, 278
  socialization of, 145–146
*Enter the Matrix* (video game), 108
Environment, 224–229, 243
Ephebiphobia, 300
Epic Games, 215
*Epic of Gilgamesh,* 111
*Erewhon* (Butler), 87
Erotics, 129–134, 258
Ethical issues and media, 15–18, 23–24
Ethnography, 266, 286
Eva (ride-sharing app), 175
Everyday life. *See also* Life in a Day
  content creation and, 190
  domestication theory, 258
  everydayness of, 300
  love in, 130
  and/in media, 7–8, 41
  mediatization of, 251
  on YouTube, 189, 191
Everything is a remix (documentary series), 14, 18–19, 262
Evolutionary theory, 87. *See also* Darwin, Charles

*Ex Machina* (Garland), 104, 109–10
Experiences as mediated, 72
Extended reality (XR), 215

Facebook, 22, 35, 70, 75, 83, 149, 159, 165, 174
Facial recognition technology, 59–60
Failure of media, 15, 17, 23
Fairmondo (global marketplace), 175
False information, 88, 97, 116–117
Family surveillance, 70
Fan Forward (organization), 180–181
Fassbinder, Rainer Werner, 277
Fast, Karin, 259
FBI, 61
Fear of computers, 17
#FeesMustFall, 152
Femen (protest group), 172
Ferguson, Kirby, 14. *See also* Everything is a remix (documentary series)
Five Eyes Intelligence Alliance, 57
Folkcommunication (Beltrão), 256
Forgiveness, 306
Forster, E. M., 94–96
  "The Machine Stops," 222
Fortnite, 114
Fortunati, Leopoldina, 265, 278
Forum on Information and Democracy (FID), 115
Foucault, Michel, 270
Foxconn, 228, 230–231
Free Brazil movement, 153
*Free Guy* (Levy), 219–223, 250–252
Freelancers, 186. *See also* Work and labor
Free Press Unlimited, 19
Freud, Sigmund, 103
Frictionless sharing, 83
#FridaysForFuture, 152, 178
Fun, 18, 150, 211. *See also* Joy; Pleasure
Functionality, 28, 300

G20 Summit, 168
Galouye, Daniel F., 277
#Gamergirls, 204
Gaming disorder, 45
"Garbage in, garbage out," 56, 79
Gauntlett, David, 263
Gebru, Timnit, 226–227
Gender, 119, 209, 300
General Data Protection Regulation (GDPR), 97
Genre conventions, 189
Ghana, 228
Ghebreyesus, Tedros Adhanom, 3
*Gibridnaya voyna* (war term), 156–157
GigaOm, 68
Glitsos, Laura, 258, 266
Global Kids Online, 267
Global Protest Tracker, 168
God, 100, 103, 111. *See also* Creation myths

# Index

**317**

Gold, Ian, 250–251
Gold, Joel, 250–251
Google. *See also* Alphabet (tech company); Android
    algorithm use by, 22
    artificial intelligence and, 226–227
    Cloud, 226
    digital commons projects and, 176
    hardware/software integration, 110–111
    Home, 71
    Maps, 46
    Street View, 105, 272
    worker organization, 231–232
    Workspace for Education, 69
Gou, Terry, 231
Governments, 46–47, 59–60
Graphic user interface (GUI), 34–35, 300
Gray, Jonathan, 259
Greenlighting, 199, 209, 300
Greenpeace, 68
Grossberg, Lawrence, 257
Group of Eight (G8), 161
Grusin, Richard, 265
GSM (global system for mobile communications)
    networks, 33
Guerrilla theater, 61

Hacking, 166, 300
Hacking Health movement, 63
Hallin, Daniel, 262
Halo effect, 119, 300
Haneke, Michael, 95
Hanich, Julian, 279–280
Hardware, 195
Harmony (sex robot), 135
*Harry Potter* (series), 180
Hartmann, Maren, 282
Harvey, Alison, 254
Hashtags, 45, 142. *See also* Social media
Hasty, Leander, 235
*Hate Poetry* (theater show), 74
Headlines Network, 236
Headspace (app), 235
Health care, 62–64, 77, 133
Hearken (firm), 202
Hepp, Andreas, 259, 282
*Her* (Jonze), 95, 104, 109–10
Heroes of Might and Magic II (New World
    Computing), 27
Hertin, Julia, 294
Hesmondhalgh, David, 289
Hjarvard, Stig, 282
Hobbes, Thomas, 99
Hoffman, Erin, 235
Hoffmann, E. T. A., 101, 103–104
Homes. *See also* Intimacy
    media as, 113–114
    as mediatopes (Quandt and Pape), 266

Honor of Kings (Tencent), 27
Horizontal integration, 192–193, 300
Hormones, 83
Houellebecq, Michel, 95–96
*How Human Is Human* (Ishiguro), 112
Hugo (award), 185
Humans
    agency of, 117
    digitization of, 112
    media as coconstituting, 308
    neo- (Itskov), 111
    sociality of, 144–145
    technology (relationship to), 87–88, 92, 100–101,
        107, 125
    as zombies, 114–115
Hybrid warfare, 156–157. *See also* War and conflict

#IAmNotAfraidToSpeak, 150
#IASolidarity, 167
IBM, 28–29, 176
Identity, 106, 196
IGN (Imagine Games Network), 204
Illich, Ivan, 98
Image modification, 117–118
Independent Media Centers, 160–162, 164
India, 180
Indiana University, 72
Indigenous media and communities, 20, 154, 156,
    161–162, 180, 256
Indirect surveillance, 74, 81. *See also* Surveillance
Individualization, 42, 207, 271
Industries. *See also* Work and labor
    convergence in, 188–189
    culture of, 234
    discrimination in, 203–204
    diversity in, 209–211
    film, 188
    heterogeneity of, 194–195
    media as, 183, 185, 188–196
    structure of, 300
    work conditions, 234
Indymedia, 160–162, 164
Influencers, 181, 191. *See also* Creator culture;
    Social media
Infodemic, 3, 116–117, 124, 247–248, 295, 300
Information
    contamination of, 63
    dis-, 224, 299
    infobesity, 44
    infodemics, 3, 116, 124
    media as carrying, 25
    misinformation, 132
    sharing of, 51
    spread of, 116
    tagging, 113
Infrastructure, 168, 263–264
Innis, Harold, 256

Instagram, 20, 114, 118, 167
Instant messaging, 165
Institutional surveillance, 68–72. *See also*
  Surveillance
Integration, 192–193
Intellectual property, 189
Intelligence community, 61
Intelligent Robotics Laboratory, 111–12
Interfaces
  definitions, 34
  design aspects, 120
  histories of, 34–36
Intermediality, 259, 300. *See also* Transmedia
International Classification of Diseases (ICD-11),
  136
International Consortium of Investigative Journalists
  (ICIJ), 61, 194
International Game Developers Association (IGDA),
  210, 235
International Panel on Social Progress, 293
Internet
  access to, 17, 59, 106
  fragmentation on, 132–133
  histories of, 13, 114
  infrastructural aspects, 168
  vs offline life, 150–151, 171, 179
  production on (by users), 247
  scholarship on, 247
  sharing on, 82
  splinternet, 174
  of things (IoT), 79, 112–113
  tracking on, 65
Internet gaming disorder, 96–97
Interpersonal surveillance, 72–76
InterPublic, 193
Intersectionality, 196, 209–210, 254
Intertextual analysis, 152, 219, 300–301
Intimacy, 7, 13, 28, 37, 146, 277, 280, 297
Investigations, 194
*Invisible Children* (documentary), 154
*Invisibles, The* (Morrison), 277–278
IPad, 29. *See also* Apple
IP addresses (Internet Protocol), 59
IPhone, 13, 28–29, 114, 228. *See also* Apple
Iraq. *See also* War and conflict
  war in, 5, 162
IS (Islamic State), 149–150
Ishiguro, Hiroshi, 111–12, 111–112
Israeli Defense Force, 179–180
Itskov, Dmitry, 111

Jackson, Michael, 15
Jackson, Peter, 203
Jansson, André, 259
Jenkins, Henry, 263
Jennnings, Humphrey, 14
Jensen, Jakob Linaa, 268

Jibo (social robot), 71
Journalism, 46–47, 121–122, 155, 164, 194, 214,
  236–237
Journalism Education and Trauma Research Group
  (JETREG), 236
Joy, 18, 250, 280. *See also* Fun; Pleasure

Karim, Jawed, 189
Kavka, Misha, 280
*Keep Running* (TV show), 73
Kennedy, John F., 122
Key performance indicators (KPIs), 199
Kidtech, 70
Klier, Michael, 272
Knowledge (self), 77
Knowledge Navigator (Apple), 29
Kony, Joseph, 154
Kony2012 campaign, 154
Korea, South, 190, 242
Koster, Raph, 243
Kurzweil, Ray, 110–11, 110–111

Labor. *See* Work and labor
*La invención de Morel* (Casares), 277
La Mettrie, Julien Offray de, 99–100
Lang, Fritz, 95
Leaver, Tama, 268, 270
Leibniz, Gottfried, 100
Lem, Stanislaw, 29
Levandowski, Anthony, 111
*Leviathan* (Hobbes), 99
LGBTQ community, 14, 167–168, 172, 234
Lievrouw, Leah, 263–265
*Life in a Day* (YouTube films), 7–10, 17–21, 261. *See
  also* Everyday life; YouTube
Life in media, 301
  everyday life and, 251–252
  mass media and, 241
  media work for a, 211–216
  mixed reality and, 215
  pervasiveness and, 13
  political response to, 105
  as private, 57
  as public, 83
  tips for, 305–312
  uncanniness and, 125
Lifelogging/lifestreaming, 53, 75–76
Lil Miquela, 191
Lion (award), 185
Literacy
  digital, 17, 20, 106
  initiatives, 97
  media, 124, 244–245, 309–310
  scholarship on, 296
  transmedia, 245–246, 304
*Little Cabin in the Woods* (tv show), 73
Live streaming, 53, 75–76

# Index

Living standards, 158
Livingstone, Sonia, 263–265, 267, 311
Logic, 198–200, 299
Lonelygirl15 (YouTube user), 191
#LongCOVID, 167. *See also* Coronavirus crisis
Los Angeles (United States), 5
Los Indignados (social movement), 153, 179
*Lost* (TV show), 277
Love
  dating apps, 5, 134, 136, 143, 280
  importance of, 147
  in life, 130
  for media, 31, 127, 143, 221
  in media, 44, 220–221
  in media studies, 130–132
  as mediating agency, 273, 278
  mediation of, 129
  mediatization of, 129
  studies, 130–131, 278
  technologies of (Madianou and Miller), 280
Loyalty programs, 65–66
Lucas, George, 95
Luddites, 302
Lundby, Knut, 282
Lupton, Deborah, 268
Lyon, David, 268

Macdonald, Kevin, 8–9, 14
Machines, 88, 94–99, 103–104
"Machine Stops, The" (Forster), 94–96
Madianou, Mirca, 259, 280
Magic of media, 307
Majal (MENA organization), 172
Making media. *See* Industries; Work and labor
Malik, Om, 68
Mancini, Paolo, 262
Manuals, 17
Maps, 46, 108–9
Marginalized individuals, 74, 206, 234, 254. *See also* Identity; Intersectionality
Market control and logic, 193, 198–199
Marlboro, 122
Martín-Barbero, Jesús, 257, 282
Martini media principle, 211–212
Marvel Cinematic Universe, 214
Mash-ups, 13
Mass communication theory, 87–88
Mass media, 2, 90–91, 97, 240, 246–247
Mass Observation project, 14
Mass self-communication, 19, 164, 309
Mastodon (social network), 175
Materiality of media, 17, 23, 27, 123, 207, 248–249, 301. *See also* Devices
*Matrix, The* series (Wachowski and Wachowski), 88, 92, 95, 107–109, 125, 214, 277–278
Mattelart, Armaud, 271
Mattoni, Alice, 281

Maturana, Humberto, 275
Maven (app), 235
Maxwell, Richard, 293
Mayer, Vicki, 289
McLuhan, Marshall, 256
McQuail, Denis, 276, 286
*Me at the Zoo* (YouTube video), 189
Media activism, 147, 149–181, 301. *See also* Activism
Media archaeology, 229, 246, 301
Media Diversity Australia award, 210
Media education, 97
Media Entertainment and Arts Alliance (union), 234
Media Excellence Award (award), 185
*Media Industries Journal,* 289
Media literacy. *See* Literacy: media
Media love, 127–147. *See also* Love
Media studies, 1–2
  communication in, 133
  as a disposition, 223
  ethical/sustainability issues in, 293
  fearful tone in, 98–99
  human/media relationship in, 106
  importance of, 22, 311
  -levels of analysis in, 291–292
  love for media in, 130–131
  love in, 130–132, 146
  media-life boundaries in, 123
  methodologies of, 242–245
  optimism, need for, 3, 249–250
  origins of media theory, 87–88
  possibilities of, 246–250
  sociological-historical scholarship in, 262
  working in media and, 237–238
Mediated monitoring, 53–54
Mediation
  about, 242, 301
  of everything, 10, 44, 47, 49, 85–86
  of existence, 88
  of experience, 72
  historical precedence of, 99–100
  of love, 129, 137–141
  mass media and, 241–242
  of reality, 99–106
  scholarship on, 257, 282
Mediatization
  about, 259, 301
  coronavirus crisis as example of, 116
  definitions, 47, 49, 129–130
  of everyday life, 251
  of fictional worlds, 221
  of institutions, 68
  of love, 129, 146
  as metaprocess, 282–283
  scholarship on, 282–283
Mediology, 121, 256, 302
Medium specificity, 27, 160, 212–213
Meikle, Graham, 265

Memes, 302
Memory and media, 30
Mental health, 233–237, 251
Mentally Healthy (initiative), 233
Meta (technology company), 35, 215. *See also* Facebook
Metacritic, 9
Metaplace (software platform), 243
Metaverse, 36, 61, 215–216, 302
#MeToo, 150, 152, 179, 204–205
*Metropolis* (Lang), 95
MeWe (social network), 175
MGM (Metro-Goldwyn-Mayer), 194
Miconi, Andrea, 264
Microsoft, 65, 76, 226
Midjourney, 239
Milano, Alyssa, 150. *See also* #MeToo
Miller, Daniel, 259, 280
Miller, Toby, 293
Mind and body dualism, 100
Mindfulness, 308–309
Mitchelstein, Eugenia, 260–261
Mitsubishi Electric, 29
Monitoring, 53–54, 82, 270. *See also* Surveillance
MonkeyShine, 194
MOOCs (massive open online courses), 69. *See also* Education
Mori, Masahiro, 103–4
Morris, Philip, 122
Morrison, Grant, 277–278
Moss, Carrie-Ann, 125
Motivation, 217
Motorola DynaTAC 8000X, 33
Movimento Brasil Livre, 153
MUDs (multi-user dungeons), 119
*Mukbang,* 190
Musk, Elon, 110
Myanmar, 224
MyLifeBits (Microsoft project), 76
MySpace, 150

Namma (Sumerian mythology), 107
Narrative (tech company), 75
Narrative techniques, 96, 107–108, 134
Nationalism, 240
Natural user interface (NUI), 35
Neohumanity, 111. *See also* Humans
Neo-Luddism, 96
Netflix, 46, 167, 187, 199
Neuralink (neurotechnological company), 110, 113
New media. *See also* Remediation
  international access to, 37–38
  monitoring enabled by, 65
  vs older media, 15, 31, 261–262, 307–308
  participatory potential of, 175
  scholarship on, 257–258
  version history of, 13

NewsCorp, 46, 193
Newsgroups, 19
News media, 3, 44, 88, 97, 116, 121–122, 164
Newspapers, 5, 31, 89, 153, 213
*New York Times,* 204, 214
*New York World* (newspaper), 189
NGOs (nongovernmental organizations), 68, 228
Nike, 15, 167–168
Nixon, Richard, 122–123
Nkemi, Mdhamiri, Á, 9
"Nobody knows" principle, 185
#Nofilter, 118
Nollywood, 188, 290
Nonhuman actors, 24, 268
Nonplayer character (NPC), 219, 222
#NotABugSplat, 61
*No Time to Die* (Fukunaga), 194
Nowism, 249, 302
Nse Asuquo, 9
NTT (telecommunications company), 71
Nüwa (Chinese mythology), 107

Obama, Barack, 72
Objectivity in journalism, 121–122. *See also* Journalism; Reality
Obsolescence, 307–308
Occupy Wall Street initiative, 153, 161
Offline life, 150–151, 171, 179
Old media, 15, 31, 127, 227, 307–308. *See also* Remediation
Omidyar Network, 173
Omnicom, 193
Omnoptic surveillance, 20, 53, 72–76, 268, 302. *See also* Surveillance
"On Exactitude in Science" (Borges and Bioy Casares), 108–9
Online. *See* internet
*On the Origin of Species* (Darwin), 86. *See also* Darwin, Charles
Ontology, 121
Open access, 176, 253, 302
OpenTable, 66
OpenUp, 19
Oramedia (Ugboajah), 256
Oreo, 167–168
Originality in media, 15
Orlowski, Jeff, 22
Oscar (award), 185, 210, 254
Osseo-Asare, Kwadwo, 229
Otomo, Katsuhiro, 278
Ouellette, Laurie, 259

Pakistan, 61
Palantir (data analytics company), 61
Palestinian community, 179–180
Pamphleteering, 153
Panama Papers, 61, 194

**Index**

Pandora Papers, 194
Panopticism, 270, 302
Panoptic surveillance, 51–52, 54–57, 64, 178. *See also* Surveillance
Papacharissi, Zizi, 260, 276
Pape, Thilo von, 266
Paradise Papers, 194
Paraodies, 13
Pardo, Ivan, 149
Parenting, 70, 127, 136, 303–304
Parks, Perry, 280
Participatory surveillance, 66
Passive data, 77, 302
PayPal, 67
Pedagogical techniques, 143
Pepsi Cola, 15
Personal assistants, 114
Personalization, 42
Pervasive, media as, 10–12
Peters, John Durham, 272–273
Philosophy, 258
Pichai, Sundar, 192
Planned obsolescence, 307–308
Platformization, 74, 82, 135, 302
Plato, 91, 222
Platon, Sara, 160
PlayerUnknown's Battlegrounds (Bluehole), 27
Pleasure, 18, 130, 211. *See also* Fun; Joy
Pluriverse, 216
PocketReporter (app), 19
Podcasting, 15
Poe, Edgar Allen, 101
POIDH, 45
Poitras, Laura, 58
*Pokémon Go* (Niantic), 105
Policing, 56, 59–60, 79
Polymedia, 259
Polysemy, 25, 128, 176, 303
 search pdf, 0
Pornography, 135–136
Positionality, 217
*Possibility of an Island* (Houllebecq), 95–96
Postman, Neil, 256
Post-reality, 124
Post-truth, 120
Power, 216–217, 239, 270
Practice, media as, 129, 301
Predictive policing, 79. *See also* Policing
*Press, The* (newspaper), 87
Prisons, 54–55. *See also* Panoptic surveillance
Privacy, 7, 19, 51
Production
 cultures, 195
 emergence as field of study, 289
 material context of, 195
 of media, 7, 123, 239
 as messy/magical, 307

 as productivity, 247
 professional, of culture, 196–198
 studies, 246
Project AWeSome (University of Amsterdam, Netherlands), 266–267
Project ecologies, 193
Propaganda, 89, 155, 225, 286
ProPublica (news organization), 57
Protests, 20, 149, 168, 173, 180
Psychology and surveillance, 82
Publicis Groupe, 193
Public life, 7, 19, 51–83, 142, 201
PublicSpaces (organization coalition), 175
Pumfrey, Alex, 234
Putin, Vladimir, 47

Quandt, Thorsten, 266
Quan-Haase, Anabel, 266–267, 311

Radio, 15, 59, 89–90, 160, 247
Rating systems, 70
Read and write technology, 159
Reality
 community and, 123
 as copies (Baudrillard), 108–109
 derealization, 239
 evaluating, 117
 lost in, 109–110
 media as distorting/losing, 91, 273
 mediation of, 99–106
 news media on, 275
 post-, 124
 as a projection, 100
 representation of, 120
 unreality, 118–119, 121
 vs virtuality, 3
Reality TV shows, 73, 280. *See also* Television
Rebellion and revolution, 157–164, 172, 302. *See also* Protests
Reeves, Keanu, 125, 278
Regulations, 106
Relational labor, 31. *See also* Work and labor
Remediation, 29–30, 34, 265, 303. *See also* New media; Old media
Remixing, 13–14
Representation, 106, 222, 224, 238–246, 294–295
Resonate (streaming service), 175
Review sites, 66
Reynolds, Ryan, 219
Rice-Edwards, Sam, 9
Ritual and belief systems, 256–257
Robots, 89, 103–4, 111–12
Romantic relationships, 134–135, 142–143. *See also* Dating apps
Rose, Jessica, 191
Roskomnadzor (media watchdog), 47
*Rossum's Universal Robots* (Èapek), 89

Rotten Tomatoes, 9
RSA Films, 19
Rusnak, Jozef, 277
Russell, Jason, 154
Russia, 46–47, 119
Russo-Ukrainian War, 45–47, 119, 155, 157, 159, 224, 286

SaaS (software-as-a-service), 59
Sampling, 13
*Der Sandmann* (Hoffmann), 101, 103
Scannell, Paddy, 273
Scher, Julia, 272
Schmidt, Eric, 64–65
Scholarship
  annotated sources, 253–296
  biases of, 254
  humanities- and social scientific approaches, 276
  open access, 253
  practice turn in, 281
  qualitative vs quanitative, 266–267
Schoolman, Carlota Fey, 21–22
Science fiction, 3, 29, 93, 110, 112, 141, 277–278
Scott, Ridley, 8, 104
Screens and screen time, 33, 96, 127
Search engines, 5, 65
Self
  branding, 189
  censorship, 83
  communication, 301
  expression, 119, 144
  knowledge, 77, 92
  love, 132
  organization, 294
  presentation, 119
  promotion, 186, 208
  surveillance, 53, 77–79 (*See also* Surveillance)
  tracking, 269–270
  true/authentic, 92
  at work, 237
Self-Investigation (educational foundation), 236
Selkirk, Jamie, 203
Serra, Marcello, 264
Serra, Richard, 21–22
Sex and sexuality, 135
Shadow profiling, 74, 303
Sharenting, 303. *See also* Parenting
Sharing, 57, 83, 120, 300, 303
Shorty (award), 185
*Shrek* (Jenson and Adamson), 104
Shteyngart, Gary, 75
Signal (app), 165
Silverstone, Roger, 258, 278, 282
Simon (mobile phone), 28–29
Simonyan, Margarita, 119
*Simulacron-3* (Galouye), 277
Singularity, 111, 303

*Singularity Is Near, The* (Kurzweil), 111
Siri, 114. *See also* IPhone
Slack, Andrew, 180
Slacktivism, 154, 166, 303
Slumdog Millionaire (Boyle), 30
Smartphones
  as charismatic devices, 37–38
  computing power, 34
  contact tracing using, 105
  as digitizing us, 112
  environmental aspects, 225
  factory conditions, 228
  histories of, 28–29
  materiality of, 33–34
  zombies and, 115
Smart technology, 29, 70–71, 78, 96, 114, 135–136
*Smombie* (smartphone/zombie), 115
*Snow Crash* (Stephenson), 277
Snowden, Edward, 57–58, 62
"Snow Fall" (*New York Times* special), 214
Social arrangements. *See* arrangements
Social change, 150–155
Social Dilemma, The (Orlowski), 22
Social media. *See also* Influencers
  appropriation of, 175
  authenticity on, 120
  creators on, 189
  election swaying, 90
  as entertainment, 189
  indirect surveillance of, 74
  influencers, 31
  popularity of, 57
  roles of, 215
  sharenting, 70
  vs traditional media, 189
Society of Editors, 236
SocioDigital Lab (Western University, Canada), 266–267
Somatechnic, 312
Sony Music, 193
Sousa, Miquela, 191
Sousveillance. *See* Omnoptic surveillance
South Africa, 168
Souza e Silva, Adriana de, 266
Speakman, Duncan, 149
Spielberg, Steven, 200
Spigel, Lynn, 265
Splinternet, 174
Sponorships, 15
Sputnik, 119
Stadler, Jane, 279–280
*Star Trek* (franchise), 110
*Star Wars* (franchise), 214
#StayHomeStaySafe, 133
Stephenson, Neal, 277
Stimpson, Catherine, 278
StoryMaker (app), 19

**Index** 323

Storytelling
  in digital activism, 172–173
  media's role in, 134, 211
  practices and methods, 213–215
  transmedia, 304
Streaming services, 187. *See also* Netflix; YouTube
Strömbäck, Jesper, 265, 282
Subtlemob, 149
Suffrage movement, 153. *See also* Women
Sundance Film Festival, 8
*Super Sad True Love Story* (Shteyngart), 75
Surrealism, 257
Surveillance, 304. *See also* Monitoring
  activism and, 102, 176–178
  algorithms and, 55–56, 78
  anti-, 52, 297
  art, 61–62, 272
  capitalism, 55–56, 73, 250, 303
  coronavirus crisis and, 20, 64, 70
  coveillance, 74
  covert, 58
  culture and, 81–82
  data and, 78–80, 299
  economics and, 81
  familiarity of, 20
  family surveillance, 70
  indirect, 74, 81
  institutional, 68–72
  interpersonal, 72–76
  life in media and, 51, 268
  omnoptic, 20, 53, 72–76, 268, 302–303
  overt, 58
  panoptic, 51–52, 54–57, 64, 178
  participation in, 55, 66, 142, 302
  psychology and, 82
  public life and, 80–83
  resistance to, 80, 270–271
  security as justification for, 270
  self-, 53, 77–79
  social media and, 74
  state/security forces, 57–62
  synoptic, 20, 52, 64, 304
  us vs them (importance of), 113
  vertical, 51–52
  at work, 69–70
  YouTube and, 61
Surveillance art, 61–62, 272
Surveillance Camera Players, 61
Surveillance capitalism, 55–56, 73, 250, 303
*Survivor* (tv show), 73
Suskind, Ron, 123
Suspicion, 271
Swift, Jonathan, 101
Swift, Taylor, 14
Synoptic surveillance, 20, 52, 64. *See also* Surveillance
Synthesia, 239
Synthetic media, 121

Tactical media, 165, 304
Talos (Greek mythology), 106–7, 106–107
Tausk, Victor, 92–93, 251
Taylor, Richard, 203
Technological specificity, 25, 192
Technology
  access to, 17–18
  anti, 96
  bias, 55–56
  charismatic, 37–38
  dependence on, 101, 103
  evolution of, 87
  freedom and, 95
  human's relationship to, 87–88, 92, 100–101, 107, 125
  moratoriums on, 56
  optimistic views on, 98
  primacy of, 256
  "read and write," 159
Technomyopia, 100, 304, 308
Technorealism, 306–307
*Tecnobrega,* 190
Telecommunications-media-technology sector (TMT), 191–192
Teledildonics (sexual technology), 135, 304
Telegram (app), 165
Teleparenting, 70, 304
Telephones, 33–34
Television
  activist media, 160
  environmental aspects, 225
  histories of, 13, 33, 35
  news on, 31
  reality television, 73
  Russo-Ukrainian War and, 155
  scholarship on, 247
  watching, 5, 130, 136
Television Delivers People (Serra and Schoolman), 21–22
*Terminator* (movie series), 85, 92, 110
*Tetsuo: The Iron Man* (Tsukamoto), 27, 278
"The Man Who Was Used Up, The" (Poe), 101
Thiel, Peter, 61
Thingness, 25, 33, 304
*Thirteenth Floor, The* (Rusank), 277
Thomas, Tanja, 282
Thunberg, Greta, 178
*THX 1138* (Lucas), 95
TikTok, 27, 45–46, 118, 159, 165, 242–243
Time, 36, 38–39, 200, 298
#Timesup, 204
Tisiot, Sandra, 73
*Tools for Conviviality* (Illich), 98
Touch screen, 34–35
Transmedia, 107–8, 213–215, 245–246, 259, 304
Transnational advocacy networks (TANs), 173
Treré, Emiliano, 281

Tripadvisor, 66
*Truman Show, The* (movie), 219–223, 250–252, 273
Truth, 120–121, 133, 155, 224
Turkle, Sherry, 57
Twitch, 114
Twitter, 20, 22, 149, 165, 174
*2001: A Space Odyssey* (Kubrick), 104
*2045 Initiative,* 111

Ubiquitous computing, 35, 264, 304
Uganda. *See* Kony2012
Ugboajah, Frank Okwu, 256
*Ultima Online* (online game), 243
Ultra HD, 30
Umbrella art, 149
Uncanniness, 103–104, 125
Union, desire for, 133
Union Creative (ad agency), 233
United Nations Educational, Scientific and Cultural Organization (UNESCO), 97, 244–245
Universality, 170
Universal Music Group, 193, 210
Universal Pictures, 194
Unreal Engine, 5, 121
Up (documentary series), 14
Usenet newsgroups, 119

Vaccination, 117, 181
*VALIS* (Dick), 277
Valkenburg, Patti, 266–267, 276, 311
Vedanta Resources (mining company), 180
Vera, Francisco, 167
Verge, The (news site), 125
Vertical integration, 192–193, 304
Vertical surveillance, 51–52. *See also* Surveillance
*Vice,* 204
Video games, 89, 96–97, 193, 204, 235, 243
Vimeo, 165
VINCI Energies Group, 63
Virtuality, 3, 191. *See also* Nonplayer character (NPC)
Visibility and vulnerability, 74. *See also* WhyIHeartMyMedia (website)
Vlogs, 19
Voice recognition technology, 114
VPNs (virtual private networks), 106. *See also* Internet

*Wall-E* (Stanton), 71
*Wall Street Journal,* 204
Walt Disney Company, 193
*Wanghong* (creators, Chinese term), 189
War and conflict. *See also* Activism
 autonomous drones, 243
 as escape, 223
 forms of, 156–157

hybrid warfare, 156, 300
 against machines, 88, 94–99
 media's role in, 46, 155–157, 286
 news coverage of, 5
Warner Brothers, 180
WarnerMedia, 193
Warner Music Group, 193
*War of the Worlds* (Wells), 89
Waste, 227–228, 230
Way of the Future (religious company), 111
Websites, 19
Weiser, Mark, 264
Wells, H. G., 89
Weta Digital (effects studio), 203–204, 206
WhatsApp, 27, 165
WHO (World Health Organization), 3
WhyIHeartMyMedia (website), 127, 143–146
Wi-Fi networks, 59, 63, 71. *See also* Internet
Wikipedia, 176
Windows, 30, 35
"Winning the hearts and minds" (saying), 156
Wireless connectivity, 18, 35
Women, 153, 172, 178
#WomenWhoCode, 204
Wonder, Stevie, 111
Wong, Roger, 203
Work and labor
 affective, 218
 atypical, 200–201, 207–208, 297
 contractors vs employees, 232–233
 demographics, 209
 descriptions of, 203–204, 206–207
 emotional, 202
 free, 8–9, 21
 as fun, 211
 life balance, 233–234
 for media, 21
 mental and health, 233–234
 overtime, 235
 policies, 217
 precarity of, 186, 303
 relational, 31
 remote, 231
 resistance of workers, 229
 robots and, 89
 scholarship on, 289–290
 surveillance at, 69–70
 unions, 231–232, 234–237, 294
 workers' rights, 224, 229–238
 working conditions, 228
World, 99–100, 239, 266, 274–275
World Federation of Advertisers (WFA), 210, 233–234
World Health Organization (WHO), 116, 136
World Values Survey, 158

Xiaoyu Zaijia (Little Fish, Baidu), 71

**Index**

Yelp, 66
#YoMeQuedoEnCasa, 133
#YoSoy132, 152
You
  as person of the year (Time magazine), 42
  use of term, 4
Young, Sherman, 265
Youth, media's impact on, 90
YouTube. *See also* Life in a Day
  everyday life videos on, 191
  Kids, 70
  medium specificity of, 27
  moderation issues, 174
  Netflix and, 187
  as protest documentation tool, 149, 165
  surveillance and, 61

Zapatista Army of National Liberation, 161–162
Zomato, 66
Zombies, 114–115
Zuboff, Shoshana, 271
Zuckerberg, Mark, 35, 51, 83, 192